TRENDS IN CONTINUUM PHYSICS

TRECOP '98

Proceedings of the International Symposium on

TRENDS IN CONTINUUM PHYSICS

TRECOP '98

Poznań, Poland 17–20 August 1998

Editors

Bogdan T Maruszewski

Poznań University of Technology
Poznań, Poland

Wolfgang Muschik

Technische Universität Berlin
Berlin, Germany

Andrzej Radowicz

Kielce University of Technology
Kielce, Poland

World Scientific
Singapore • New Jersey • London • Hong Kong

Published by

World Scientific Publishing Co. Pte. Ltd.

P O Box 128, Farrer Road, Singapore 912805

USA office: Suite 1B, 1060 Main Street, River Edge, NJ 07661

UK office: 57 Shelton Street, Covent Garden, London WC2H 9HE

Library of Congress Cataloging-in-Publication Data

International Symposium on Trends in Continuum Physics (1998 : Poznań, Poland)

Proceedings of the International Symposium on Trends in Continuum Physics, TRECOP '98, Poznań, Poland, 17–20 August 1998 / editors, Bogdan T. Maruszewski, Wolfgang Muschik, Andrzej Radowicz.

p. cm.

Includes bibliographical references.

ISBN 981023760X

1. Field theory (Physics) -- Congresses. I. Maruszewski, Bogdan. II. Muschik, W. (Wolfgang), 1936– . III. Radowicz, Andrzej. IV. Title.

QC173.68.I585 1998

530--dc21 98-53243

British Library Cataloguing-in-Publication Data

A catalogue record for this book is available from the British Library.

This book is printed on acid-free paper.

Printed in Singapore by Uto-Print

SCIENTIFIC COMMITTEE

- A.M. Anile, Universitá di Catania, Catania, Italy,
- D. Bedeaux, University of Leiden, Leiden, The Netherlands,
- Z. Bilicki, Institute of Fluid-Flow Machinery, PAN, Gdańsk, Poland,
- J. Engelbrecht, Estonian Academy of Sciences, Tallinn, Estonia,
- K.-H. Hoffmann, Technische Universität Chemnitz-Zwickau, Chemnitz, Germany,
- R.K.T. Hsieh, The Royal Institute of Technology, Stockholm, Sweden,
- H. Irschik, Johannes-Kepler Universität, Linz, Austria,
- D. Jou, Universitat Autonoma de Barcelona, Bellaterra, Spain,
- R. Kotowski, Institute of Fundamental Technological Research, PAN, Warszawa, Poland,
- G. Lebon, Université de Liege, Liege, Belgium
- G.A. Maugin, Université Pierre et Marie Curie, Paris, France,
- L Restuccia, Universitá di Messina, Messina, Italy,
- Cz. Rymarz, Military University of Technology, Warszawa, Poland,
- S. Sieniutycz, Warsaw University of Technology, Warszawa, Poland,
- J. Stefaniak, Poznań University of Technology, Poznań, Poland,
- C. Trimarco, Universitá di Pisa, Pisa, Italy,
- A.A.F. van de Ven, Eindhoven University of Technology, Eindhoven, The Netherlands,
- J. Verhás, Technical University of Budapest, Budapest, Hungary,
- H. Zorski, Institute of Fundamental Technological Research, PAN, Warszawa, Poland.

PREFACE

The International Symposium on Trends in Continuum Physics took place on August 17-20, 1998 at the Polish-German Academic Centre of Poznań University of Technology in Poznań, Poland. It was jointly planned and organized by B.T.Maruszewski (Poznań University of Technology and Poznań Society of Friends of Arts and Sciences), by W. Muschik (Technische Universität Berlin) and by A. Radowicz (Kielce University of Technology). More than 55 participants from 11 countries attended the symposium. One of the main aims of the meeting was to bring together scientists from Eastern Europe working in different fields of widely understood continuum physics with those of Western and Central Europe for extending their existing co-operation and for creating new connections.

During the symposium 9 invited and 27 contributed lectures were held out of the field of continuum physics, mainly on material properties of

* electro-magnetic
* heat conducting
* thermo-viscoplastic
* superconducting
* porous
* 2-phase flow
* biological

materials. Also non-material items such as

* non-linear differential equations
* waves and vibrations
* dynamical systems
* films
* fiber bundles
* mesoscopic mechanics
* extended thermodynamics
* dissipation inequality and hyperbolicity

were treated in lectures. As a prominent non-continuum item

* endoreversible thermodynamics

is mentioned. Special emphasis has been put on the representation of different concepts applied in the different fields of continuum physics. Parallel to the lectures 16 posters were discussed in a poster session of a vivid and friendly scientific atmosphere.

The organizers thank the Scientific Committee for its supporting scientific and organizational advice. Special thanks should be expressed to G. A. Maugin, Université Pierre et Marie Curie, Paris, for his idea, how to create the final form of the logo of the symposium which you can find at the beginning of these proceedings. The scientific and local organizers thank warmly, also on behalf of all participants, the Stiftung für deutsch-polnische Zusammenarbeit, Warsaw, the Batory Foundation, Warsaw, the State Committee for Scientific Research, Warsaw, and the Poznań Society of Friends of Arts and Sciences, Poznań, for their generous support and outstanding care which made the symposium possible.

Because of its informative character in this wide field of continuum physics and because of this highly suitable meeting place in Poznań generating a creative scientific atmosphere in lectures and round table discussions the organizers and participants have proposed to repeat this symposium each three years in Poznań.

B.T.Maruszewski
W. Muschik
A.Radowicz

CONTENTS

SOME REMARKS ON THERMODYNAMICAL TERMINOLOGY

W. MUSCHIK *
INSTITUT FÜR THEORETISCHE PHYSIK
TECHNISCHE UNIVERSITÄT BERLIN
HARDENBERGSTR. 36
D-10623 BERLIN GERMANY

Often difficulties in understanding thermodynamics arise from different terminology or semantics. Therefore here remarks should be made for clarifying some expressions which appear in thermodynamical terminology.

First of all one has to distinguish between two different description of thermodynamical systems: Such a system can be described as a ***discrete*** (or lumped) ***system*** or by a ***field formulation.*** Discrete systems are separated from its environment by a partition. The interaction between the discrete system and its environment can be described by exchanges of heat, of power, and of material. The discrete system is characterized by in general time-dependent quantities belonging to the whole system [1]. The exchanges between system and its environment depend on both their states. Because no gradients appear in the description of discrete systems, this concept is used in thermostatics. But also complex machines in engineering sciences are described by discrete systems, if mainly exchange quantities are of interest. Thus for a reduced description of complex situations, the concept of a discrete system is a very useful one. The theoretical background is complicated [2].

If more information is neccessary, we have to describe the thermodynamical system by a field formulation. Here the quantities describing the system are in general n time-dependent fields $\boldsymbol{a}(\boldsymbol{x},t)$ [3] which satisfy the following balances

$$\partial_t(\varrho\boldsymbol{a}) + \nabla\cdot(\boldsymbol{v}\varrho\boldsymbol{a}+\boldsymbol{\Phi}) = \boldsymbol{\Sigma}. \tag{1}$$

Here $\varrho(\boldsymbol{x},t)$ is the field of the mass density, $\boldsymbol{v}(\boldsymbol{x},t)$ the field of the material velocity, $\boldsymbol{\Phi}(\boldsymbol{x},t)$ the fields of the conductive fluxes belonging to $\boldsymbol{a}$, and $\boldsymbol{\Sigma}(\boldsymbol{x},t)$ the sum of the production and supply terms of $\boldsymbol{a}$. In general the $\boldsymbol{a}(\boldsymbol{x},t)$ are not the ***wanted fields.*** Usually, that means in not-extended thermodynamics [4], these five wanted fields are

$$(\varrho,\epsilon,\boldsymbol{v})(\boldsymbol{x},t). \tag{2}$$

Here $\epsilon(\boldsymbol{x},t)$ is the field of the internal energy.

*E-MAIL: MUSCHIK@PHYSIK.TU-BERLIN.DE

The balance equations (1) are underdetermined: That are n equations for more than n quantities. Consequently we need additional equations for obtaining a determined system of differential equations. These additional equations are the *constitutive equations* which are describing the considered material. For performing the derivatives in (1) we need a *state space* which is defined as the domain of the constitutive equations. In general the state space is different from that spanned by the wanted fields [5]. It includes also derivatives of the wanted fields

$$\boldsymbol{z} = (\varrho, \epsilon, \boldsymbol{v}, \nabla\varrho, \nabla\epsilon, \nabla\boldsymbol{v}, \partial_t\epsilon, ...). \tag{3}$$

In *extended thermodynamics* [6,7,9] the state space does not contain derivatives of the wanted fields, but instead of them the stress tensor and dissipative fluxes

$$\boldsymbol{z} = (\varrho, \epsilon, \boldsymbol{v}, \mathbf{P}, \boldsymbol{q}, ...)(\boldsymbol{x}, t) \tag{4}$$

for which one needs additional balances equations. The constitutive equations are now

$$\boldsymbol{a}(\boldsymbol{z}(\boldsymbol{x}, t)), \boldsymbol{\Phi}(\boldsymbol{z}(\boldsymbol{x}, t)), \boldsymbol{\Sigma}(\boldsymbol{z}(\boldsymbol{x}, t)), \tag{5}$$

and the derivatives in (1) become

$$\partial_t\boldsymbol{a} = \frac{\partial\boldsymbol{a}}{\partial\boldsymbol{z}}\cdot\partial_t\boldsymbol{z}, \qquad \nabla\cdot\boldsymbol{a} = \frac{\partial\boldsymbol{a}}{\partial\boldsymbol{z}}\cdot\nabla\boldsymbol{z}. \tag{6}$$

If we now introduce (6) into the balances (1), we obtain by taking the mass balance into consideration

$$\varrho\frac{\partial\boldsymbol{a}}{\partial\boldsymbol{z}}\cdot\partial_t\boldsymbol{z} + \varrho\boldsymbol{v}\frac{\partial\boldsymbol{a}}{\partial\boldsymbol{z}} : \nabla\boldsymbol{z} + \frac{\partial\boldsymbol{\Phi}}{\partial\boldsymbol{z}}\cdot\partial_t\boldsymbol{z} = \boldsymbol{\Sigma}(\boldsymbol{z}). \tag{7}$$

These balance equations are called *balances on the state space.* Dependent on the special constitutive equations this system of differential equations is highly non-linear, but often, it is not sufficient to describe complex materials. For describing complex materials one need *additional fields* $\boldsymbol{b}(\boldsymbol{x}, t)$ beyond the $\boldsymbol{a}(\boldsymbol{x}, t)$. Additional fields demand additional equations for them

$$\partial_t(\varrho\boldsymbol{b}) + \nabla\cdot(\boldsymbol{v}\varrho\boldsymbol{b} + \boldsymbol{\Psi}) = \boldsymbol{\Xi}, \tag{8}$$

and the original balances (1) contain different fluxes, production and supply terms

$$\partial_t(\varrho\boldsymbol{a}) + \nabla\cdot(\boldsymbol{v}\varrho\boldsymbol{a} + \boldsymbol{\Phi}') = \boldsymbol{\Sigma}'. \tag{9}$$

The usual way to describe *complex materials* is to include additional fields and their governing equations (8). The difficulty which arises is the often missing knowledge of the additional equations (8) and of the material equations for $\boldsymbol{\Phi}'$ and for $\boldsymbol{\Sigma}'$.

Another method of describing complex materials is the *mesoscopic theory*. Instead of introducing additional fields we introduce additional variables, called *mesoscopic variables* which are extending space-time [8]

$$(\boldsymbol{x},t) \longrightarrow (\boldsymbol{x},t,\boldsymbol{b}) \equiv (\cdot). \tag{10}$$

The mesoscopic variable $\boldsymbol{b}$ is not an additional macroscopic field, but a variable. The fields are now defined on the *mesoscopic space* (10)

$$\varrho(\cdot), \boldsymbol{a}(\cdot), \boldsymbol{v}(\cdot), \boldsymbol{\Phi}(\cdot), \boldsymbol{\Sigma}(\cdot), \tag{11}$$

and are called the *mesoscopic fields*. Their balance equations are obtained by additional derivation with respect to the mesoscopic variables $\boldsymbol{b}$

$$\partial_t(\varrho\boldsymbol{a})(\cdot) + \nabla\cdot(\boldsymbol{v}\varrho\boldsymbol{a}+\boldsymbol{\Phi})(\cdot) + \nabla_{\boldsymbol{b}}\cdot(\boldsymbol{v}\varrho\boldsymbol{a}+\boldsymbol{\Phi})(\cdot) = \boldsymbol{\Sigma}(\cdot). \tag{12}$$

These *mesoscopic balance equations* have to be solved by methods which are also applied to the balances (1). Examples for mesoscopic variables are the dryness fraction in two-phase flow, the microscopic director in liquid crystal theory, damage parameters in crack theory, and the orientation of microcrystallites in microstructured media. The mesoscopic theory introduces a statistical element, the *mesoscopic distribution function* [9]

$$f(\cdot) := \frac{\varrho(\cdot)}{\int \varrho(\cdot)d\mathcal{M}}. \tag{13}$$

Here the integration is performed on the mesoscopic variables of the mesoscopic space (10). The macroscopic fields are obtained by integrating over the mesoscopic variables

$$\varrho(\boldsymbol{x},t) = \int \varrho(\cdot)d\mathcal{M}, \tag{14}$$

$$\boldsymbol{v}(\boldsymbol{x},t) = \frac{1}{\varrho(\boldsymbol{x},t)}\int \varrho(\cdot)\boldsymbol{v}(\cdot)d\mathcal{M}, \tag{15}$$

$$\epsilon(\boldsymbol{x},t) = \int \left\{\epsilon(\cdot) + \varrho(\cdot)[\boldsymbol{v}(\cdot)-\boldsymbol{v}(\boldsymbol{x},t)]^2\right\} d\mathcal{M}, \ldots.. \tag{16}$$

From the mesoscopic distribution function (13) we obtain a *family of macroscopic order parameter* fields which describe the complex material additionally to the usual fields

$$\int f(\cdot)d\mathcal{M} = 1, \tag{17}$$

$$\int \boldsymbol{b} f(\cdot) d\mathcal{M} = \boldsymbol{B}(\boldsymbol{x}, t), \tag{18}$$

$$\int \boldsymbol{b}\boldsymbol{b} f(\cdot) d\mathcal{M} = \mathcal{B}(\boldsymbol{x}, t), \tag{19}$$

$$\int \boldsymbol{b}\text{...m times...}\boldsymbol{b} f(\cdot) d\mathcal{M} = \underline{\mathcal{B}}_m(\boldsymbol{x}, t). \tag{20}$$

This is a system of differential equations for determining the macroscopic order parameter family.

Besides the balance equations (1) or (12) we have to take into consideration the *dissiption inequality* by which the *second law* is introduced

$$\partial_t(\varrho s) + \nabla \cdot (\boldsymbol{v}\varrho s + \boldsymbol{\phi}) = \sigma \geq 0. \tag{21}$$

There are different possibilities of *exploiting the dissipation inequality*. The most simple one is first to solve the balances on the state space (7), and then to determine the parameters in the constitutive equations in such a way, that the dissipation inequality is satisfied. The second method of exploitation is those of *Coleman and Noll* [11]. This exploitation method does not take into consideration the balance equation as constraints which is done by the *Liu exploitation* [12].

With these remarks we want to finish the short course on thermodynamical terminology. The original literature can be found in the references cited below.

References

1. W. Muschik: Fundamentals of Non-Equilibrium Thermodynamics, in: Non-Equilibrium Thermodynamics with Application to Solids (W. Muschik Ed), Springer Wien, 1993, chap.1
2. W. Muschik, R. Domínguez-Cascante: On extended thermodynamics of discrete systems, Physica A 233 (1996) 523
3. see [1] chap.6 ff
4. W. Muschik: Aspects of Non-Equilibrium Thermodynamics, World Scientific Singapore, 1990, sect.6.3
5. W. Muschik, C. Papenfuss, H. Ehrentraut: Concepts of Continuum Thermodynamics, Kielce University of Technology 1996, ISBN 83-905132-7-7, chap.4
6. D. Jou, J. Casas-Vázquez, G. Lebon: Extended Irrebersible Thermodynamics, Springer Berlin, 1993
7. I. Müller, T. Ruggeri: Extended Thermodynamics, Springer Berlin, 1993

8. see [5] chap. 5.3
9. see [5] chap. 6.3
10. see [5] chap. 5.4
11. see [4] sect. 6.4
12. W. Muschik, H. Ehrentraut: An amendment to the second law, J. Non-Equilib. Thermodyn. 21 (1996) 175

SIMULATION OF PERCOLATION THROUGH POROUS MEDIA USING PARTICLE DYNAMICS

W. ALDA, W. DZWINEL, J. KITOWSKI
Institute of Computer Science, University of Mining and Metallurgy,
al. Mickiewicza 30, 30-059 Cracow, Poland
E-mail: alda@uci.agh.edu.pl

L. FIORETTI
Nuova Simonelli, Via Madonna d'Antenagio 62-031 Belforte (MC) Italy
E-mail: n.simonelli@wnt.it

J. RYBICKI
Department of Physics, Technical University of Gdansk,
ul. Narutowicza 11/12, 80-952 Gdansk, Poland
E-mail: ryba@sunrise.pg.gda.pl

We present two computer models based on particle approach, which can be suitable for the simulation fluid flow through deformable and undeformable porous media. We take under consideration Molecular Dynamics of Lennard-Jones liquid Dissipative Particle Dynamics. We include sample results obtained with 2-D simulations.

1 Introduction

Numerical solution of fluid mechanics problems is a challenging task due to its wide theoretical and practical implications[1]. Classical approach is a continuum-based model described by Navier-Stokes equations, which can by numerically solved using Finite Elements Method (FEM) or Finite Differences Method (FDM). This model, however powerful, may be extremely difficult to use with complex systems, time-dependent boundary conditions and/or sediment granules added. Flow through porous media can be such a case, especially when granules forming the medium can have different shapes and sizes, can be deformable, can split or glue together. It may also be assumed that they act in some sort of chemical reactions with the fluid. With the above assumptions classical approach may face problems practically impossible to solve.

A particle representation of fluid can overcome many difficulties found in classical methods. This new approach got a very strong impact only in the last few years both in computational and theoretical aspects. Topics, such as convection, sedimentation, thermodynamic instabilities, developing of turbulence, flow in porous media, two-component flows, shock waves propagation and many others are of particular interest of particle models. In the paper we describe briefly two models based on this alternative method: Molecular Dynamics (MD) and Dissipative Particle Dynamics (DPD).

Molecular Dynamics is most straightforward and most accurate, but also most time-consuming method. As it is pure particle approach governed only by interactions and Newtonian equations of motion, complex deformations of porous material are easy to simulate. One of MD drawbacks is its limitation to microscopic space and time scales. However, it is possible to rescale the system and treat it as mesoscopic. One of the most promising approaches is DPD, which merges benefits of MD and Lattice Gas Automata [2] methods.

In such investigations it is necessary to deal with large systems [3]. For Molecular Dynamics simulations at least 10^5 particles in 2D are needed.

2 Particle methods for flow through porous media

2.1 *Molecular Dynamics*

Molecular Dynamics has been first introduced for equilibrium systems. It is defined entirely by the properties of particles and their mutual conservative interactions. The particles are responsible for energy and momentum transport according to the rules of conservation. Each particle of a N-particle system interacts with all neighbours within a sphere with r_{cut} radius. Usually the interaction is defined as a two-body, distance dependent potential function $U(r_{ij})$. An effective force $\vec{F_i}$ acting on particle i is defined as a sum of additive pair forces $\vec{f_{ij}}$:

$$\vec{F_i} = \sum_{j \in S(i, r_{cut})} \vec{f_{ij}}, \qquad \vec{f_{ij}} = \nabla U(r_{ij}). \tag{1}$$

Particles motion is described by the Newtonian equations:

$$m_i \frac{d\vec{v_i}}{dt} = \vec{F_i}, \qquad \frac{d\vec{r_i}}{dt} = \vec{v_i}. \tag{2}$$

The interaction applied in the MD program is the short-range Lennard-Jones potential in the form:

$$U(r_{ij}) = 4\epsilon_{ij} \left[\left(\frac{\sigma_{ij}}{r_{ij}} \right)^{12} - \left(\frac{\sigma_{ij}}{r_{ij}} \right)^{6} \right] \tag{3}$$

where ϵ is the well depth of the potential, σ can be roughly treated as a nearest distance of particles, and r is the actual distance between particles i and j.

The MD simulations have been originally applied for computations of different parameters of gases, fluids and solids (such as diffusion coefficient, specific heat, electrical conductivity) and for investigation of material properties in micro-scale. The MD approach to a fluid flow modelling is a relatively recent development, however. It seems to be a viable approach to study flows under extreme conditions, to analyse the microscopic origin of hydrodynamic instabilities and to find out what happens inside a boundary layer. First non-equilibrium MD simulations of fluid flows started in late 80's [4,5]. They concerned the flow past an obstacle and eddy formation. A lot of work has been done to simulate formation of patterns of Rayleigh-Benard convection [6,7].

2.2 Dissipative Particle Dynamics

Disssipative particle dynamics (DPD) is a relatively new particle approach introduced in 1992 by Hoogerbrugge and Koelman [8] and developed significantly in the last few years [9,10]. The basic idea of the method is similar to that of MD. However the particles here are not considered as atoms, but rather as mesoscopic clusters of physical particles. In this scale the total force acting on DPD particles consists of conservative (MD-type) interactions, dissipative term responsible for viscosity and stochastic component which controls internal pressure and the temperature.

Similarly to the MD method particles move in continuum space defined by positions r_i and momenta p_i for every particle ($i = 1, \ldots, N$). On the other hand system is updated in discrete time steps δt in two phases: collision and propagation. This is similar to Lattice Gas Automata approach as well as to event driven particle dynamics [4], where free time between collisions is greater then that in MD. In the collision phase the momenta change as:

$$\vec{p}_i' = \vec{p}_i + \sum_j \Omega_{ij} \vec{e}_{ij} \tag{4}$$

where $\vec{e}_{ij} = (\vec{r}_i - \vec{r}_j)/|\vec{r}_i - \vec{r}_j|$ is the unit vector from particle j to i, while Ω_{ij} describes the momentum transferred from j to i. During the propagation phase particles move freely according to:

$$\vec{r}_i' = \vec{r}_i + \frac{\delta t}{m_i} \vec{p}_i', \tag{5}$$

The interactions are limited to neighbours within a distance r_{cut} by assuming $\Omega_{ij} = 0$ for $|\vec{r}_i - \vec{r}_j| > r_{cut}$. Ω_{ij} depends not only on distance between particles, but also on relative momenta expressesd with $(\vec{r}_i - \vec{r}_j)(\frac{\vec{p}_i}{m_i} - \frac{\vec{p}_j}{m_j})$. The expression for Ω_{ij} is usually applied in the form:

$$\Omega_{ij} = W(|\vec{r}_i - \vec{r}_j|)\{\Pi_{ij} - \omega_{ij}(\vec{p}_i - \vec{p}_j)\vec{e}_{ij}\}. \tag{6}$$

In Eq. (6) $\omega_{ij} = 1 - \frac{r_{ij}}{r_{cut}}$ is a positive coefficient responsible for dissipation of energy dependent on relative momenta of neighbour particles. $W(r)$ is a dimensionless non-negative weight function. In the mesoscale it is sufficient to apply linear form of $W(r)$ decreasing and vanishing at r_{cut}: $W(r) = (3/\pi r_{cut}^2 n))(1 - r/r_{cut})$, where n is number density equal to N/V. $W(r)$ is a repulsive, MD-type component, while Π_{ij} and $\omega(\vec{p}_i - \vec{p}_j)\vec{e}_{ij}$ are stochastic and dissipative terms respectively.

2.3 Other particle methods

Among other particle models suitable for simulation of complex fluid flows we should mention Lattice Gas Automata (LGA) and Lattice Boltzmann (LB) models.

LGA can be treated as a maximal simplification of molecular dynamics, where particles of unit mass can be placed only in the nodes on a hexagonal mesh and can move from one node to another with unit velocity fulfilling certain rules of collisions that conserve momentum and number of particles [2]. By introducing nodes which cannot be occupied by particles, unmovable obstacles forming porous medium are created. LGA method has been intensively used for simulation of particle fluid flow through porous material [11]. Lattice Boltzmann (LB) method is an extension to LGA approach in that sense that it uses probabilistic particle distributions, rather than discrete particles [12]. Thus the velocities are represented by floating-point values and the method is considerably less noisy than LGA. LB becomes now more popular in simulation of complex flows, in particular flows through porous media [13,14].

2.4 MD and DPD programs for simulation of flow through porous media

The program written for the simulation of flow through porous media has some characteristic features:

1. The program accepts different types of particles, i.e. particles with different masses and different interction parameters.

2. External gravitational force is applied only to particles representing fluid.

3. To avoid growth of energy in presence of gravitational field, damping coefficient is used in the MD program. We use the *leap-frog* scheme in

the form:

$$\vec{v}_i^{n+1/2} = \frac{1-d}{1+d}\vec{v}_i^{n-1/2} + \frac{1}{m_i(1+d)}\vec{F}_i^n \cdot \Delta t \tag{7}$$

$$\vec{r}_i^{n+1} = \vec{r}_i^n + \vec{v}_i^{n+1/2}\Delta t \tag{8}$$

where n is a timestep number of Δt, and d is a damping coefficient.

4. In DPD case the resultant force $\vec{F}_i^n$ depends on $(\vec{v}_i^n - \vec{v}_i^n)$ terms which are a part of Ω_{ij} expression, and which are not available from Eqs. (7) and (8). Thus the numerical scheme is changed to the form:

$$\vec{v}_i^{n+1/2} = \vec{v}_i^{n-1/2} + \frac{\Delta t}{m_i} \sum_{j \in S(i, r_{cut})} \Omega_{ij}^n \tag{9}$$

$$\vec{v}_i^n = \frac{1}{2}(3\vec{v}_i^{n-1/2} - \vec{v}_i^{n-3/2}) \tag{10}$$

$$\vec{r}_i^{n+1} = \vec{r}_i^n + \vec{v}_i^{n+1/2}\Delta t \tag{11}$$

3 Simulation and Results

The results presented have been obtained from MD and DPD simulations. The system consists of three types of particles. Light particles, which model a liquid phase, are placed in the upper part of the box. Porous material is represented by granules formed of particles 100 times heavier than "fluid" particles. It is initially placed in the middle of the box between obstacles made of very heavy (10^{10} times heavier then "fluid" particles) and practically unmovable particles. Number of moving particles in the simulations is about 10^5. Qualitative results are shown in subsequent figures in a form of snapshots with current coordinates of particles. Figures 1 and 2 come from two MD runs, while figures 3 and 4 show results from single DPD run. In the run C007 (Fig. 1) the ϵ value in the Lennard-Jones potential of the fluid is such, that the reduced temperature of the system is 0.7 which makes the behaviour of the particles more like a liquid. In the second run C107, shown in Fig. 2, the reduced temperature value is 100 times higher, which makes the fluid behave more like steam. In both cases moving of granules due to "fluid" particles interaction is clearly seen. Results from DPD run can be seen in Fig. 3 and Fig. 4. Compared with MD, DPD simulation can be completed in much fewer timesteps. Also due to different potential, DPD results can be interpreted as mesoscopic.

These results, however only qualitative, show the flexibility and potential power of the method. We expect to increase the number of particles for 3D simulations with the DPD model and try to obtain quantitative results in macro-scale.

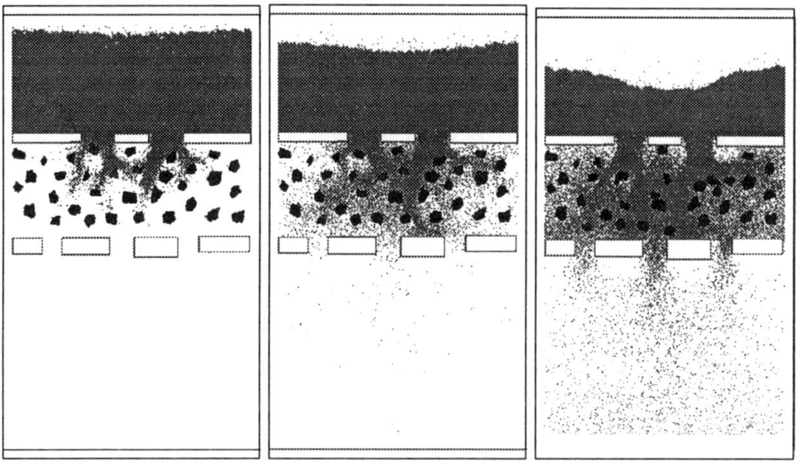

Figure 1: Snapshots from the run C007 after 6000, 12400 and 26800 timesteps.

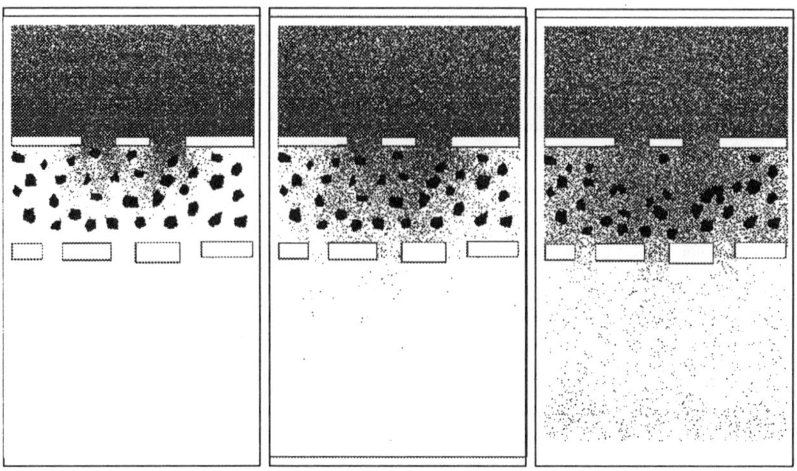

Figure 2: Snapshots from the run C107 after 6000, 12400 and 26800 timesteps.

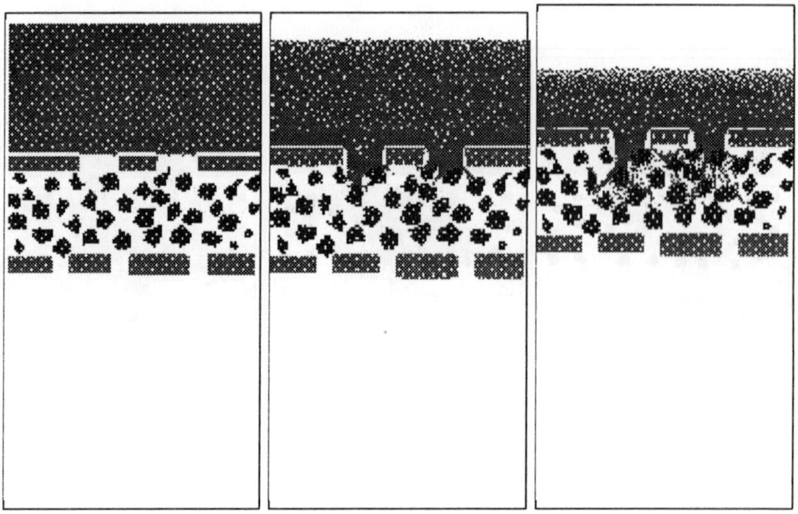

Figure 3: Snapshots from the run DPD7 after 10, 200 and 300 timesteps.

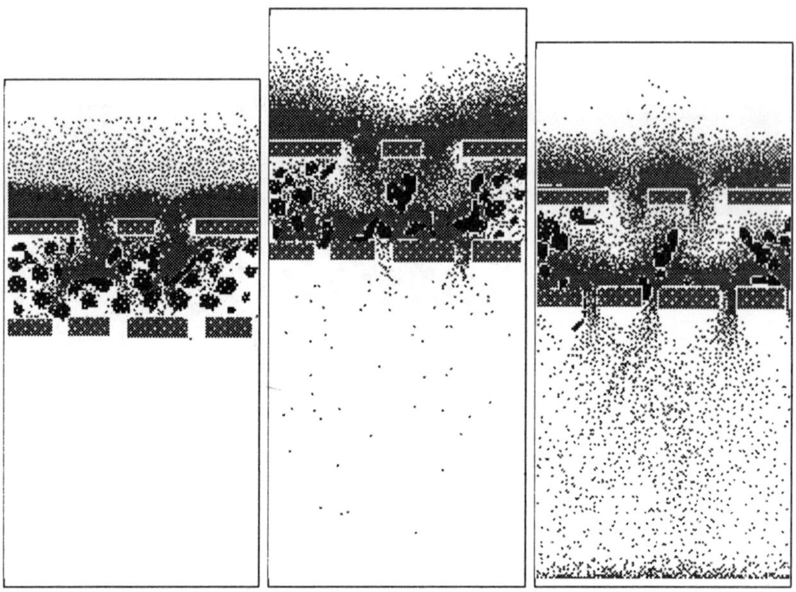

Figure 4: Snapshots from the run DPD7 after 400, 500 and 600 timesteps.

Acknowledgments

The authors are grateful to Academic Computer Centres: CYFRONET – Kraków and TASK – Gdańsk for computer time and to Nuova Simonelli for support.

References

1. U. Frisch and S.A. Orszag, *Physics Today* **1**, 24 (1990).
2. U. Frisch *et al*, *Complex Systems* **1**, 649 (1987).
3. M. Mareschal and B.L. Holian in *Microscopic Simulations of complex hydrodynamic phenomena*, ed. M. Mareschal and B.L. Holian (Plenum Press, New York, 1992).
4. D.C. Rapaport, *Phys. Rev. Lett.* **60**, 2480 (1988).
5. A. Puhl *et al*, *Phys. Rev.* A **40**, 1999 (1989).
6. D.C. Rapaport, *Phys. Rev.* A **43**, 7046 (1991).
7. D.C. Rapaport, *Phys. Rev.* A **46**, 1971 (1992).
8. P.J.Hoogerbrugge and J.M.V.A. Koelman, *Europhys. Lett.* **19**, 155 (1992).
9. P.V. Coveney and K.E. Novik, *Phys. Rev.* E **54**, 5134 (1996).
10. P. Español, *Phys. Rev.* E **57**, 2930 (1998).
11. D.H Rothman and S. Zalesky, *Rev. Modern Phys.* **66**, 1417 (1994).
12. R. Benzi *et al*, *Phys. Rep.* **222**, 145 (1992).
13. A.K. Gunstensen, D.H. Rothman, *J. Geophys. Res.* **98**, 6431 (1993).
14. N. Martys, H. Chen, *Phys. Rev.* E **53**, 743 (1996).

MAGNETOINDUCED MICROPLASTICITY OF NON-MAGNETIC CRYSTALS

V.I.Alshits, E. V. Darinskaya, O.L.Kazakova, M. V. Koldaeva, E. Yu. Mikhina, E.A. Petrzhik
Institute of Crystallography, Russian Academy of Sciences, Leninsky Pr. 59,117333 Moscow, Russia
E-mail: alshits@mechan.incr.msk.su

The main features of the magnetoplastic effect manifesting itself in dislocation displacements in non-magnetic crystals under the action of external magnetic field are discussed. The influence of external electric field, mechanical loading and preliminary X-ray irradiation on the magnetoplastic effect are demonstrated. The threshold value of the magnetic field below which the magnetoplastic effect does not exist is predicted and experimentally observed. It is shown that the threshold field is sensitive to temperature, X-ray irradiation of the samples and Ca impurity concentration in NaCl crystals. The limiting magnetic field above which dislocation mobility does not depend of the value of the magnetic induction is found out.

The magnetic properties of non-magnetic crystals with dislocations are manifested in different phenomena [1-3]. However it was rather surprising when so-called magnetoplastic effect was observed for the first time [4]. It was found that exposing of non-magnetic crystals to a static magnetic field without any mechanical loading causes dislocation displacements. Now the existence of magnetoplastic effect has been reliably established and many experimental and theoretical investigations have been performed by a number of independent groups [5-14].

The experiments were carried out for NaCl, CsI, LiF, Zn and A1 single crystals at the magnetic field B=(0.02÷2)T at three temperatures T=4.2, 77 and 293K during the time t ranging from several seconds up to hours. The dislocation displacements were revealed by selective etching. The alternating magnetic field was created by a specimen rotation in the static magnetic field at various frequencies ν=(0÷200)Hz. The electrical contacts were provided by silver glue on the (100) surfaces of NaCl and LiF crystals to investigate the influence of the electric field E=(0.2÷20)kV/m on the magnetoplastic effect. NaCl crystals were irradiated by X-ray ($Mo_{k\alpha}$, I=35mA, U=45kV) during the time t_{ir} ranging from 5s up to 1 min. Experimental details were published in [4, 7,10].

The main properties of magnetoplastic effect are the following. The average displacement of dislocations l in a magnetic field is proportional to the magnetic treatment time t (Fig.la), but at a long enough time and a sufficiently high density ρ of dislocations the dependence $l(t)$ reaches saturation at a level corresponding to an average distance between "forest" dislocations (Fig.lb). Fig.2a demonstrates that the dependence $l(B)$ has the form $l = l_o + \alpha B^2$ where l_o is the background mean dislocation path, which also exists at $B = 0$ and is associated with the etching out of near-surface pinning centers. Magnetoplastic effect appears to be very sensitive to the impurity concentration in the samples. The mean dislocation path decreases with the growth of Ca impurity concentration C in NaCl crystals as $1/\sqrt{C}$ (Fig.2b). But a slight addition of Ni impurity to NaCl crystals leads on the contrary to the increase of dislocation mobility. Magnetoplastic effect is temperature independent in the range T=(4.2÷77)K and grows only by 15÷20% with further increase T up to room temperature (Fig.3). The magnitude of mean dislocation path depends on the mutual orientation of the magnetic field, the dislocation line and its Burgers vector. The magnetic field applied parallel to the dislocation line (i.e. in the [100] direction) does not cause any displacements. The density of mobile dislocations ρ_m also increase under the action of magnetic field with t and B and it can reach almost 100% of the freshly induced dislocation density. The study of dislocation mobility in an alternating over orientation magnetic field, created by rotating of the samples in a static magnetic field, has shown that when the frequency ν exceeds some critical value ν_c the mean dislocation path abruptly decreases (Fig.4a). The critical frequency is proportional to the square of the magnetic induction (Fig.4b): in normalized coordinates both dependences for different values of magnetic induction map onto a universal curve which is insensitive to B. Within the limits of the experimental error, ν_c is independent of the duration of sample rotation in a magnetic field and of the temperature (Fig.5, curves 1, 2). The critical frequency remains the same when temperature is lowered from the room temperature to that of liquid nitrogen. The variation of Ca impurity concentration in NaCl crystals (Fig.5) does not change ν_c. But the critical frequency turned out to be sensitive to the impurity type. Addition of Ni impurity to NaCl crystals leads not only to the mean dislocation path increase, but also to the essential increase of ν_c. The critical frequency appears to be different for edge and screw dislocations.

The experimental data of $l(t)$ saturation (Fig.1) allow to suppose that the driving force for dislocation motion in the magnetoplastic effect is a long-range internal stress field of other dislocations. The role of the magnetic field is reduced to the unlocking of dislocations from paramagnetic centers. The experimentally established temperature independence of the effect (Fig.3), proportionality of the dislocation path to the square of the magnetic induction (Fig.2), its inverse proportionality to the square root of the paramagnetic impurity concentration (Fig.3) and the existence of the critical frequency of the sample rotation in magnetic field above which dislocations do not move (Fig.4, 5) allow to suppose that dislocation is unlocked from paramagnetic centers because of spin-dependent electron transitions which are stimulated in the system impurity + dislocation by the magnetic field, whose role reduces to removing the spin forbiddenness on the processes which break down local barriers. The phenomena which are determined by the spin evolution of systems in a magnetic field are well known in physics [15-17].

The driving force from internal stresses is rather small in our crystals. This explains why the effect turns out to be very sensitive to a weak external electric field providing an additional force on charged dislocations in NaCl and LiF crystals (Fig.6). It is essential that in the absence of a magnetic field even the highest of the electric fields used in our experiments did not cause a islocation motion. The electric influence is found only for edge (charged) dislocations and only then the **E** vector has a nonvanishing projection both on the slip plane and on the velocity of dislocation. No difference in the effect was observed at 300K and 77K. Experiments in an alternating magnetic field have shown that the critical frequency is not sensitive to the electric field. The grows of mean dislocation path with B is usually changed by saturation at a level, corresponding to an average distance between "forest" dislocations (Fig.6a, curves 3, 4). However, at sufficiently high E one can see that saturation level is defined by the value of the electric field and its magnitude appears to be multiple to the initial one (Fig.6a, curves 1, 2). The dependences $l(B^2)$ begin to saturate at the same value of the magnetic induction. The existence of such limiting magnetic field B_{lim} is also well seen from Fig.6b: there is no dislocation mobility increase after $B = 1$ T for LiF crystals.

The similar effect of dislocation path saturation on the multiple levels is observed in NaCl crystals under the combined action of magnetic field and mechanical loading (Fig.7a). The special setup has been designed to investigate crystal microplasticity by uniaxial compression in a static magnetic field. A possibility of programming crystal loading by mechanical triangular and trapeziform pulses of various amplitude front inclinations was realised. It was found out that the mechanical impulse form influences the dislocation mobility when $B \neq 0$ (Fig.7a) but the mean dislocation path appears to be almost insensitive to the impulse form when $B = 0$ (Fig.7b).

The critical frequency being the fundamental characteristic of an elementary act of dislocation detachment from a point defect must be sensitive to such external influences which modify the spectrum of the point defects in the crystals. Indeed, the dislocation mobility under the external magnetic field in NaCl and LiF single crystals has proved to be sensitive to rather weak preliminary X-ray irradiation $(\sim 10^2 \text{rad})$ Fig.8a shows that after irradiation a one-step curve $l(\nu)$ for edge dislocations is replaced by two-step function characterized by two critical frequencies ν_{c1} and ν_{c2}. With the dose of irradiation increasing the first critical frequency goes up as high as the level of the second critical frequency. The frequency ν_{c2} is insensitive to the irradiation dose (Fig. 8b). After exposing of the irradiated crystals to the light of the tungsten lamp over (1÷2)h the second step disappears and the position of the first one returns to the initial value. But the mean dislocation path in NaCl crystals after such de-excitation never returns to the initial level. This indicates that the irradiation provides not only the new states of initial paramagnetic complexes (the first step) and the additional magnetosensitive defects (the second step), but also the new non-magnetic obstacles which determine the decrease of l. Variation of Ca impurity concentration in NaCl crystals does not change the frequencies ν_{c1} and ν_{c2}. Whereas addition of small amount of Ni makes the Ca complexes radiation resistant and the critical frequency ν_{c1}, corresponding to them, initially equals ν_{c2} for crystals without Ni. In NaCl:Ni crystals both critical frequencies become insensitive to the irradiation dose. Screw dislocations in LiF crystals react to irradiation in another way: dependence $l(\nu)$ remains a single-step curve, the frequency ν_{c1} in the irradiated crystals is less than ν_c in non-irradiated ones and it is insensitive to the irradiation dose. Thus, the investigations demonstrate that magnetoplastic effect is more sensitive to low irradiation doses then the optical measurements with the help of spectrometers.

The critical frequency of an alternating magnetic field determine the time τ_{dp} of dislocation depinning from a paramagnetic center. As we have shown in [8, 9] the depinning time can be estimated as: $\tau_{\text{dp}} = \left(10^{-2} \div 10^{-3}\right)/\nu_c$. The magnetoplastic effect can be observed as long as the depinning time does not exceed the spin-lattice relaxation time τ_{sl} in the given dislocation + impurity system. Otherwise thermal oscillations, mixing the spin states in the system, will virtually eliminate spin-dependent transitions and thus the depinning of the dislocations. According to experimental data $\tau_{\text{dp}} \propto B^{-2}$, so one can expect the existence of a threshold value of the magnetic field B_c (when $\tau_{\text{dp}}(B_c) = \tau_{\text{sl}}$) below which the magnetoplastic effect disappears. The existence of the above-predicted B_c is demonstrated in Fig. 9,10,11 for NaCl and LiF crystals. The effect of the temperature on the threshold field is shown in Fig.10. The observed leftward shift of the step in $l(B)$ dependence as the temperature decreases from room one to 77K is connected with the spin-lattice time increasing when τ_{dp} remains the same. As the preliminary irradiation of alkali halide crystals converts the single-step curve $l(\nu)$ into a two-step one (Fig.8), it is natural to expect a second step to appear in the curve $l(B)$ after irradiation. This is indeed observed in Fig.11. The comparing of $l(\nu)$ and $l(B)$ dependences gives the estimation of spin-lattice time: τ_{sl} for an impurity complex $\approx \left(10^{-3} \div 10^{-4}\right)$s and τ_{sl} for a new pinning centers produced under irradiation $\approx \left(10^{-2} \div 10^{-3}\right)$s. Thus, the magnetoplastic effect provides a very simple new method for measuring the spin-lattice relaxation time in a dislocation + paramagnetic center system.

Along with the spin-lattice relaxation time measurements, the discovery of the threshold B_c of magnetoplastic effect allows to estimate the average size of paramagnetic complexes. As it was shown in Fig.9, the threshold magnetic field appear to be sensitive to the Ca impurity concentration in NaCl crystals. The impurity concentration increase leads to B_c increase as it is demonstrated on Fig.12. On the other hand, the critical frequency of rotation is identical for NaCl crystals having different Ca concentration C (Fig.5). So the depinning time does not depend on C. Taking into account that B_c is governed by the relation $\tau_{\rm dp}(B_c)=\tau_{\rm sl}$, one can suggest that $\tau_{\rm sl}$ is sensitive to the concentration C. It is well known that the average size of the impurity complexes in annealed crystals grows with C increasing. The dependence $\tau_{\rm sl}(C)$ characterizes the response of spin-lattice relaxation in the dislocation + impurity atom system to variation of the size of the impurity complexes. The response of the spin system to thermal lattice vibrations must increase as the vibration amplitude increases. In particular, this is why the time $\tau_{\rm sl}$ decreases as the temperature is raised. The same thing happens at a fixed temperature when the number of vacancies among the nearest neighbors of each impurity atom increase. Fig.12 demonstrates the theoretical dependence of average number of cation n_v vacancies among the nearest neighbors of Ca impurity on the number N of Ca atoms in the complex [11] in the comparison with the experimental dependence $B_c(C)$. Fig.12 allows to determine the average size of Ca complexes in the given NaCl crystals.

The authors would like to thank V.L.Indenbom, B.N.Grechushnikov and A.A.Urusovskaya, for useful discussions of the results, V.P.Kisel, E.B.Rudneva, I.L.Smol'skii and V.M.Chernov for technical assistance and L.M.Soifer for providing the LiF crystals. This work was partialy supported by the Russian Foundation for Basic Research, grant 97-02-16327.

References:

1. E.J.Sharp, D.A.Avery, *Phys. Rev.*,158 (1967) 511.
2. V.A.Pavlov, I.A.Pereturina, I.L.Pecherkina, *Phys.Status Solidi A,* 57 (1980) 449.
3. V.P.Lebedev, V.S.Krylovskii, *Fiz. Tverd Tela,* 27 (1985) 1285.
4. V.I.Alshits, E.V.Darinskaya, T.M.Perekalina, A.A.Urusovskaya, *Fiz. Tverd. Tela,* 29 (1987) 467.
5. V.I.Alshits, E.V.Darinskaya, E.A.Petrzhik, *Mat. Sci. Eng.*, A164 (1993) 332.
6. V.I.Alshits, E.V.Darinskaya, O.L.Kazakova, E.Yu.Mikhina, E.A.Petrzhik, *J. Alloys Comp.*, 211/212 (1994) 548.
7. V.I.Alshits, E.V.Darinskaya, E.Yu.Mikhina, E.A.Petrzhik, *Fiz. Tverd. Tela,* 38 (1996) 2426.

8. V.I.Alshits, E.V.Darinskaya, O.L.Kazakova, E.Yu.Mikhina, E.A.Petrzhik, *JETP Lett.,* 63 (1996) 628.
9. V.I.Alshits, E.V.Darinskaya, O.L.Kazakova, E.Yu.Mikhina, E.A.Petrzhik, *Mat. Sci. Eng.,* A234-236 (1997) 617.
10. V.I.Alshits, E.V.Darinskaya, O.L.Kazakova, *JETP,* 111 (1997) 615.
11. V.I.Alshits, E.V.Darinskaya, O.L.Kazakova, *Fiz. Tverd Tela,* 40 (1998) 81.
12. Yu.I.Golovin, R.B.Morgunov, *JETP Leg.,* 61 (1995) 596.
13. M.I.Molotskii, R.E.Kris, V.Fleurov,. *Phys. Rev. B,* 51 (1995) 12531.
14. A.A.Urusovskaya, V.I.Alshits, A.E.Smirnov, N.N.Bekkauer, *JETP Lett.,* 65 (1997) 470.
15. Ya.B.Zel'dovich, A.L.Buchachenko, E.L.Frankevich, *Usp. Fiz. Nauk,* 155 (1988) 3.
16. K.M.Salikhov, Yu.N.Molin, R.Z.Sagdeev, A.L.Buchachenko, *Spin Polarization and Magnetic Effects in Radical Reactions* (Elsevier, Amsterdam,1984; Nauka, Novosibirsk,1978).
17. V.V.Kveder, Yu.A.Osip'yan, A.I.Shalynin, *JETP,* 83 (1982) 699.

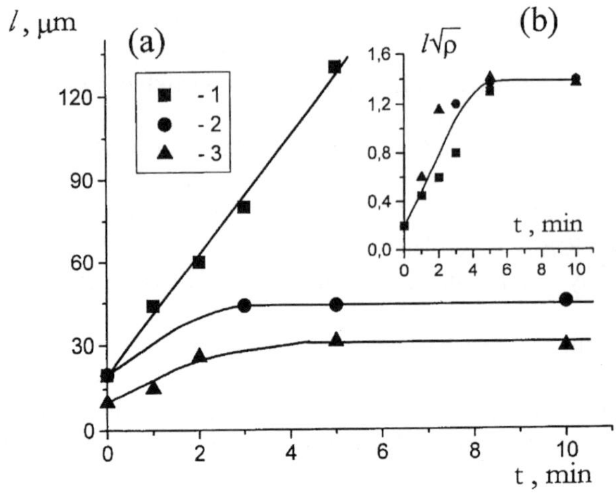

Fig.1 (a) - Dependence of the mean dislocation path l in NaCl crystals on the time t of exposure to the magnetic field $B = 0.5T$ for specimens with various densities ρ of dislocations: 1-1x10^4 cm^{-2}, 2 -1x10^{5}cm^{-2}, 3 - 4x10^{5}cm^{-2}; (b) - dependences (a) replotted in normalized coordinates.

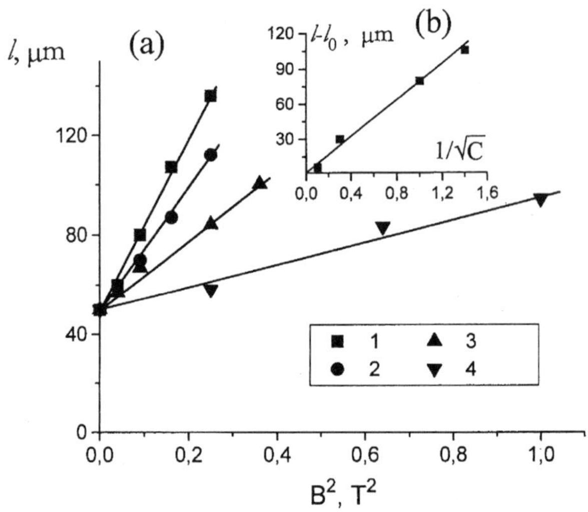

Fig.2 Dependence of the mean dislocation path l in NaCl:Ca crystals: (a) - on the magnetic induction B for different impurity concentrations: 1- 0.5ppm, 2 - lppm, 3-10ppm, 4-100ppm; (b) - on the impurity concentration, $B = 0.5T$, t = 5min.

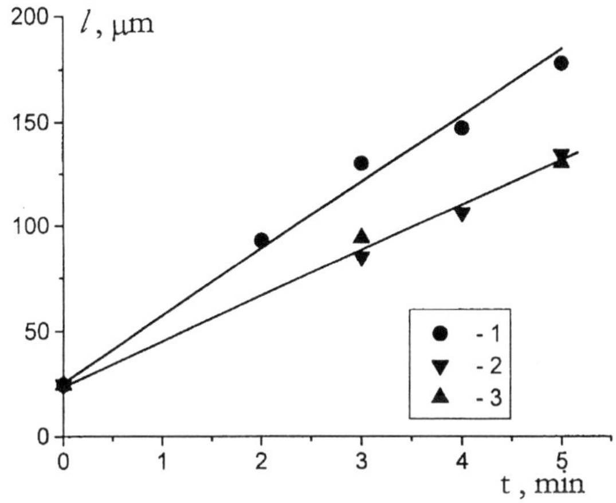

Fig.3 Dependence of the mean dislocation path in NaCl:Ni crystals on the time t of exposure to the magnetic field $B = 0.5\text{T}$ for different temperatures: 1 - 293K, 2 - 77K, 3 - 4.2K.

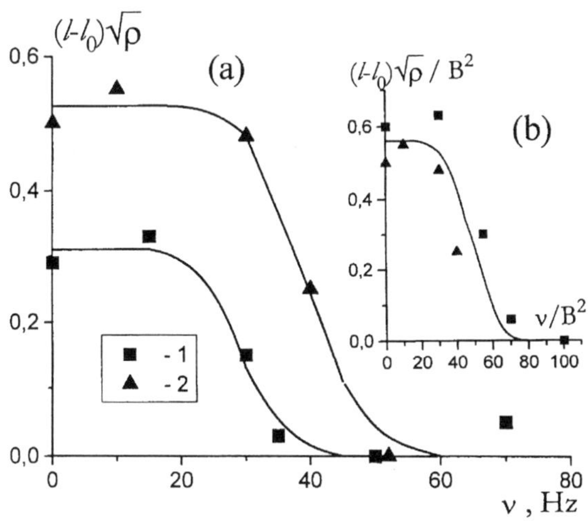

Fig.4 (a) - Dependence of the mean dislocation path l in Al crystals on the frequency ν of specimen rotation in a magnetic field, obtained for different values of magnetic induction B: 1- 0.7T, 2 -1T; (b) - dependences (a) replotted in normalized coordinates.

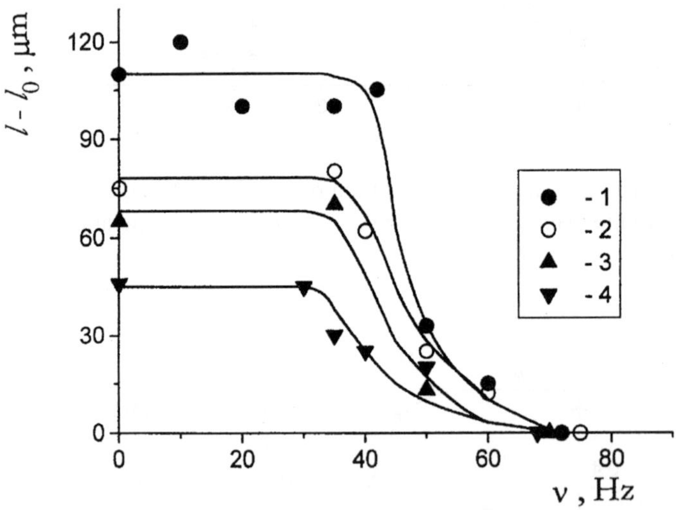

Fig. 5 Dependence of the mean dislocation path l on the frequency ν of specimen rotation in the magnetic field $B = 0.5$T for NaCl crystals with various impurity concentration C for different temperatures T: 1, 3, 4 - T=293K, 2 - T=77K, 3 - C=1ppm, 4 - C=10ppm.

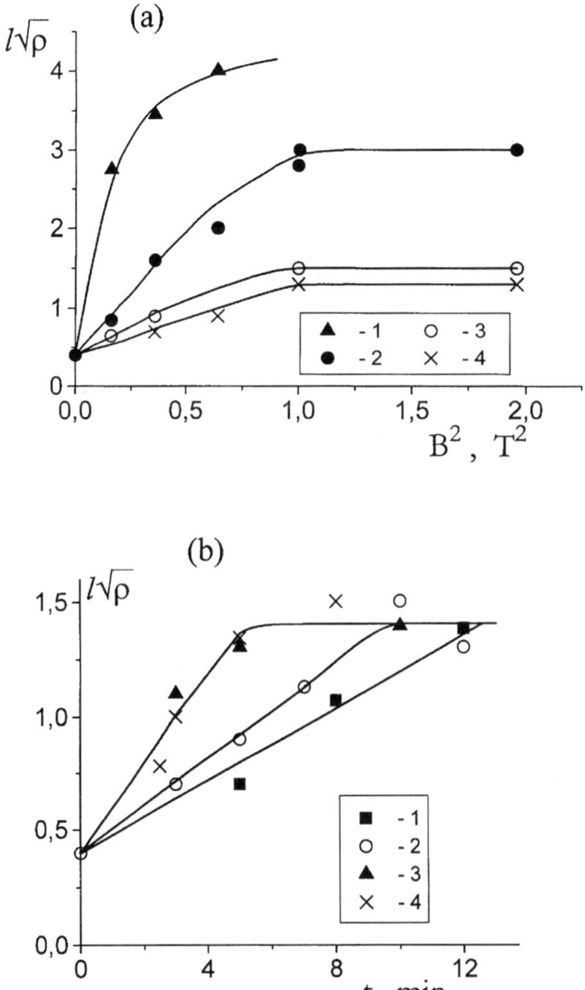

Fig.6 Dependence of the mean dislocation path l in LiF crystals: (a) - on the magnetic induction B for different electric fields: 1 - 9kV/m, 2 - 6kV/m, 3 - 3kV/m, 4 – $E=0$, $t = 5\,\text{min}$; (b) - on the time t of exposure to the magnetic field B for different B: 1- 0.6T, 2 - 0.8T,3-1T,4-1.4T.

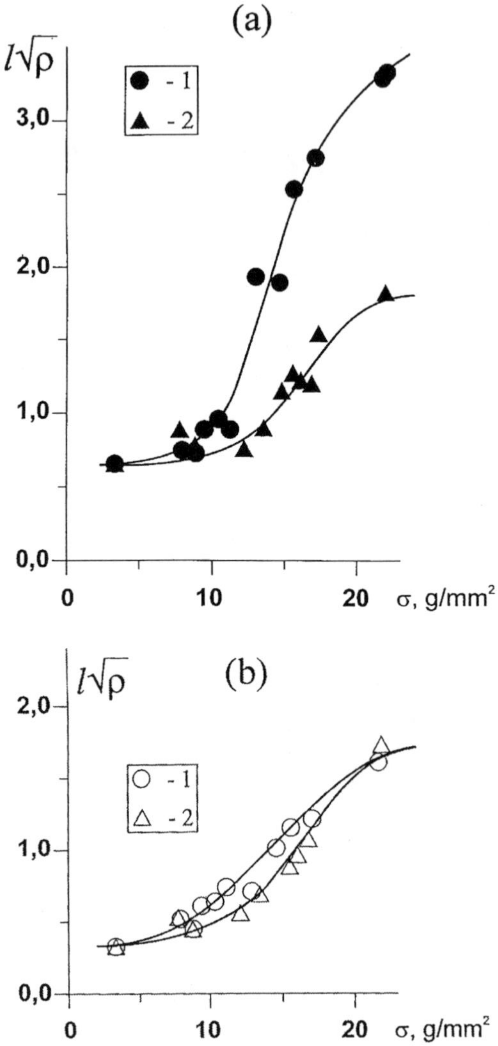

Fig.7 Dependence of the mean dislocation path in NaCl crystals on the external mechanical stress for trapeziform (1) and triangular (2) impulses; (a) - combined action with magnetic field $B = 0.3\text{T}$, (b) $- B = 0$.

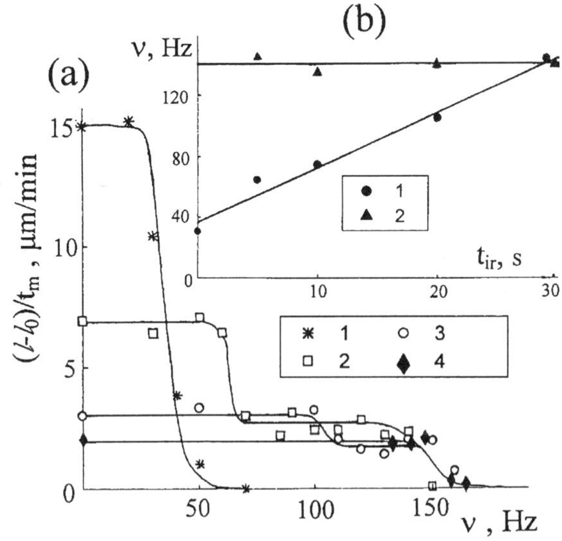

Fig.8 (a) - Dependence of the effective dislocation velocity in NaCl crystals on the frequency ν of specimen rotation in a magnetic field $B = 0.5$T for dif«erent irradiation times: 1-Os, 2 - 5s, 3 - 20s, 4 - 30s; (b) - critical frequencies v_{c1} (1) and v_{c2} (2) versus the irradiation time t_{ir}.

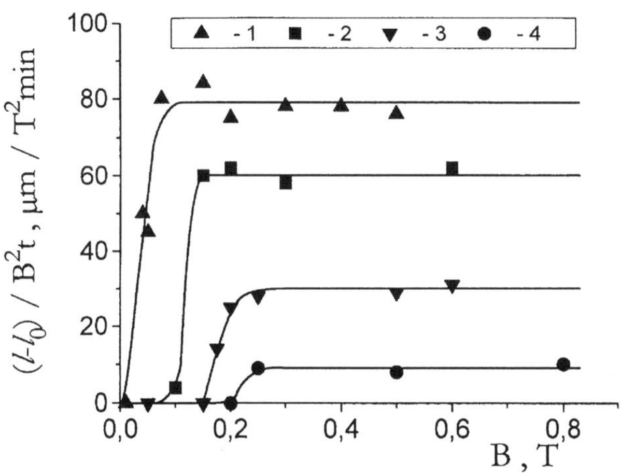

Fig.9 Dependence of the normalized mean dislocation path on the magnetic induction B for different Ca impurity concentration in NaCl crystals: 1- 0.5ppm, 2 - 1ppm, 3 - 10ppm, 4 - 100ppm.

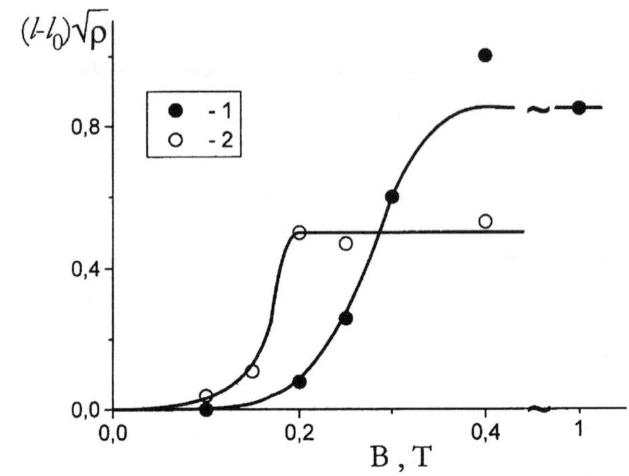

Fig.10 Dependence of the normalized mean dislocation path in LiF crystals on the magnetic induction B for two temperatures: 1- 293K, 2 - 77K.

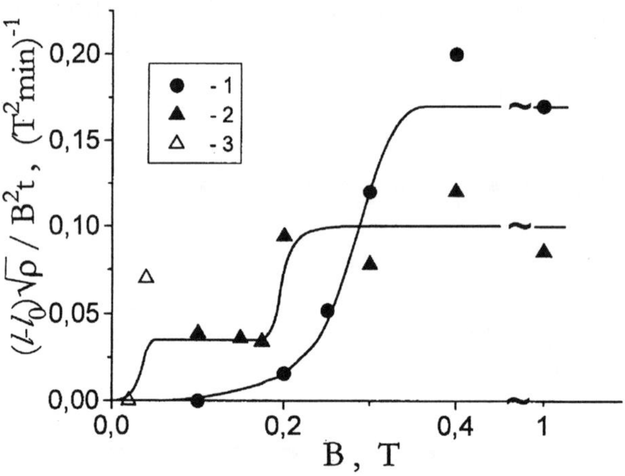

Fig.11 Dependence of the normalized mean dislocation path in LiF crystals on the magnetic induction B for non-irradiated crystals (1) and irradiated for 5s crystals (2, 3) without the application of an electric field (1, 2) and with the electric field $E = 10\text{kV/m}$ (3).

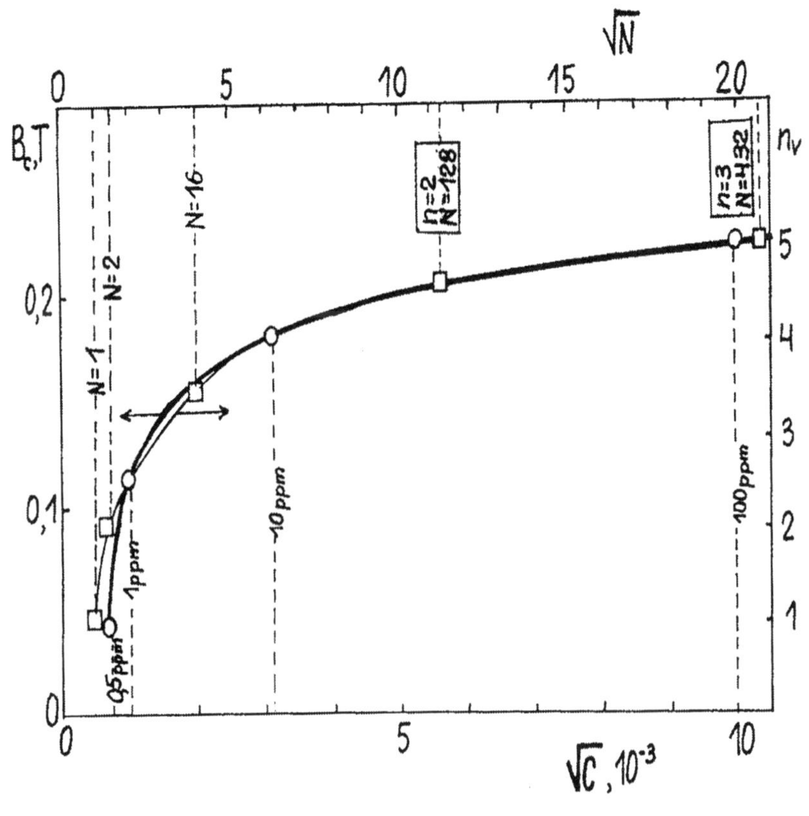

Fig.12 Threshold magnetic field B_c versus Ca impurity concentration C (experimental data), in coordinates $(B_c, \sqrt{C})$ and average number n_v of vacancies in the nearest neighborhood of a Ca impurity versus the number N of Ca atoms in the complex (theoretical curve), in coordinates $(n_v, \sqrt{N})$

THE METHODS OF RESONANT EXCITATION OF SURFACE WAVES IN CRYSTALS

V. I. AISHITS, A. S. GORKUNOVA, V. N. LYUBIMOV
Institute of Crystalography Academy of Sciences of Russia, Leninski pr. 59, 117333 Moscow, Russia
E-mail: alshits@mechan.incr.msk.su

W. GIERULSKI, A. RADOWICZ
Kielce University of Technology, Al. 1000-lecia Państwa Polskiego 7, 25-314 Kielce Poland
E-mail: gierulsk@eden.tu.kielce.pl, radowicz@eden.tu.kielce.pl

R. KOTOWSKI
Institute of Fundamental Technological Research Polish Academy of Sciences, ul. Świętokrzyska 21, 00-049 Warszawa Poland
E-mail: rkotow@ippt.gov.pl

A general method of resonant excitation of acoustic and electromagnetic surface waves is theoretically studied for a bicristalline system – a layer on a substrate. The idea of this method is related to the fact that for sufficiently short-wave-length comparing with the layer thickness, the surface wave on a free boundary of the bicrystal has a structure of a leaky wave which is characterized by a small energy leakage from the surface. This fact is due to the bulk wave propagation into the substrate with an exponentially small amplitude. Therefore at the reflection in the substrate, when the bulk partial wave of the leaky eigenmode, a resonance must arise with exponentially high amplitude. The physical parameters of this kind of resonance are analyzed both for acoustics and for optics.

1. Two Types of Surface Waves in Solids

Dynamic steady-state processes in acoustics and optics are usually described in terms of stationary plane waves or built on their basis wave packets, beams etc. In physical media with surfaces or interfaces such waves are customarily devided on bulk and surface elastic, acoustoelectric or electromagnetic waves. *A* surface wave is mathematically understood as an eigenmode satisfying definite boundary conditions, carrying energy flux parallel to the surface and localized close to that due to an exponential decay of the wave amplitude A to the depth of the semi-infinite medium. The corresponding wave field can be formed by a superposition of several partial waves of the type

$$A_\alpha(y) = A_{s\alpha}\exp(-\gamma_\alpha ky), \tag{1}$$

where k is the length of the wave vector projection on the surface, γ_α is a dimensionless constant of order of unity. These are typical features of the Rayleigh elastic waves or the Bleustein-Gulyaev acoustoelectric waves in piezoelectrics with metallized surfaces (Fig. 1). The other type of surface waves is represented by the Bleustein-Gulyaev waves at nonmetallized surfaces and by the surface electromagnetic waves (surface polaritons) which include also additional localized components in the adjoining vacuum

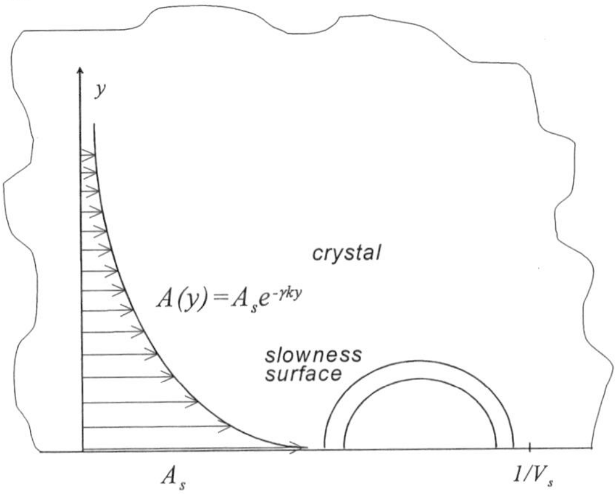

Fig.1. Scheme of the surface wave propagation in the half infinite medium

Both types of surface waves are usually more slow than corresponding bulk waves in a sense that the reverse phase speed of a surface wave $1/v_s$ proves to be as a rule outside the slowness (or refraction) surface for the bulk waves (Fig. 1).

2. Leaky Wave Eigenmodes in Layered Structures

Let us consider now the above surface wave solutions as a zero approximation of the wave field in the thick enough layer on the substrate (Fig.2). The exponentially small field on the interface, $A(h) \sim A_s \exp(-\gamma kh), (\gamma = \min \gamma_\alpha)$, must create in the substrate the wave field of the same order of magnitude. If to choose the substrate material so that the surface wave phase speed v_s in the layer belongs to the region of bulk waves in the substrate (Fig.2) then the accompanying wave in the substrate will be a bulk mode of a small amplitude $A_i \sim A_s \exp(-\gamma kh)$ propagating from the surface to the depth of the substrate. For the open boundary conditions at the interface crystal - vacuum which are specific for surface polaritons and acoustoelectric surface waves in piezoelectrics with nonmetallized surface the analogous consideration leads to completely the same physical picture. In the both cases a small leakage of energy occurs from the surface wave to the substrate, which must cause a corresponding decrease of the amplitude A_s along the surface.

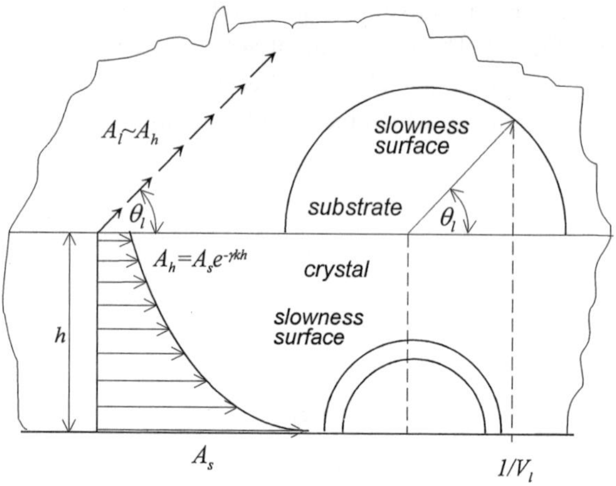

Fig.2. Leaky wave in the structure: layer on a substrate

In its turn this means that the phase speed of such wave construction, known as a leaky wave, must have a small negative imaginary part

$$v_s \to v_l - iv_l', \; v_l' > 0, \tag{2}$$

where obviously $v_l' \propto |A_l|^2 \propto \exp(-2\gamma kh)$. One can show that the same estimate is valid for $\Delta v = |v_l - v_s|$.

3. Resonant Excitation of Surface Waves

A resonant reflection of bulk acoustic waves in the vicinity of the leaky eigenmode was first described in [1] for the situation when the leaky wave branch arises from the supersonic Rayleigh wave in anisotropic semi-infinite elastic medium at a small perturbation of the direction of its propagation. In series of further publications [2 - 6] analogous resonances were studied for other different acoustic and acoustoeletric systems. In this paper we consider a new variant of the resonance which is specific in two aspects:

1). it is universal for any types of surface waves in acoustics and optics;

2). it provides an exponential dependence $v_l' \propto \exp(-2\gamma h)$ which allows to control parameters of the resonance in very wide region and to obtain (at least in principle) exponentially high coefficients of the surface wave excitation.

The idea of the resonance is illustrated by Fig.3: the resonant conditions arise if to choose the incidence angle of the bulk wave in the substrate so that the reflected wave would propagate along the direction $\theta = \theta_l$ of the bulk partial wave of the leaky eigenmode. As a result, in the layer there must be excited the surface wave of exponentially large amplitude. Taking into account that the amplitudes (but not the phases) of the incident wave (A_i) and the reflected wave (A_r) must be equal, the incident wave must completely compensate the leakage of the energy from the surface wave in the layer. Accordingly, the amplitude A_s of the latter will not decrease along the surface and the phase speed v of the whole wave superposition must be real.

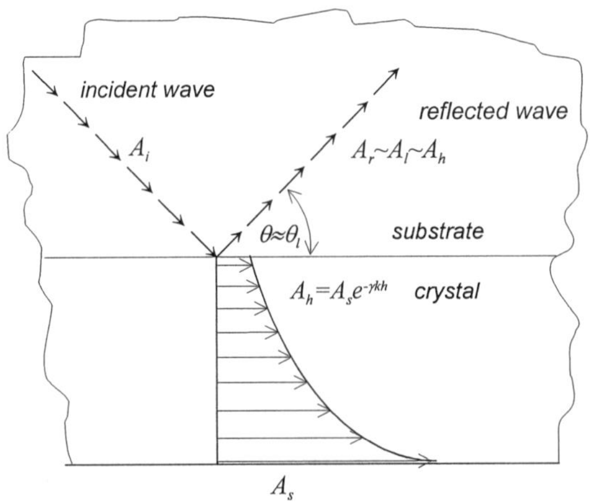

Fig.3. Geometry of the resonant reflection

From this consideration the following estimate for the coefficient of resonant excitation of the surface wave is expected

$$S_{exc}^{res} = A_s / A_i = A_s / A_r \sim A_s / A_l \sim \exp(\gamma kh) \tag{3}$$

Explicit calculations show that in all cases the form of the resonance is described by the universal function

$$S_{exc}(v) = \frac{A_s}{A_i} = C\frac{\sqrt{v_l v_l'}}{v - v_l + iv_l'}, \tag{4}$$

where C is a dimensionless constant of order of unity. Taking into account that $v_l' \sim v_l \exp(-2\gamma kh)$, the resonance proves to be very narrow and very sharp (Fig.4).

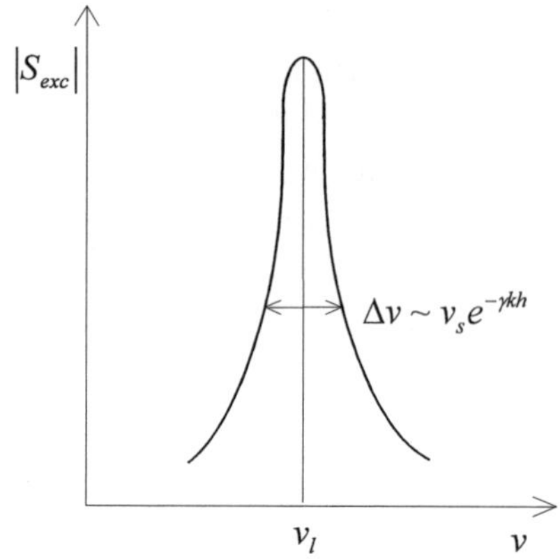

Fig.4. Schematic form of the resonant curve $|S_{exc}(\nu)|$ of the surface wave excitation

One can see that at the resonant incidence angle corresponding to $\nu = \nu_l$ the maximum value of the excitation coefficient is indeed determined by the estimate (3).

4. Physical Limitations for the Parameters of Resonance

Thus, we have come to the mechanism of excitation of a strong surface wave in the crystal plate by means of a special reflection of a weak bulk pumping wave in the adjoined substrate. Certainly, this result does not contain any contradiction to the energy conservation law. Indeed, the above consideration and eqn. (4) are based on the stationary description which excludes any questions about the energy sources of the surface wave. On the other hand, it is obvious that at a nonstationary statement of the problem, when the time origin corresponds to the switching on the pumping incident wave in the substrate, the time of coming the system to a stationary state would be the bigger the more is the value $S_{exc} = A_s / A_i$ in order to provide the necessary energy density in the wave field propagating in the crystal plate.

Nevertheless, there are definite physical limitations which prevent unlimited exponential growth of the coefficient of excitation of the surface wave with increasing parameter kh. The first limiting factor is related to the natural divergence of the wave beams which are used in practice instead of plane waves [2,3]. In view of this divergence, the less is ν_l' the narrower is the resonant zone over incidence angles and therefore the less part of the beam belongs to the resonant zone.

The second physical limitation for a narrowing resonance and a corresponding increase of its amplitude at $kh \to \infty$ is associated with a natural dissipative damping of the waves at their propagation [6]. Phenomenologically this leads to the replacement $\nu_l \to \nu_l - i\nu'_d$, where ν'_d is determined by the damping decrement of the wave. As a result, the denominator in eqn. (4) is transformed to $\nu - \nu_l + i(\nu'_l + \nu'_d)$. This leads to the dependence

$$\left|S_{exc}^{res}\right| = C\frac{\sqrt{\nu_l \nu'_l}}{\nu'_l + \nu'_d} \qquad (5)$$

which goes to zero both at $\nu'_l = 0$ and at $\nu'_l \to \infty$ and has a maximum at $\nu'_l = \nu'_d$ (Fig.5)

$$\max\left|S_{exc}^{res}\right| = \frac{1}{2}C\sqrt{\frac{\nu_l}{\nu'_d}} >> 1. \qquad (6)$$

Thus, the considered mechanism of resonant excitation of surface waves must be the more effective the less are divergence and damping of the incident wave beams.

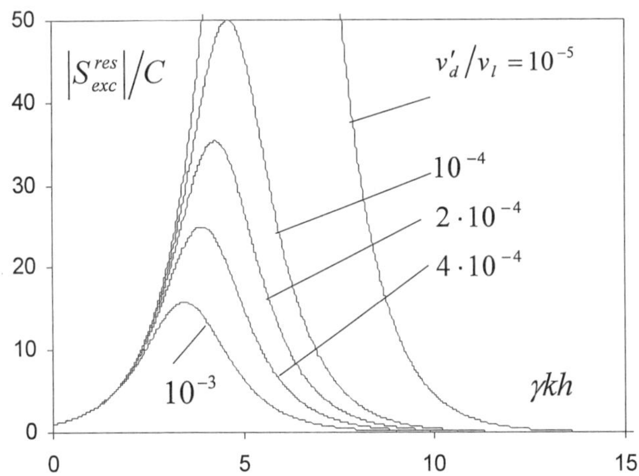

Fig.5. The dependence $\left|S_{exc}^{res}(\gamma kh)\right|/C$ for different magnitudes of the parameter ν'_d/ν_l

In acoustics the damping is usually rather weak and the main physical limitation for practical realization of the above mechanism reduces to a divergence of acoustic beams. In optics with an appropriate choice of the materials and the wave length region the damping also can be done very small.

On the other hand, in this case due to lasers the beam divergence can be made extremely small as well. That is why we would expect first practical successes on this way at resonant excitation of surface polaritons.

Acknowledgements

This work, supported by the Russian Foundation for Basic Research (grants 98-02-16077, and 98-02-16096) was completed during a visit of V.I.A. to the Kielce University of Technology.

References

1. V.I. Alshits, J.Lothe, Wave Motion 3, 297 (1981)
2. V.I. Alshits, et al., Sov.Phys.Crystallography 33, 318 (1988)
3. V.I. Alshits, et al., Kristallografija 35, 812 (1990)
4. V.I. Alshits, et al., Kristallografija 36, 284 (1991)
5. V.I. Alshits, et al., Phys.Solid.State 38, 604 (1996)
6. V.I. Alshits, et al., Proceedings of Second Workshop on Dissipation in Physical Systems, ed. A.Radowicz (Mechanika 66, Kielce, 1998)

DYNAMICAL STABILITY OF THIN SHELLS WITH THERMAL CONVECTION

JAN AWREJCEWICZ

Technical University of Łódź, Division of Automatics and Biomechanics, 1/15 Stefanowskiego St., 90-924 Łódź, POLAND

VADIM A. KRYSKO

Saratov State University, Department of Mathematics, Saratov State Technical University, Saratov, 410054 RUSSIA

The nondimensional partial differential equations governing dynamics of the geometrically nonlinear shells with thermal convection represented by the heat transfer equation and with different boundary conditions are analysed. The used finite difference method yields the algebraic and ordinary differential equations. Examples and discussion of the vibrational processes and stability loss during heating or cooling and for different boundary conditions are given.

1 Introduction

It is obvious that an investigation of geometrically nonlinear shells when a surrounding medium temperature changes sufficiently has an important meaning for nuclear power plants and aircraft engineering. The mechanical properties (including stability phenomena) can be changed due to the thermal action and therefore many of researchers have been attracted by the mentioned problems of heat transfer and thermal stresses. The moderately thick thermally induced plates have been considered in [1] within the frame of the Mindlin theory. The obtained class of solutions is restricted to simply supported plates. The multilayered composite plates and shells with decoupled steady-state heat transfer and a geometrically nonlinear thermoelastic problem have been considered in [2]. The applied numerical approaches solved the defined thermophysical problem. A two-dimensional thermo-visco-elastic problem of an infinitely thin plate containing a circular hole has been defined and solved in [3]. The free vibrations of a truncated conical thin shell subjected to thermal gradients have been analyzed in [4]. Simply supported and clamped boundary conditions have been considered and a temperature loading due to a supersonic flow has been taken into account. The obtained differential equations have been solved using the collocation method.

In this work we investigate a thin shell with thermal convection and with different boundary conditions in the frame of heating and cooling processes. The finite difference method has been applied and then the Gauss technique as well as the Runge-Kutta method have been used to solve the obtained equations.

2 Governing equations

We consider a flexural conical shell, rectangular in the plane. Its average surface is related to the orthogonal system of coordinates x, y. The coordinate normal to the average surface is denoted by z, which is defined as positive with respect to the curvature centre direction.

The following hypotheses are used:

1. The material is isotropic and fulfils the Hook's principle.
2. The Kirchhoff-Love hypothesis is valid.
3. A relation between the deformations and displacements is of the second power (it is widely used in the geometrically nonlinear theory of shells).
4. Thermal and mechanical material characteristics depend on temperature in general, but they are constant in the calculation examples given further.

The heat flow density is defined as the heat amount per time unit and related to the surface unit [J/m^{2}s].

The shells deformations have the form of:

- flexural deformations

$$e_{xx} = \varepsilon_{xx} + z\kappa_{xx}, \quad e_{xy} = \varepsilon_{xy} + z\kappa_{xy}, \quad (x \leftrightarrow y) \tag{1}$$

$$\text{where}: \kappa_{xx} = -\frac{\partial^2 w}{\partial x^2}, \quad (x \leftrightarrow y), \quad \kappa_{xy} = -2\frac{\partial^2 w}{\partial x \partial y}, \quad (x \leftrightarrow y) \tag{2}$$

- deformations of the average surface

$$\begin{aligned} \varepsilon_{xx} &= \frac{\partial u}{\partial x} + \frac{1}{2}\left(\frac{\partial w}{\partial x}\right)^2 - K_x w, \\ \varepsilon_{xy} &= \frac{\partial u}{\partial y} + \frac{\partial \upsilon}{\partial x} + \frac{\partial w}{\partial x}\frac{\partial w}{\partial y}, \qquad (x \leftrightarrow y, \upsilon \leftrightarrow u). \end{aligned} \tag{3}$$

The thermoelastic stresses, according to the Hook's principle and the Duhamel-Neuman hypothesis concerning disjoining of the elastic and thermal stresses, have the form:

$$\sigma_{xx} = \frac{E}{1-\nu^2}\left[\left(e_{xx} + \nu e_{yy}\right) - \left(e^t_{xx} + \nu e^t_{yy}\right)\right], \quad \sigma_{xy} = \frac{E e_{xy}}{\left(2(1+\nu)\right)}, \quad (x \leftrightarrow y), \tag{4}$$

where E and ν are the Young's and Poisson's coefficients, respectively.

The deformations caused only by temperature are as follows:

$$e^t_{xx} = \alpha_t \theta(x, y, z, t), \quad (x \leftrightarrow y), \tag{5}$$

and e^t_{xx}, e^t_{yy} are the deformations occurring in the shell due to thermal loads.

Let us consider the shell element being in equilibrium with the bending action (M_1, M_2) and the rotational (M_{12}) moments, the longtidunal (T_1, T_2), transversal (N_1, N_2) and tangent (S) forces.

The forces applied to the shell, taking into account the thermal and mechanical characteristics (α_t and E) versus temperature, are given by the following formulae:

$$T_1 = \int_{-\frac{h}{2}}^{\frac{h}{2}} \sigma_{xx} dz = \frac{F_t}{1-\nu^2}\left(\varepsilon_{xx} + \nu\varepsilon_{yy}\right) + \frac{S_t}{1-\nu^2}\left(\kappa_{xx} + \nu\kappa_{yy}\right) - N_t, \tag{6}$$

$$S = \int_{-\frac{h}{2}}^{\frac{h}{2}} \sigma_{xy} dz = \frac{1}{2(1+\nu)}[F_t\varepsilon_{xy} - S_t\kappa_{xy}],\ M_{12} = \int_{-\frac{h}{2}}^{\frac{h}{2}} \sigma_{xy} z dz = \frac{S_t}{2(1+\nu)}\varepsilon_{xy} + D_t(1-\nu)\kappa_{xy},$$

$$M_1 = \int_{-\frac{h}{2}}^{\frac{h}{2}} \sigma_{xx} z dz = D_t\left(\kappa_{xx} + \nu\kappa_{yy}\right) + \frac{S_t}{1-\nu^2}\left(\varepsilon_{xx} + \nu\varepsilon_{yy}\right) - M_t,\ (1 \leftrightarrow 2; x \leftrightarrow y),$$

where: $D_t = \int_{-\frac{h}{2}}^{\frac{h}{2}} \frac{Ez^2}{1-\nu^2} dz,\quad M_t = \int_{-\frac{h}{2}}^{\frac{h}{2}} \frac{\alpha_t \theta E_z}{1-\nu} dz,\quad S_t = \int_{-\frac{h}{2}}^{\frac{h}{2}} Ezdz,$

$$N_t = \int_{-\frac{h}{2}}^{\frac{h}{2}} \frac{\alpha_t \theta E}{1-\nu} dz,\quad F_t = \int_{-\frac{h}{2}}^{\frac{h}{2}} Edz,\quad \Theta = T - T_0,$$

and T is the temperature.

A consideration of the shell element equilibrium leads to the following equations:

$$\frac{\partial T_1}{\partial x} + \frac{\partial S}{\partial y} - K\left(\frac{\partial M_1}{\partial x} + \frac{\partial M_{12}}{\partial y}\right) = 0, \qquad (1 \leftrightarrow 2),\ (x \leftrightarrow y),$$

$$\frac{\partial^2 M_1}{\partial x^2} + 2\frac{\partial^2 M_{12}}{\partial x \partial y} + \frac{\partial^2 M_2}{\partial y^2} + K_x T_1 + K_y T_2 + \frac{\partial T_1}{\partial x}\frac{\partial w}{\partial x} + T_1\frac{\partial^2 w}{\partial x^2} + \frac{\partial S}{\partial x}\frac{\partial w}{\partial y} +$$

$$+ 2S\frac{\partial^2 w}{\partial x \partial y} + \frac{\partial S}{\partial y}\frac{\partial w}{\partial x} + \frac{\partial T_2}{\partial y}\frac{\partial w}{\partial y} + T_2\frac{\partial^2 w}{\partial y^2} - \frac{\gamma h}{g}\left(\frac{\partial^2 w}{\partial t^2} + \xi\frac{\partial w}{\partial t}\right) + q = 0, \tag{7}$$

where: q - transversal load, γ - weight density, h - shell thickness, ξ - damping.

Substituting the expressions for forces and moments (6) to equations (7), we obtain the following differential thermoelastic equations for displacements:

$$\frac{\partial}{\partial x}\left[A_{1x}A_{7x}-A_{2x}A_{5x}\right]+\frac{\partial}{\partial y}\left[\frac{1}{2}A_{1x}A_{8}-(1-\nu)A_{2x}\frac{\partial^{2}w}{\partial x\partial y}\right]=\frac{\partial A_{3x}}{\partial x},$$

$$\frac{\partial}{\partial x}\left[\frac{1}{2}A_{1y}A_{8}-(1-\nu)A_{2y}\frac{\partial^{2}w}{\partial x\partial y}\right]+\frac{\partial}{\partial y}\left[A_{1y}A_{7y}-A_{2y}A_{5y}\right]=\frac{\partial A_{3y}}{\partial y}, \tag{8}$$

$$-D_t\nabla^4 w-2\frac{\partial D_t}{\partial x}A_{4x}-2\frac{\partial D_t}{\partial y}A_{4y}-2\frac{\partial^2 D_t}{\partial x^2}A_{5x}-\frac{\partial^2 D_t}{\partial y^2}A_{5y}-2(1-\nu)\frac{\partial^2 D_t}{\partial x\partial y}\frac{\partial^2 w}{\partial x\partial y}-$$

$$-\frac{S_t}{1-\nu^2}\left[(1-\nu)L(w,w)+\left(\nabla^2 w\right)^2+K_xA_{5x}+K_yA_{5y}-\frac{\partial^3 u}{\partial x}-\nu\frac{\partial^3 \upsilon}{\partial x^2\partial y}-\nu\frac{\partial^3 u}{\partial x\partial y^2}-\frac{\partial^3 \upsilon}{\partial y^3}+\right.$$

$$\left.+\frac{\partial w}{\partial x}A_{4x}+\frac{\partial w}{\partial y}A_{4y}\right]+\frac{1}{1-\nu^2}\left[\frac{\partial S_t}{\partial x}A_{6x}+\frac{\partial S_t}{\partial x^2}A_{6y}\right]+\frac{\partial^2 S_t}{\partial x^2}A_{7x}+\frac{\partial^2 S_t}{\partial y^2}A_{7y}+\frac{\partial^2 S_t}{\partial x\partial y}\frac{A_8}{1-\nu^2}+$$

$$+\frac{F_t}{1-\nu^2}\left[K_xA_{7x}+K_yA_{7y}+\frac{\partial w}{\partial x}\left(\frac{\partial^2 u}{\partial x}+\frac{1}{2}(1+\nu)\frac{\partial^2\upsilon}{\partial y\partial x}+\frac{1}{2}(1-\nu)\frac{\partial^2 u}{\partial y^2}\right)+\right.$$

$$\left.+\frac{\partial w}{\partial y}\left(\frac{\partial^2\upsilon}{\partial y^2}+\frac{1+\nu}{2}\frac{\partial^2 u}{\partial y\partial x}+\frac{1-\nu}{2}\frac{\partial^2\upsilon}{\partial x^2}\right)+\frac{\partial^2 w}{\partial x^2}A_{7x}+\frac{\partial^2 w}{\partial y^2}A_{7y}+\frac{\partial^2 w}{\partial y\partial x}A_8\right]+$$

$$+\frac{1}{1-\nu^2}\left[\frac{\partial F_t}{\partial x}\left(\frac{\partial w}{\partial x}A_{7x}+\frac{1}{2}\frac{\partial w}{\partial y}A_8\right)+\frac{\partial F_t}{\partial y}\left(\frac{\partial w}{\partial y}A_{7y}+\frac{1}{2}\frac{\partial w}{\partial x}A_8\right)\right]-$$

$$-\left(\nabla^2 M_t+N_t\left(K_x+K_y\right)+\nabla^2 wN_t+\frac{\partial N_t}{\partial x}\frac{\partial w}{\partial x}+\frac{\partial N_t}{\partial y}\frac{\partial w}{\partial y}-q\right)=$$

$$=\frac{\gamma h}{g}\left(\frac{\partial^2 w}{\partial t^2}+\xi\frac{\partial w}{\partial t}\right),$$

where: $A_{1x}=\left(F_t-K_xS_t\right)$, $A_{2x}=\left(S_t-\left(1-\nu^2\right)K_xD_t\right)$, $A_{3x}=\left(N_t-K_xM_t\right)\left(1-\nu^2\right)$,

$$A_{4x}=\left(\frac{\partial^3 w}{\partial x^3}+\frac{\partial^3 w}{\partial x\partial y^2}\right),\ A_{5x}=\left(\frac{\partial^2 w}{\partial x^2}+\nu\frac{\partial^2 w}{\partial y^2}\right),\ A_{7x}=\left(\frac{\partial u}{\partial x}+\nu\frac{\partial\upsilon}{\partial y}\right),$$

$$A_{6x}=\left(2\frac{\partial^2 u}{\partial x^2}+(1+\nu)\frac{\partial^2\upsilon}{\partial x\partial y}+(1-\nu)\frac{\partial^2 u}{\partial y^2}-\frac{\partial w}{\partial y}\frac{\partial^2 w}{\partial x\partial y}+\frac{\partial w}{\partial x}A_{5x}\right),$$

$$A_8=\left(\frac{\partial u}{\partial y}+\frac{\partial\upsilon}{\partial x}+\frac{\partial w}{\partial x}\frac{\partial w}{\partial y}\right)(1-\nu),\quad (x\leftrightarrow y),\quad (\upsilon\leftrightarrow u),$$

$$L(\alpha,\beta)=\frac{\partial^2\alpha}{\partial x^2}\frac{\partial^2\beta}{\partial y^2}+\frac{\partial^2\alpha}{\partial y^2}\frac{\partial^2\beta}{\partial x^2}-2\frac{\partial^2\alpha}{\partial x\partial y}\frac{\partial^2\beta}{\partial x\partial y},$$

$$\nabla_k^2=K_x\frac{\partial^2}{\partial y^2}+K_y\frac{\partial^2}{\partial x^2}.$$

For a uniformly distributed temperature field, equations (8) are simplified and take the form:

$$A_{1x}\left(\frac{\partial^2 u}{\partial x^2}+\frac{1-\nu}{2}\frac{\partial^2 u}{\partial y^2}+\frac{1+\nu}{2}\frac{\partial^2 \upsilon}{\partial x\partial y}\right)-A_{2x}A_{4x}=0,$$

$$A_{1y}\left(\frac{1-\nu}{2}\frac{\partial^2 u}{\partial x^2}+\frac{1+\nu}{2}\frac{\partial^2 u}{\partial x\partial y}+\frac{\partial^2 \upsilon}{\partial y^2}\right)-A_{2y}A_{4y}=0,$$

$$D_t\nabla^4 w-\frac{S_t}{1-\nu^2}\left[(1-\nu)L(w,w)+\left(\nabla^2 w\right)^2+K_xA_{5x}K_yA_{5y}-\frac{\partial^3 u}{\partial x^3}-\nu\frac{\partial^3\upsilon}{\partial x^2\partial y}-\right.$$

$$\left.-\nu\frac{\partial^3 u}{\partial x\partial y^2}-\frac{\partial^3\upsilon}{\partial y^3}+\frac{\partial w}{\partial x}A_{4x}+\frac{\partial w}{\partial y}A_{4y}\right]+\frac{F_t}{1-\nu^2}\left[K_xA_{7x}+K_yA_{7y}+\right.$$

$$+\frac{\partial w}{\partial x}\left(\frac{\partial^2 u}{\partial x^2}+\frac{1}{2}(1+\nu)\frac{\partial^2 v}{\partial x\partial y}+\frac{1}{2}(1-\nu)\frac{\partial^2 u}{\partial y^2}\right)+\frac{\partial w}{\partial y}\left(\frac{\partial^2\upsilon}{\partial y^2}+\frac{1+\nu}{2}\frac{\partial^2 u}{\partial x\partial y}+\frac{1-\nu}{2}\frac{\partial^2\upsilon}{\partial x^2}\right)+$$

$$\left.+\frac{\partial^2 w}{\partial x^2}A_{7x}+\frac{\partial^2 w}{\partial y^2}A_{7y}+\frac{\partial^3 w}{\partial x\partial y}A_8\right]-\left[N_t\left(K_x+K_y\right)+\nabla^2 wN_t-q\right]=$$

$$=\frac{\gamma h}{g}\left(\frac{\partial^2 w}{\partial t^2}+\xi\frac{\partial w}{\partial t}\right). \tag{9}$$

The given systems of differential equations with variable coefficients for a nonuniformly (8) and uniformly (9) distributed temperature field can be used for the computation of flexural conical shells with boundary conditions for the displacements u,υ,w.

In addition, we obtain a system of differential equations in the hybrid form in relation to the deflection and stress function F satisfying the following relations:

$$T_1=\frac{\partial^2 F}{\partial y^2}-P_x(y,t),\ T_2=\frac{\partial^2 F}{\partial x^2}-P_y(x,t),\ S=-\frac{\partial^2 F}{\partial x\partial y}. \tag{10}$$

2. Formulation of the problem

We consider dynamical stability of a shell with thermal convection, when the Young's modulus E of the shell material as well as the linear thermal expansion coefficient α do not depend on temperature. Proceeding in a similar way and taking into account (10), the following dimensionless equations are obtained:

$$\frac{1}{12(1-\nu^2)}\left(\lambda^{-2}\frac{\partial^4 w}{\partial x^4}+2\frac{\partial^4 w}{\partial x^2\partial y^2}+\lambda^2\frac{\partial^4 w}{\partial y^4}\right)- \\ -\nabla_K{}^2F-L(w,F)+\left(\frac{\partial^2 w}{\partial \tau^2}+\xi\frac{\partial w}{\partial \tau}\right)=q, \tag{11}$$

$$\lambda^{-2}\frac{\partial^4 F}{\partial x^4}+2\frac{\partial^4 F}{\partial x^2\partial y^2}+\lambda^2\frac{\partial^4 F}{\partial y^4}+\nabla_K{}^2\,w+\frac{1}{2}L(w,w)=0, \tag{12}$$

where: $\lambda = a/b$ (a, b are the length and width of a shell, respectively), dimensionless time $\tau=(\alpha t)/h^2$, $\kappa=a^2b^2\gamma\alpha/(gEh^6)$ is the parameter characterising the physical and geometrical shell properties, $\overline{w}=w/h$, $\overline{x}=x/a$, $\overline{y}=y/b$, $\overline{K}_x=(K_x a^2)/h$, $\overline{K}_y=(K_y b^2)/h$, $\overline{\xi}=(\xi h^2)/\alpha$, $\overline{q}=(qa^2b^2)/Eh^4$ (bars are omitted in equations (11) and (12)).

The nonstationary convection process occurs due to the Newton's law on the shell internal surface (other shell surfaces are isolated).

A heat transfer equation has the form:

$$\frac{\partial T}{\partial \tau}=\frac{\partial^2 T}{\partial z^2} \tag{13}$$

with the boundary conditions

$$\begin{aligned} &\frac{\partial T}{\partial z}+Bi(T-T_s)=0 \quad &\text{for}\quad z=\frac{1}{2},\\ &\frac{\partial T}{\partial z}=0 \quad &\text{for}\quad z=-\frac{1}{2}, \end{aligned} \tag{14}$$

and the initial conditions

$$T=T_0 \quad \text{for}\quad \tau=0. \tag{15}$$

Above $Bi=\alpha h/\lambda_g$, where α is the heat expansion coefficient and λ_g is the thermal transfer coefficient on the surface $z=1/2$; T is the temperature, whereas T_S denotes the surrounding medium temperature.

Further we consider a spherical shell with $K_x = K_y = 24$, $\xi = 0$ and $\kappa = 0.393$. It corresponds, for instance, to a real shell with the following parameters: a =b=1m, h=0.005m, ν=0.3, E=2.1·10³MPa, $\alpha = 1.2 \cdot 10^{15}\,\mathrm{grad}^{-1}$, $\lambda_g = 45.4\,\mathrm{W/mK}$ and the specific thermal capacity c =460J/kgK.

The nondimensional solution to the formulated heat transfer problem has the following analytical solution [5]:

$$T = T_s - (T_s - T_0)\sum_{n=1} C_n \cos\left[\mu\left(z + \frac{1}{2}\right)\right]\exp\left(-\mu_n^2\tau\right), \tag{16}$$

where: $C_n = \dfrac{2\mathrm{Bi}}{[\mathrm{Bi}(1+\mathrm{Bi}) + \mu_n^2]\cos\mu_n}$ and μ_n are the roots of the equation $\mu \operatorname{tg}\mu = \mathrm{Bi}$.

The thermal moments and forces have the form

$$M_t = -\theta\sum_{n=1}\frac{C_n}{2\mu_n}\left[\mu_n \sin\mu_n + 2\cos\mu_n - 2\right]\exp\left(-\mu^2\tau\right), \tag{17}$$

$$N_t = \theta\left[1 - \sum_{n=1}\frac{C_n}{\mu_n}\sin\mu_n \exp\left(-\mu^2\tau\right)\right]. \tag{18}$$

The series occurring in the terms of the thermal moments (17) and forces (18) are convergent.

3. Finite-difference method

In order to solve the obtained equations for arbitrary boundary conditions, we use the finite-difference method with the approximation $0(h^2)$ along the space coordinates. The shell area has been covered by a rectangular mesh with a step $a_x = a/n$, $b_x = b/n$ along the *x* and *y* axes (*n* denotes the number of the surface parts).

The governing equations in relation to the stress and deflection functions have the form:

$$\Phi_{ij}^1(F) = \Phi_{ij}^2(w), \tag{19}$$

$$\left(\frac{d^2 w_{ij}}{dt^2} + \xi\frac{dw_{ij}}{dt}\right) = \Phi_{ij}^3(w, F), \tag{20}$$

where: Φ_{ij}^k $(k = 1,2,3)$ are the finite-difference operators for each (i, j) point (i, j = 1, 2...N); N^2 denotes the number of equations ([6]).

The following algorithm has been used:

1. The values of deflection in each i, j point from the previous step (in the first step the initial deflection values are taken) are substituted to the right-hand side of linear algebraic equations (19). They are solved in relation to F_{ij} using the method of the main element removal in relation to the Gauss rows.
2. The obtained values of the force functions F_{ij} and the deflection in each i, j point (w is taken from the previous step) are substituted to the right-hand side of system (20) and the integration is carried out using the Runge-Kutta method. As a result, we get the w_{ij} values on the step of calculations.
3. The next step in time is made, and the described procedure is repeated.

The numerical results have been compared with the analytical ones, showing a very good agreement. The calculations have been stopped when $\left|a_n / \Sigma a_{n-1}\right| \leq 10^{-8}$, where Σa_{n-1} denotes the sum of the n-1 series components.

4. Results and conclusions

The following boundary conditions are taken into account:

1. Free support

$$\begin{aligned} & w = M_1 = T_1 = S = 0 \quad \text{for } x = 0, a, \\ & w = M_2 = T_2 = S = 0 \quad \text{for } y = 0, b. \end{aligned} \tag{21}$$

2. Slip support

$$\begin{aligned} & w = 0, \ \frac{\partial w}{\partial x} = 0 \quad T_1 = \varepsilon_{yy} = 0 \quad \text{for } x = 0, a, \\ & w = 0, \ \frac{\partial w}{\partial y} = 0 \quad T_2 = \varepsilon_{xx} = 0 \quad \text{for } y = 0, b. \end{aligned} \tag{22}$$

3. Hybrid type

$$\begin{aligned} & \text{for } x = 0, a \ \text{ - free support - (21)} \\ & \text{for } y = 0, b \ \text{ - slip support - (22)} \end{aligned} \tag{23}$$

4. Free support on the flexural nonstretching (noncompressed) ribbs in their tangent surfaces

$$\begin{aligned} & w = M_1 = T_1 = \varepsilon_{yy} = 0 \quad \text{for } x = 0, a, \\ & w = M_2 = T_2 = \varepsilon_{xx} = 0 \quad \text{for } y = 0, b. \end{aligned} \tag{24}$$

We consider a vibrational process for the given boundary conditions for shells working either in cooling or heating conditions. The vibrational process (dependence $w(\tau)$) is presented in Fig. 1.

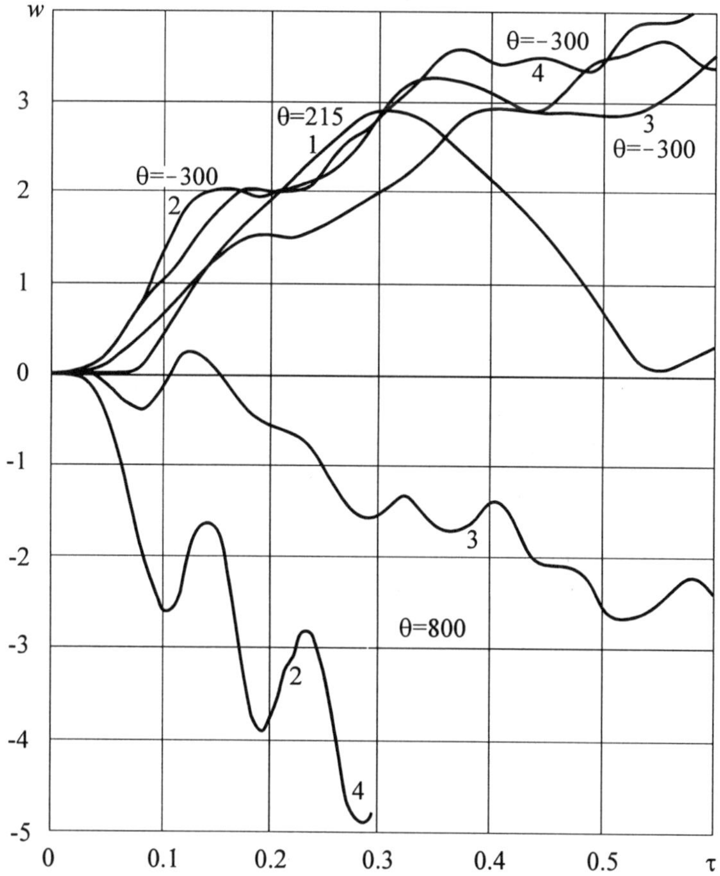

Figure 1. The shell centre vibrations (the digits correspond to the boundary condition number).

Only for the shells with number 1 boundary conditions the stationary vibrational process with positive values of deflections is observed during heating. The axis of vibrations is always parallel to the initial deflection state. With an increase in temperature the shell loses its stability. For other boundary conditions in the heating regime, the shell subjected to the thermal forces deflects in the negative direction, and a stability loss behaviour cannot occur. The vibrational process is observed around the axis sloped with respect to the initial deflection location and the deflections increase infinitely with time.

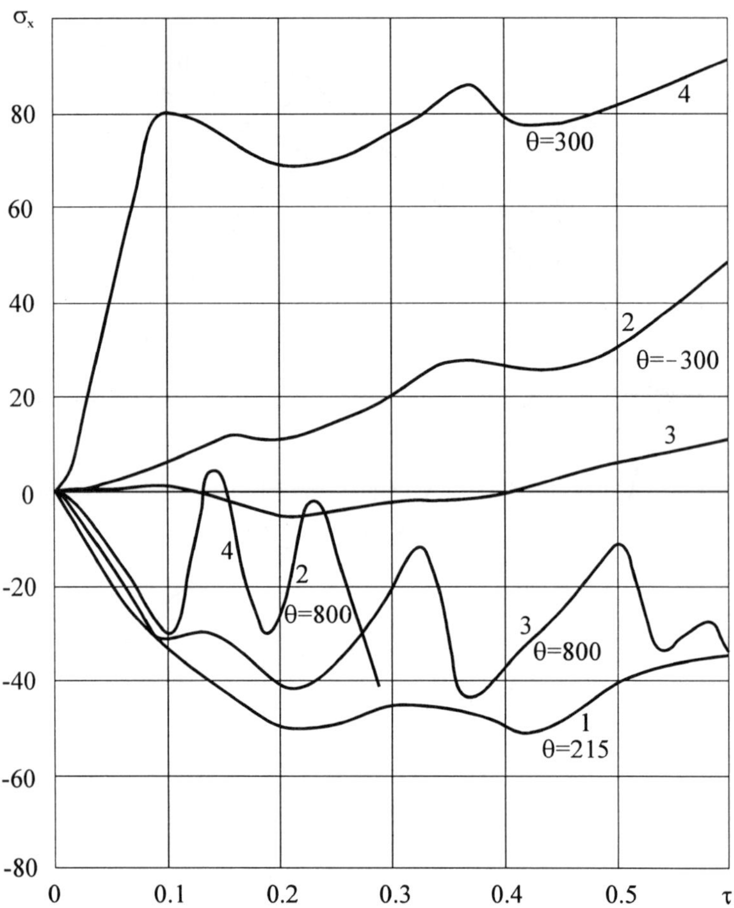

Figure 2. The shell centre internal stresses versus time.

Comparing the vibrational process for boundary conditions 2 and 3, it has been observed that the vibrational process of the shell deflection with boundary conditions 3 is characterized by complexity and aperiodicity. An angle of the slope of the vibration axis is sufficiently smaller than in the case of boundary condition 2 for the same temperature values. The action of thermal moments and thermal forces for heating (cooling) is opposite. In the case of heating the thermal moment is going to deflect the shell inside, along the curvature direction, contrary to the thermal force action. As similar consideration can be carried out for the cooling process.

We analyse the shell stress state for the cases given in Figure 1. In dependence $\sigma_x(\tau)$ presented in Fig. 2 a vibrational process of full stresses is governed by equations (4), where e_{xx}^t, e_{yy}^t are temperature deformation components in the shell centre on its internal surface. Comparing the vibrations of stresses (Fig. 2) and deflection (Fig. 1), a qualitative similarity has been observed. The deflection maximum in most cases corresponds to the stress maximum. For boundary conditions 2, 3 and 1 for both heating and cooling regimes the stresses unboundedly increase. The largest stresses have been observed in the cooling process for shells with boundary conditions 4, subjected to both temperature moments and forces. The stresses occurring in the shells (except for boundary condition 1) have the same sign as the corresponding deflection.

References

1. Heuer R. and Irschik H. and Ziegler F., Thermally forced vibrations of moderately thick polygonal plates, *Journal of Thermal Stresses,* Vol. 15, 2 (1992) pp. 203-210.
2. Argyris J. and Tenek L., Combined steady-state nonlinear heat transfer/thermal postbuckling computations in unstiffened and stiffened laminated composite plates and shells, *Computer Methods in Applied Mechanics and Engineering,* Vol. 138, 1-4 (1996) pp. 131-185.
3. Banerjee S. and Roychoudhuri S. K., Generalized thermo-visco-elastic problem of an infinitely extended thin plate containing a circular hole, *Journal of Thermal Stresses,* Vol. 19, 5 (1996) pp. 465-480.
4. Mecitoglu Zahit, Free vibrations of a conical shell with temperature-dependent material properties, *Journal of Thermal Stresses.* Vol. 19, 8 (1996) pp. 711-729.
5. Kovalenko A. D. Basis of thermoelasticity (Naukova Dumka, Kiev, 1970) in Russian.
6. Awrejcewicz J. and Krysko V. A. Numerical Analysis of Shells Vibrations with Thermal Load (WNT, Warsaw, 1998) in Polish.

TWO-PHASE ONE-COMPONENT SYSTEM AS A CONTINUUM WITH AN INTERNAL STRUCTURE

ZBIGNIEW BILICKI
Institute of Fluid-Flow Machinery, Polish Academy of Sciences, IMP PAN,
ul. Gen J. Fiszera 14, PL-80952 Gdansk
E-mail: zb@imppan.imp.pg.gda.pl

The two-phase flow system is described by a set equations with respect to average parameters on the macroscopic level. It refers to the homogeneous model as well as to the two-fluid model. However, some significant features of two-phase flow originate from the phenomena occurring at the interface and its vicinity, which are not taken into account in the equations describing the two-phase system on the macroscopic level. Therefore, it is proposed to analyse the two-phase system as a system of internal structure where the transport phenomena are taken into consideration on the mesoscopic level. The analysis is carried out based on thermodynamics of nonequilibrium processes. The analysis of thermodynamic nonequilibrium in the two-phase medium yields theoretical determination of some characteristic quantities. Among them are the relaxation time corresponding to the dryness fraction or to the temperature of liquid, the operative heat conductivity, the operative shear viscosity, the bulk viscosity. All these quantities characterise the internal structure of the two-phase system and the transport processes at the interface.

1. Introduction

Two-phase medium is a heterogeneous medium. It is a mixture of two-phases - the liquid phase and the gas phase. The latter can be either an inert gas - then we have a two-phase two-component flow, or vapour of the liquid making the two-phase flow one-component. The phases are separated by an interfacial surface. The configuration of the interfacial surface is a factor that makes a distinction between several two-phase structures, including bubbly flow, slug, churn, froth, annular, annular-mist or mist flow. The properties of two-phase systems vary with their structure. This is due to a decisive effect of processes of mass, momentum and heat transfer at the interface. All these processes feature in one-component two-phase flows. Therefore, we will limit ourselves to this most general class of two-phase flow. The dominant role of processes of mass, momentum and heat transfer at the interface, that is in the mesoscale, explains why the two-phase system is considered at the mesoscopic level and treated as a system remains in mechanical and thermodynamic non-equilibrium. The mechanical non-equilibrium is understood here as a property of the medium pronounced by different mean velocities of the vapour and the liquid phase. The thermodynamic non-equilibrium manifests itself in different temperatures, pressures and chemical potentials of the phases.

In modelling, the two-phase system is regarded as a continuum. Two approaches can be distinguished. In the first approach, it is assumed that the mean velocities of both phases in certain volumes into which the two-phase system is divided do not considerably differ from each other [1]. This assumption is at the foundation of a model homogeneous with respect to the velocity - the so-called homogeneous model. In the second approach, two continuum fields of the state vector are superposed at each point of the system. This is the so-called two-fluid model. It should be realised, however, that both approaches which transform a real two-phase system into an equivalent continuum tend to lose some important features that follow from the processes at the mesoscopic level, that is in the vicinity of, for example, vapour bubbles. This paper is meant to present a method that takes into account the said processes of mass, momentum and heat transfer in two-phase structures. For the sake of simplification, let us assume that the two-phase liquid-vapour system is described with the aid of the homogeneous model.

2. Continuum model

The most general continuum model is described by equations of conservation of mass, momentum and energy in the following form, see Nigmatulin [2]:

- mass conservation equation, separately for each phase , $i = 1,2$, where $i = 1$ refers to the liquid 'l' and $i = 2$ to the gas or vapour 'g':

$$\frac{\partial \rho^i \varphi^i}{\partial t} + \frac{\partial \rho^i \varphi^i w_k^i}{\partial z_k} = \Gamma^i , \tag{1}$$

where Γ^i is the production of the phase i and

$$\Gamma^i = -\Gamma^j ; \tag{2}$$

- momentum conservation equation, separately for each phase, $i = 1,2$

$$\frac{\partial \rho^i \varphi^i w_k^i}{\partial t} + \frac{\partial \rho^i \varphi^i w_m^i w_k^i}{\partial z_m} = \rho^i \varphi^i f_k^i - \frac{\partial \varphi^i P_{mk}^i}{\partial z_m} + \sigma_k^i , \tag{3}$$

where f_k denotes the mass force, P_{mk}^i the pressure tensor prescribed to the component or phase i and σ_k^i is here to express the rate of momentum transfer between the components, or phases, j and i that increases the momentum of the phase i in a recoverable way, that is

$$\sigma^i = -\sigma^j ; \tag{4}$$

- energy conservation equation, separately for each phase, $i = 1,2$

$$\frac{\partial \rho^i \varphi^i E^i}{\partial t} + \frac{\partial \rho^i \varphi^i E^i w_k^i}{\partial z_k} = \rho^i \varphi^i f_k^i w_k^i - \frac{\partial}{\partial z_k}\left(\varphi^i P_{km}^i w_m^i + \varphi^i q_k^i\right) + \varepsilon^i , \tag{5}$$

where ε^i is a total energy transferred in a recoverable way to the component, or phase, i at the expense of the same amount of energy of the component, or phase, j

$$\varepsilon^i = -\varepsilon^j . \tag{6}$$

The remaining quantities that appear in the conservation equations of mass, momentum and energy are: ρ - density, φ - volumetric fraction of the phase, w_k^i - velocity of i phase, E - total energy, that is a sum of internal and kinetic energy $E^i = u^i + \frac{w_k^{i\,2}}{2}$, q_k - heat flux vector. In this way, we have formulated the conservation equations for the two-fluid model.

The operation of summation of the respective conservation equations referring to each phase, taking into account Eqs. (2), (4) and (6), yields the following system of mass, momentum and energy conservation equations for the two-phase mixture model:

- mass conservation equation for the mixture

$$\frac{\partial \rho}{\partial t}+\frac{\partial \rho w_k}{\partial z_k}=0\,, \tag{7}$$

where ρ is the density of the mixture

$$\rho=\sum_{i=1}^{2}\rho^i\varphi^i, \qquad\qquad \rho=\rho^l\left(1-\varphi\right)+\rho^g\varphi\,, \tag{8}$$

$$\varphi^i=\frac{V^i}{V},\quad \varphi=\varphi^g=\frac{V^g}{V}\Rightarrow\varphi^l=1-\varphi\,, \tag{9}$$

and w_k denotes the barycentric velocity

$$w_k=\frac{1}{\rho}\sum_{i=1}^{2}\rho^i\varphi^i w_k^i,\quad w_k=\frac{\rho^l\left(1-\varphi\right)w_k^l+\rho^g\varphi w_k^g}{\rho}\,; \tag{10}$$

- momentum conservation equation for the mixture

$$\rho\frac{dw_k}{dt}=\rho f_k-\frac{\partial P_{mk}}{\partial z_m}-\frac{\partial}{\partial z_m}\left(\rho^l\varphi^l\widetilde{w}_m'^{\,l}\widetilde{w}_k'^{\,l}+\rho^g\varphi^g\widetilde{w}_m''^{\,g}\widetilde{w}_k''^{\,g}\right), \tag{11}$$

where $\frac{d}{dt}=\frac{\partial}{\partial t}+w_k\frac{\partial}{\partial z_k}$ and $\widetilde{w}_k^i \overset{def}{=} w_k^i-w_k$ is the diffusion velocity that determines the velocity of one phase with respect to the barycentric velocity with $\sum_{i=1}^{2}\rho^i\varphi^i\widetilde{w}_k^i=0$;

- energy conservation equation for the mixture

$$\rho\frac{dE}{dt}+\sum_{i=1}^{2}\frac{\partial}{\partial z_k}\left(\rho^i\varphi^i\widetilde{w}_k^iE^i\right)=-\sum_{i=1}^{2}\frac{\partial}{\partial z_k}\left(\varphi^iP_{km}^iw_m^i-\varphi^iq_k^i\right)+\rho f_kw_k\,. \tag{12}$$

None of the above equations that characterise the two-fluid or mixture model, including the homogeneous model, substantiates two-phase structures. In other words, these models are not capable of describing strong deformations of single-connected fields in the vicinity of the interfaces as the equations do not incorporate the description of local properties. For example, significant changes of the velocity in the hydraulic boundary layer surrounding vapour bubbles as well as temperature variations there can be successfully described only with the aid of local equations. Local velocity and temperature fields give rise to more intensive dissipation of thermal and kinetic energy than that following from the macroscopic description of these fields. Therefore, two-phase flow models, like the two-fluid model, mixture model or homogeneous model must be subject to modifications to embody real two-phase structures.

The most simple structure to consider is bubbly flow. We will limit ourselves to present a problem of modelling the effects that take place at a mesoscopic level, that is at the level of two-phase structure, into the macroscale. An example is a one-dimensional model where the physical properties are averaged across the channel and the diffusion velocity is neglected in the conservation equations as negligibly small, compared to other terms. This model was applied to describe the structure of shock waves in a two-phase mixture of water and vapour [3,4] and reads as below:

- mass conservation equation

$$\frac{d\rho}{dt}+\rho\frac{\partial w}{\partial z}=-\frac{\rho w}{A}\frac{dA}{dz}, \tag{13}$$

- momentum conservation equation

$$\rho\frac{dw}{dt}=-\frac{\partial P}{\partial z}+\frac{\partial}{\partial z}\left[\tilde{\mu}\frac{\partial w}{\partial z}\right]-\tau_w\frac{C}{A}+\rho g cos\alpha\,, \tag{14}$$

- energy conservation equation

$$\rho\frac{d}{dt}\left(h+\frac{w^2}{2}\right)=\frac{\partial P}{\partial t}+\frac{\partial}{\partial z}\left(\tilde{\mu}w\frac{\partial w}{\partial z}\right)+\frac{\partial}{\partial z}\left(\tilde{\lambda}\frac{dT_S}{dz}\right)+q_w w\frac{C}{A}+\rho w g\, cos\alpha\,. \tag{15}$$

This system of equations is supplemented with a state equation that takes into account the fact that the temperature of the vapour phase T^l is different than that of the liquid

$$h=h\left(P,x,T^l\right)=xh_S\left(P\right)+\left(1-x\right)h^l\left(P,T^l\right), \tag{16}$$

where $h_s(P)$ corresponds to the enthapy of liquid at the saturation line x=0. The mean liquid temperature introduced here, T^l, undergoes a relaxation process to reach the saturation temperature of the vapour $T_s(P)$

$$\frac{DT^l}{Dt}=-\frac{T^l-T_s}{\theta_T}, \tag{17}$$

where θ_T is the relaxation time that can be evaluated based on the analysis of heat transfer between the liquid and an expanding or condensing vapour bubble in bubbly flow [6]. Therefore, the relaxation time θ_T itself includes the two-phase structure. The two-phase structure is also incorporated in the concept of the operative viscosity $\tilde{\mu}$ and heat conductivity $\tilde{\lambda}$.

3. Relaxation time θ_T

Let us assume that vapour bubbles of initial radius R_o are evenly distributed in the two-phase medium. The vapour bubbles are at the saturation temperature $T_s(P)$, whereas at the initial moment the liquid has a higher constant temperature T^l. With a lapse of time, the temperature of the liquid will redistribute and flow of heat to the vapour bubbles will take place. The vapour bubbles will grow at the expense of heat flux from the liquid. The temperature field in the liquid can then be described by an equation put forward by Scriven [5]

$$\frac{\partial T^l}{\partial t}+\varepsilon\dot{R}\frac{R^2}{r^2}\frac{\partial T^l}{\partial r}=a^l\left(\frac{\partial^2 T^l}{\partial r^2}+\frac{2}{r}\frac{\partial T^l}{\partial r}\right), \tag{18}$$

where

$$\varepsilon = 1 - \frac{\rho^g}{\rho^l}, \tag{19}$$

$$\dot{R} = \frac{dR}{dt}. \tag{20}$$

This equation with the following initial and boundary conditions

$$T^l(r,0) = T_\infty, \quad T^l(R,t) = T_s, \quad T^l(\infty,t) = T_\infty, \tag{21}$$

was solved analytically by Scriven. The solution of this equation with the boundary conditions as below

$$\left(\frac{\partial T^l}{\partial r}\right)_{r=R} = \frac{\alpha}{\lambda^l}\left[T^l(R,t) - T^l(r_k,t)\right], \quad \left(\frac{\partial T^l}{\partial r}\right)_{r=r_k} = 0, \tag{22}$$

was found numerically by Bilicki et al. [6]. As a result of calculations of the transient temperature field in the liquid, a relaxation time θ_T was determined and found to agree well with that obtained from the experiments of Moby Dick, see [6, 13].

4. Operative conductivity $\tilde{\lambda}$

The operative heat conductivity should be treated, similar to the turbulent heat diffusivity, not as a physical property of the medium alone, but also a feature dependent on the flow parameters. The operative conductivity can be estimated from the comparison of effects of energy dissipation at the mesoscopic level due to heat transfer in the thermal boundary layer around the vapour bubbles $\dot{\sigma}_{Tloc}$ and at the macroscopic level $\dot{\sigma}_{Tmac}$ as found from the equations that describe the assumed model of two-phase flow, see Bilicki et al. 1998 [7]

$$\dot{\sigma}_{Tmac} = \dot{\sigma}_{Tloc}, \tag{23}$$

where

$$\dot{\sigma}_{Tmac} = \tilde{\lambda}\left(\frac{\nabla \hat{T}^{lg}}{\hat{T}_{lg}}\right)^2, \tag{24}$$

$$\dot{\sigma}_{Tloc} = 3\lambda^l \varphi \frac{\dfrac{\Theta^2}{\delta_T R}}{T^{l2}}. \tag{25}$$

In the above equations, $\tilde{\lambda}$ is the operative conductivity, $T^{lg} = T^l - T^g$ - temperature difference in the channel section, $\Theta = T^l - T_s(P)$ - temperature difference in the thermal boundary layer in the liquid around the vapour bubbles, δ_T - thermal boundary layer thickness, λ^l - heat conductivity of the liquid, φ - void fraction ($\varphi = \frac{V^g}{V}$), R - bubble radius. The comparison of Eqs. (24) and (25) yields

$$\tilde{\lambda} = \frac{3\lambda^l \varphi \Theta^2 \hat{T}^{lg^2}}{R^2 T^{l^2} \left(\nabla T^{lg}\right)^2} \left(\frac{R}{\delta_T}\right). \tag{26}$$

As $\tilde{\lambda} >> \lambda$ during expansion of the superheated liquid, it is assumed that

$$\frac{\tilde{\lambda}}{\lambda^l} = \left[1 + \frac{3\varphi \Theta^2 \hat{T}^{lg^2}}{R^2 T^{l^2} (\nabla T^{lg})^2} \left(\frac{R}{\delta_T}\right)\right]. \tag{27}$$

5. Operative viscosity

The operative viscosity can be determined similar to the operative conductivity from the comparison of effects of dissipation of mechanical energy in the macroscale $\dot{\sigma}_{Mlmac}$ and at the mesoscopic level due to the difference in the velocity of the vapour bubble and the liquid in the hydraulic boundary layer $\dot{\sigma}_{Mmac}$

$$\dot{\sigma}_{Mmac} = \dot{\sigma}_{Mloc}, \tag{28}$$

where

$$\dot{\sigma}_{Mmac} = \frac{\tilde{\mu}}{T} \left(\frac{\partial w}{\partial z}\right)^2, \tag{29}$$

$$\dot{\sigma}_{Mloc} = \frac{V_h}{T^l} \left(\frac{4}{3}\mu^l + \varsigma^l\right) \left(\frac{w^g - w^l}{\delta_h}\right)^2, \tag{30}$$

$$\frac{\partial w}{\partial z} \approx \frac{w^g - w^l}{\delta_h}. \tag{31}$$

Assuming that

$$\delta_h = \frac{R}{\text{Re}}, \tag{32}$$

$$\text{Re} = \frac{2\rho^l \left|w^g - w^l\right| R}{\mu^l}, \tag{33}$$

and that the slip between the phases follows from the momentum balance for the vapour bubble with the liquid phase in the form of the so-called added mass

$$\left(\rho^{g}+\frac{\rho^{l}}{2}\right)w^{g}\frac{dw^{g}}{dz}-\frac{\rho^{l}}{2}w^{g}\frac{dw^{l}}{dz}=-\frac{dP}{dz}-\frac{3}{8}\frac{C_{D}}{R}\rho^{l}\left(w^{g}-w^{l}\right)\left|w^{g}-w^{l}\right|, \tag{34}$$

where C_D is the drag coefficient for a sphere. The comparison of Eqs. (29) and (30) gives

$$\widetilde{\mu}=\mu^{l}\frac{T^{l}}{T}\frac{3\varphi\delta_{h}}{R}\left(\frac{w^{g}-w^{l}}{\delta_{h}}\right)^{2}\left(\frac{\partial w}{\partial z}\right)^{-2}. \tag{35}$$

Assuming that the reference temperature $T=T^{l}$ and $\widetilde{\mu}>>\mu^{l}$, a relationship can be found that ,asymptotically for w^{g} - $w^{l}=0$, yields the following solution

$$\frac{\widetilde{\mu}}{\mu^{l}}=\left[1+\frac{3\varphi\delta_{h}}{R}\left(\frac{w^{g}-w^{l}}{\delta_{h}}\right)^{2}\left(\frac{\partial w}{\partial z}\right)^{-2}\right]. \tag{36}$$

The bulk viscosity that appears in Eq. (30) is a complex quantity dependent on the frequency of perturbations of the two-phase system [1]. Because the presented model that consists of equations (13-17) does not incorporate an equation describing these perturbations, the effect of ς^{l} is neglected in the analysis at a high level of simplification. This effect can be taken into account by supplementing the model with the Rayleigh equation, or extending the model to include the viscosity of the liquid around the vapour bubble and the surface tension [8].

6. Bulk viscosity

One reason for the existence of thermodynamic nonequilibrium in the two-phase medium is a difference between the real pressure P^{h} measured by manometers or other gauges and the thermodynamic equilibrium pressure P. This difference, called also the viscous pressure P^{V}, occurs when the fluid volume of the medium varies with time, $V(t)$. It can take significant values if the velocity divergence for the medium, which is the measure of volume changes, $\left(\nabla\cdot\mathbf{w}\equiv\frac{1}{V(t)}\frac{dV(t)}{dt}\right)$, as well as the bulk viscosity (known also as the second viscosity) are significant. This can be seen from the following relationship derived with the aid of classical thermodynamics of irreversible processes, [9], [10],

$$P^{v}\equiv P^{h}-P=-\varsigma\nabla\cdot\mathbf{w}. \tag{37}$$

It follows from the above equation that the pressure difference P^{V} can be of particular importance for flows with spontaneous evaporation or condensation, that is for flashing flow or for shock waves where the fluid volume, and at the same time $\nabla\cdot\mathbf{w}$, undergo radical changes. This leads to the entropy production as below:

$$\dot{\sigma}=-\frac{1}{T}P^{v}\nabla\cdot\mathbf{w}>0. \tag{38}$$

It is clear then that during expansion ($\nabla \cdot \mathbf{w} > 0$) the inequality $P > P^h$ should be satisfied whereas during compression ($\nabla \cdot \mathbf{w} < 0$) $P < P^h$. At this stage it is worth noticing that the Newton hypothesis, widely used in mechanics of continua, especially in fluid mechanics, for evaluation of the pressure tensor P_{ij} considers the thermodynamic pressure P and two kinds of viscosity - the first (shear) viscosity μ , μ_l for example, and the already mentioned second (bulk) viscosity ς

$$P_{ij} = (P - \varsigma \nabla \cdot \mathbf{w})\delta_{ij} - \mu\left(\frac{\partial w_i}{\partial z_j} + \frac{\partial w_j}{\partial z_i} - \frac{2}{3}\delta_{ij}\nabla \cdot \mathbf{w}\right). \tag{39}$$

Consequently, the equations describing the flow contain the thermodynamic pressure P considerably different from the real pressure P^h under certain conditions mentioned before. One advantage of placing P instead of P^h in Eq. (39) is the fact that the physical properties appearing in basic equations of the flow can be found directly from the thermodynamic tables where they are referred to the thermodynamic pressure. However, when the theory is compared with the experiment it should be remembered that the measured pressure P^h is not the equilibrium pressure P .

It arises from the above presented analysis that an essential question in describing the thermodynamic nonequilibrium is the determination of the bulk viscosity ς . This question was studied in detail by Badur et al. 1997 [11]. Following the way of reasoning initially conceived by Mandelsztam & Leontowicz [12] and developed further by Landau & Lifshic [9], one can assume that the intensity of dissipative processes, and also ς , depend on the relation between the intensity of compression or expansion processes and the relaxation time associated with volume changes *V(t)*. If the compression or expansion is caused by a shock wave, then the bulk viscosity should be a dispersive quantity, that is dependent of the shock wave frequency ω . The following expression was derived by Landau and Lifszic standing on the grounds of fenomenological thermodynamics

$$\varsigma = \frac{\theta_T \rho}{1 - i\omega\theta_T}\left(a_f^2 - a_e^2\right), \tag{40}$$

where a_f and a_e are the frozen and equilibrium phase velocity referring to the disturbance frequency $\omega \to \infty$ and $\omega \to 0$, respectively, see Bilicki & Kestin [14]. The real part of the bulk viscosity is equal to

$$\Re e\, \varsigma = \frac{\theta_T \rho}{1 + (\omega\theta_T)^2}\left(a_f^2 - a_e^2\right). \tag{41}$$

For relatively small *Deborah Number,* which corresponds to disturbance frequencies, that is $\omega\theta_T << 1$, the bulk viscosity takes its maximum real value

$$\varsigma = \theta_T \rho\left(a_f^2 - a_e^2\right). \tag{42}$$

On the other hand ς vanishes for infinitely large frequencies, ensuing that the two-phase system does not follow the higher frequency disturbing oscillations, which remains in keeping with the intuitive, physical meaning of the bulk viscosity. It is clear from Eqs. (41) and (42) that the bulk viscosity depends on the internal structure of the two-phase system because it depends on the relaxation time θ_T characterising the microstructure, as shown in Chapter 3. In the "Moby Dick" experiments [13] over a water/vapour mixture under flashing flow, it was found that at the beginning of evaporation when the void fraction φ is very low and the density of the mixture ρ is near that of water $\rho \approx 10^3\ \mathrm{kg/m}^3$, the frozen velocity a_f is larger than the equilibrium velocity a_e by at least one order of magnitude and equal to about 100 m/s. At the same time the relaxation time θ_T is of the order of $0(10^{-1})$s and the bulk viscosity can have unexpectedly large values $0(\varsigma) = 10^6$, larger by nine orders of magnitude than the shear viscosity of water which is $0(\mu) = 10^{-3}$. The bulk viscosity decreased from the above value in the course of the "Moby Dick" experiments [13], down to $0(\varsigma) = 10^2$. However, it is still considerably higher than the shear viscosity of water.

6. Conclusions

As shown in the paper, the main problem in modelling two-phase flows is the appropriate representation of processes of mass, momentum and energy transfer that take place at the interfacial surface. The macroscopic description, including the two-fluid model, mixture model or homogeneous model, does not include these processes. In the paper, an idea has been put forward how to incorporate mesoscopic processes into the macroscale description. This idea draws on the introduction of operative parameters - the operative heat conductivity and the operative viscosity and also the bulk viscosity. Both operative quantities are not physical properties, but depend on the mutual motion of both phases and the degree of thermodynamic non-equilibrium.

References

1. Bilicki, Z., Thermodynamic nonequilibrium in the two-phase system - a continuum with an internal structure, *Archives of Thermodynamics* **17**, 109-134 (1996).
2. Nigmatulin, R.I., *Dynamics of Multiphase systems*, Vol. 1, Hemisphere, Washington (1991).
3. Bilicki, Z. & Kwidziński, R., Thermodynamic and mechanical nonequilibrium in flashing flow with a shock wave in liquid-vapour mixture, *Journal of Theoretical and Applied Mechanics* **35**, 263-284 (1997).
4. Kwidziński, R, An experimental and theoretical study of the stationary shock wave in the flow of a water-vapour mixture, *Ph. D. Thiesis*, Institute of Fluid-Flow Machinery, P. A. Sci., Gdańsk, Poland (1998).
5. Scriven, L.E., On the dynamics of phase growth, *Chem. Eng. Sci.* **10**, 1-13 (1959).
6. Bilicki, Z., Kwidziński, R. & Mohammadein, S.A., Evaluation of the relaxation time of heat and mass exchange in the liquid-vapour bubble flow, *Int. J. Heat Mass Transfer* **39**, 753-759 (1996).

7. Bilicki, Z., Kardas, D. & Michaelides, E.E., Relaxation models for wave phenomena in liquid-vapor bubble flow in channels, *ASME J. Fluids Engineering* **120,** 369-377 (1998).
8. Plesset, M.S. & Prosperetti, A., Bubble dynamics and cavitation, *Ann. Rev. Fluid Mech.*,145-185 (1997).
9. Landau, L.D. & Lifshic, E.M., *Continuum Mechanics* (in Polish), PWN, Warszawa (1958).
10. Bilicki, Z., The concept of the hidden variables theory on the background of conterporary trends in thermodynamics of irreversible processes applied to two-phase flow (in Polish), *Zeszyty Naukowe IMP PAN* **421**/1379/94 (1994).
11. Badur, J., Bilicki, Z. & Kwidziński, R., Operative bulk viscosity in the process of momentum transport of the flashing of water and shock wave of condensing bubble mixture (in Polish), *Zeszyty Naukowe IMP PAN* **479**/1428/97 (1997).
12. Mandelsztam, L.I. & Leontowicz, M.A., A theory of sound absorption in liquids, *J. Exp. Theor. Phys.* **7**, 438-449, (English trans. by R.T. Beyer) (1937).
13. Reocreux, M., Contribution a l'etude des debits critiques en ecoulement diphasique eau-vapeur. *Ph.D. thesis*, Université Scientifique et Médicale de Grenoble, France (1974).
14. Bilicki, Z. & Kestin, J., Physical aspects of the relaxation model in two-phase flow, *Proc. Royal Society London A* **428**, 379-397 (1990).

MOLECULAR DYNAMICS SIMULATIONS OF CATION SUBSTITUTIONS IN $YBa_2Cu_3O_7$

J. BOŚKO, J. RYBICKI

Department of Solid State Physics, Faculty of Technical Physics and Applied Mathematics, Technical University of Gdansk, Narutowicza 11/12, 80-952 Gdansk, Poland

E-mail: jbosko@task.gda.pl, ryba@pg.gda.pl

The structure of the $YBa_2Cu_3O_7$ crystal has been studied using the molecular dynamics (MD) technique. A two-body model potentials proposed by Zhang and Catlow have been used in MD simulations performed in wide ranges of parameters as temperature or pressure. The temperature-induced orthorhombic-to-tetragonal phase transition is observed. Starting from the model of the pure $YBa_2Cu_3O_7$, several series of simulations of cation substitutions have been performed using modified classical Abrahamson parametrisations. Rear earth elements were substituted for Y with doping level = 1/27. Replacement of Y or Ba ions with Ca^{2+} cation is also studied. The structural changes due to the substitutions and defect formation energies are described. The preferential dopant positions are discussed.

1 Introduction

The changes of various physical properties of $YBa_2Cu_3O_7$ due to the cation substitutions may be explained in terms of variation of either electron sub-system or structure of the crystal. In this paper we present some structural results of the MD simulations of $YBa_2Cu_3O_7$ (YBCO). These were performed in a wide range of temperature and pressure in order to check whether the used model is able to reproduce experimental data. Starting from the model of undoped material, we have examined structural changes obtained in the MD simulations of YBCO doped with the rear earth elements and calcium.

The YBCO substituted with Ca^{2+} has been studied experimentally [7,8]. Because of its radius and charge, the Ca^{2+} cation can replace either Y^{3+} or Ba^{2+}. Contradictory observations are reported in the literature. For low concentration of Ca (up to 20%), according to reference [7] dopant atoms occupy the Y sites, whereas according to the reference [8] they occupy the Ba sites.

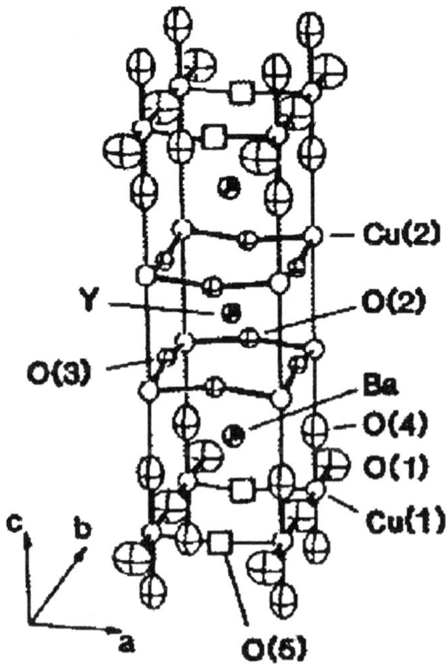

Figure 1. Structure of $YBa_2Cu_3O_7$

In the case of the YBCO substituted with rear earth (RE) elements (except for Pr which strongly depresses superconductivity), it seems that due to the similar electron configurations and cation radii, the properties of the (Y or RE)$Ba_2Cu_3O_7$ are mainly correlated with the size of RE cation.

2 Simulation method

The atomistic simulation method was chosen as a tool in our study. The method accuracy is determined by the quality of the interatomic potentials describing the system, and the assignment of charges to each ion in the solid. In our model the charge states of 3+, 2+ and 2+ were assigned to Y, Cu and Ba ions respectively. To achieve electroneutrality, all the O ions, except for those in the CuO chains, were given a charge of 2-. The O anions in chains have a charge of 1-. The system was treated as an assembly of rigid ions. Interatomic potentials were separated into Coulombic, and short-range terms. The long-range Coulomb interactions were handled using the Ewald technique. The short-range potentials were of the two-body Buckingham form:

$$\phi_{ij} = A \cdot \exp(-r/\rho) - C/r^6 .$$

The potential parameters proposed by Zhang and Catlow [9] were used in our study (Table 1).

Table 1: Potential parameters for short-range interactions.

	A (eV)	ρ (Å)	C (eV·Å^6)
O^{2-}-O^{2-}	22764.3	0.14900	25.0
O^{2-}-O^{-}	22764.3	0.14900	25.0
O^{2-}-Cu^{2+}	3799.3	0.24273	0.0
O^{2-}-Ba^{2+}	3115.5	0.33583	0.0
O^{2-}-Y^{3+}	20717.5	0.24203	0.0
O^{-}-O^{-}	22764.3	0.14900	25.0
O^{-}-Cu^{2+}	1861.6	0.25263	0.0
O^{-}-Ba^{2+}	29906.5	0.27238	0.0
Cu^{2+}-Ba^{2+}	168128.6	0.22873	0.0
Ba^{2+}-Ba^{2+}	2.663.7	0.25580	0.0

The simulations of the pure YBCO were performed using a box containing 975 atoms which corresponded to 75 crystal unit cells. The system was maintained in the isothermal-isobaric regime with the Nose'-Hoover thermostat. In the time of studying the substituted crystal, the simulation box consisted of 351 atoms in 27 crystal unit cells with one substituted atom, what resulted in the doping level = 1/27. During the simulation the system remained in NVT ensemble (isochoric-isothermal regime). Volume V corresponded to the volume of the pure YBCO in the same temperature and under pressure of 0 Pa obtained in the simulation. The short-range potential parameters, used to describe the interactions of the dopant atoms with neighbouring atoms, were taken from [10,11]. In the case of Ca^{2+} - Y^{3+} exchange, an additional hole has to be incorporated to other ions. We distributed it on 8 neighbouring Cu cations. The charge assigned to each of them was 2.25+.

The equations of motion were integrated with a 1 fs time step. Initially the Maxwell-Boltzmann distribution of velocities was assigned to all the atoms in the simulation box. Iterative velocity scaling was employed to achieve a stable state (configuration). The calculations were performed on the SGI Power Challenge 8xR10000 computer using a MOLDY computer code.

3 Simulation results

In the five subsequent sections we present the equilibrium structure, the temperature-induced and pressure-induced structural changes of the pure YBCO, and finally the energy and the structure of the cation substituted crystal.

3.1 Equilibrium structure

The potential applied in our model reproduces the structure of YBCO with a maximum error of 4% in the most structural parameters (see Table 2). The essential feature of the crystal is its anisotropy and layered structure. The bond lengths, obtained with less accuracy, correspond to the distances between those layers. Within layers the bond lengths are in a good agreement with the experimentally observed values [12]. In a further part of the paper, in order to avoid the inaccuracy of results, only the relative changes of the structure due to the variation of temperature, pressure or composition will be discussed.

Table 2: Comparison of the calculated and experimental [12] crystal constants and bond lengths.

	Calculated length (Å)	Experimental length (Å)	Difference (Å)
a	3.842	3.820	0.022
b	3.908	3.885	0.023
c	12.823	11.683	1.140
Y-O(2)	2.392	2.418	0.026
Y-O(3)	2.392	2.399	0.070
Ba-O(1)	3.247	2.891	0.356
Ba-O(2)	3.034	2.980	0.054
Ba-O(3)	3.034	2.948	0.086
Ba-O(4)	2.856	2.750	0.106
Cu(1)-O(1)	1.947	1.947	0.000
Cu(1)-O(4)	1.819	1.834	0.015
Cu(2)-O(2)	1.943	1.929	0.014
Cu(2)-O(3)	1.943	1.961	0.018
Cu(2)-O(4)	2.927	2.341	0.586

3.2 *The temperature-induced structural changes*

A series of simulations was performed for the system held in the temperature varying from 300 K up to 1500 K. Figure 2 shows how the calculated lattice parameters *a* and *b* vary as a function of temperature. They expand linearly up to 800 K. Above this value, the *a* axis expands much faster, and finally the *a* and *b* axes become equal, and the crystal becomes tetragonal. The temperature at which the orthorhombic-to-tetragonal transition occurs amounts approximately to 1100 K.

This phase transition was observed experimentally [12]. It occurs due to the diffusion of the O(1) ions. Moreover, above 700K the oxygen loss takes place. In our model the number of atoms is conserved, so the structural changes are caused only by the redistribution of O anions over O(1), and O(5) sites (see Figure 1). Because of a constant stoichiometry, the simulated crystal corresponds to the one which is held in a high oxygen partial pressure. The temperature at which the transition occurs for the oxygen pressure equal 1 ATM, is ~ 950 K [12]. It is in a good agreement with the value obtained in our calculation.

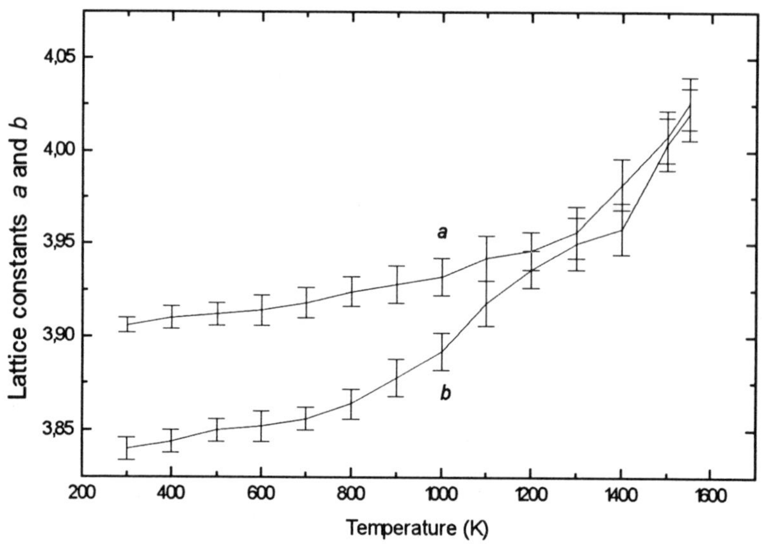

Figure 2: The lattice constants *a* and *b* versus temperature.

3.3 *The pressure-induced structural changes*

Another series of simulations corresponds to a crystal, remaining in the constant temperature equal to 300 K and pressures in the range of 0 ÷ 10 GPa. Using the simulation data, we calculated the compressibility coefficients along *a*, *b* and *c* axes and the volume compressibility. The obtained values are compared to the experimental ones [5] in Table 3.

As it is seen, the Zhang-Catlow potential gives rather poor agreement between the calculated values of the compressibility coefficients and corresponding experimental values. However, the general tendencies of changes of the crystal structure, due to the applied external pressure, are similar to those observed experimentally. The crystal is less compressive along the *a* axis than along the *b* axis, and the compressibility coefficient along the *c* axis is the biggest one.

The YBCO crystal undergoes the pressure-induced orthorhombic-to-tetragonal phase transition at the pressure of ~ 6 GPa. In our study the *a* and *b* lattice constants converge much slower with pressure, and the crystal would become tetragonal at the pressure much higher than for a real crystal.

Table 3: Comparison of the coefficients of compressibility observed experimentally [5] and obtained in the simulations.

	Experimental value	**Simulation** (compression)	**Simulation** (decompression)
a_0^{-1} **da/dp**	1.30 ± 0.41	0.73	0.62
b_0^{-1} **db/dp**	2.80 ± 0.45	1.08	1.13
c_0^{-1} **dc/dp**	3.89 ± 0.55	4.22	4.27
v_0^{-1} **dv/dp**	7.82 ± 1.20	6.00	5.93

3.4 *Ca^{2+} substitutions*

The introduction of Ca into YBCO caused the structural changes of the crystal, which we describe in terms of relative changes of the atom positions and the bond lengths within the simulation box of constant volume. These displacements are due to the differences in sizes of Ca and substituted cations, and due to new potential parameters describing interaction with the neighbouring atoms, and in the case of Ca^{2+} occupying Y^{3+} site also due to a new charge distribution. The main structural changes are:

- The distance Cu(2) - O(2,3) decreases, which is reflected in an increase of the Cu(2) - O(2,3) - Cu(2) angle.
- The Cu(2) cations move toward the plane containing Y cations and consequently the distance between CuO_2 planes decreases.

- The distance between Ca - Ba or Y - Ca, respectively, is shorter in comparison to the Y - Ba distance in the undoped crystal.

These displacements of atoms, within the crystal unit cell, are common for both configurations with Ca substituting either Y or Ba ions. The observed changes are in the agreement with the experimental data [7]. The conclusion is that the preferential positions of Ca cannot be determined only from the structural changes of the crystal.

We have also studied the influence of a distribution of the dopant atoms in a crystal on the energy of the system. With a constant doping level we varied the sites replaced by calcium. Every configuration, in which the dopant atoms occupy the neighbouring sites, is energetically beneficial in comparison to the energy of the system with Ca uniformly incorporated in the crystal. This observation is similar for Ca replacing either Y or Ba atoms. This tendency of the system to having the inclusions of the phases with high concentration of Ca rather than having one homogenous phase, can be explained in terms of stresses which appear at the boundaries of the crystal unit cells: pure and the deformed by the dopant atom. When the Ca atoms occupy sites in the neighbouring unit cells the accompanying stress in minimised. In the literature the results of the experimental study of the substituted YBCO crystal are interpreted under the assuming the homogenous distribution of dopant atoms in the system.

3.5 Rear earth elements substitutions

In our study of the cation substituted YBCO we also took into consideration the crystal doped with the rear earth elements (RE). We replaced Y^{3+} with Tb^{3+}, Gd^{3+} and Pr^{3+}. Substitution of each of these elements was associated with the introduction to the crystal the cation with the larger radius ($r(Y^{3+}) = 1.04$ Å, $r(Tb^{3+}) = 1.063$ Å, $r(Gd^{3+}) = 1.078$ Å, $r(Pr^{3+}) = 1.130$ Å). We observed the changes in the energy of the system, and in the structure as a function of the radius of the dopant atom. With the increase of the radius of the RE element the changes are as follow:

- The cohesive energy of the crystal increases.
- The distance RE - Cu(2) increases, and RE - O(2,3) decreases. It is accompanied by a decrease of the Cu(2) - O(2,3) - Cu(2) angle. Also the distance between the CuO_2 planes increases.
- The distance RE - Ba increases.

According to [1] the similar structural changes are related directly to the observed changes of T_c.

4 Conclusions

In our study of the YBCO crystal we employed the molecular dynamics technique. The advantages and disadvantages of the two-body interaction potential parametrised by Zhang and Catlow have been analysed. The potential parametrisation reproduces well the experimentally observed structural features of pure YBCO in their dependence on temperature in the range of 300 ÷ 1500 K. The pressure-induced changes of the crystal structure are reproduced with less accuracy, but the qualitative tendencies agree with the experimental ones. In our simulations of the substituted crystal we used the classical Abrahamson potential parametrisation. Such a straightforward approach can not lead to the results which are in quantitative agreement with experimental data. However, the qualitative tendencies, as the dependence of the structure on the doping atom radius, are qualitatively correct. In order to obtain more reliable description of the structure and properties of doped YBCO crystals, the potential parameters have to be tuned. Such work, based on the ab initio calculations, is in progress.

Acknowledgement

The opportunity to perform our numerical calculations at the TASK Computer Centre (Gdansk, Poland) is kindly acknowledged.

References

[1] Rao C N R and Ganguli A K 1995 *Chemical Society Reviews* 1
[2] Gupta R P and Gupta M 1995 *Physical Review* B **51(21)** 15617
[3] Li J and Ladik J 1997 *Solid State Commun.* **102(10)** 769
[4] Lubritto C, Rościszewski K and Oleś M 1996 *J.Phys.: CM* **8** 11053
[5] Akhtar M J, Akhtar Z N and Catlow C R A 1990 *J.Phys.: CM* **2** 3231
[6] Zhang X, Yip K W and Ong C K 1995 *J.Phys. Chem. Solids* **56(2)** 153
[7] Böttger G, Mangelschots I, Kaldis E, Fischer P, Krüger Ch and Fauth F 1996 *J.Phys.CM* **8** 8889
[8] Chattopadhyay M K, Ghatak S K and Dey T K 1997 *Physica C* **281** 211
[9] Zhang X and Catlow C R A 1992 *Phys. Rev.* B **46(1)** 457
[10] Abrahamson A A 1969 *Phys. Rev.* **178(1)** 76
[11] Sangster M J L and Stoneham A M 1981 *Phil. Mag.* B **43** 597
[12] Beyers R and Shaw T M *Solid State Physics* **42** 135

SOLUTION OF INVERSE HEAT CONDUCTION PROBLEM WITH THE USE OF NEW TYPE OF FINITE ELEMENT BASE FUNCTIONS

M. J. CIAŁKOWSKI
Poznań University of Technology, ul.Piotrowo 3, 60 965 Poznań
E-mail:cialkows@sol.put.poznan.pl

The paper presents application of heat polynomials for purposes of solving a heat conduction inverse problem. Due to unfavourable properties of high degree polynomials a finite element approximation of the solution was made, using heat polynomials for formulating the base functions. A new functional was introduced, minimization of which leads to the solution. The inverse problem is stable also in the case of 2nd- and 3rd-degree time approximation. The paper presents a comparison of the new approach to classical finite element method.

1. Introduction

In spite of development of computer methods analytical and numerical methods do not become less important. Heat polynomials, i.e. the polynomials that identically satisfy the heat conduction equation, provide also such methods. Generally, the methods of solution of differential equations may be divided into three groups:

- 1st Group

This group is characterized by:

- analytical solution of differential equation,
- accurate fulfillment of initial and boundary conditions.

The group includes the methods basing on spectral and integral analysis.

- 2nd Group

This group is characterized by:

- approximate solution of differential equation
- approximate satisfaction of boundary and initial conditions.

Such a solution may be achieved e.g. by means of finite element method, boundary element method, or finite difference method (or combination of above mentioned approaches).

- 3rd Group

This group ensures:

- accurate satisfaction of differential equation, in discussed case by heat polynomials,
- approximate fulfillment of initial and boundary conditions.

Hereinafter the use of heat polynomials will be shown [1,2] for purposes of local approximation of finite element solution [4]. Table 5 of the paper [2] presents the polynomials satisfying the equation.

2. Governing equation and solution of direct problem

Let us consider a one-dimensional heat conduction equation for a layer

$$\rho \cdot c \frac{\partial T}{\partial \tau} = \lambda \frac{\partial^2 T}{\partial x^2}, \quad \begin{array}{l} \tau \in (0,\infty), x \in (0,l) \\ T \in C^2(0,l) \cap C^1(0,\infty) \end{array} \tag{1}$$

with constant coefficients ρ, c, λ and the following conditions

- initial condition

$$T(x,0) = f(x), \quad x \in \langle 0,1 \rangle \tag{2}$$

- boundary condition

$$\begin{array}{l} T(x=0,\tau) = g(\tau), \quad \tau \in (0,\infty) \\ T(x=1,\tau) = h(\tau), \quad \tau \in (0,\infty) \end{array} \tag{3}$$

Substitution

$$\xi = \frac{x}{l}, \quad t = \frac{\lambda}{\rho \cdot c} \cdot \frac{\tau}{l^2} \tag{5}$$

transforms the (1) equation to the following dimensionless form

$$\frac{\partial T}{\partial t} = \frac{\partial^2 T}{\partial \xi^2}, \quad \xi \in (0,1), \; t \in (0,\infty) \tag{6}$$

In order to solve the equation (6) the heat polynomials [1] will be used. They are obtained by expansion in series of generating function

$$w = e^{p\xi + p^2 t} = \sum_{n=1}^{\infty} v_n(\xi,t) \cdot p^{n-1} \tag{7}$$

that leads to the relations:

$$v_1(\xi,t)=1\,, \quad v_2(\xi,t)=\xi\,, \quad v_3(\xi,t)=\frac{1}{2}\xi^2+t\,, \quad v_4(\xi,t)=\frac{1}{6}\xi^3+t\cdot\xi$$

$$v_5(\xi,t)=\frac{1}{24}\xi^4+\frac{1}{2}\xi^2 t+\frac{1}{2}t^2\,, \qquad v_6(\xi,t)=\frac{1}{120}\xi^5+\frac{1}{6}\xi^3 t+\frac{1}{2}\xi t^2$$

$$v_8(\xi,t)=\frac{1}{5040}\xi^7+\frac{1}{120}\xi^5 t+\frac{1}{12}\xi^3 t^2+\frac{1}{6}\xi t^3$$

with recurrence formula

$$v_{n+1}(\xi,t)=\frac{\xi}{n}\cdot v_n(\xi,t)+\frac{2t}{n}\cdot v_{n-1}(\xi,t), \;\; n\geq 2$$

and properties for derivatives

$$\frac{\partial v_n(\xi,t)}{\partial \xi}=v_{n-1}(\xi,t), \;\; n\geq 1, \qquad \frac{\partial v_n(\xi,t)}{\partial t}=v_{n-2}(\xi,t), \;\; n\geq 2\,, \; v_o=0 \tag{8}$$

The polynomials identically satisfy the (6) equation. Solution of the (6) equation may be presented in the form

$$T(\xi,t)=\sum_{n=1}^{\infty} A_n \cdot v_n(\xi,t) \tag{9}$$

if the system $\{v_n(\xi,t)\}$ is complete.

The above equation enables to determine the A_n coefficients by minimizing mean square deviation from the boundary-initial condition. The form of the functional is as follows

$$J(\{A\})=\int_0^1 [T(\xi,0)-f(\xi)]^2 d\xi+\int_0^{t_k}[T(0,\tau)-g(\tau)]^2 d\tau+\int_0^{t_k}[T(1,\tau)-h(\tau)]^2 d\tau \tag{10}$$

It can be easily shown that approximation of the $f(\xi)$ function allows to generate the matrix elements of the system of equations defining coefficients A_i and including the elements of very badly conditioned Hilbert's matrix. On the other hand, the approximation performed with high degree polynomials results in manifestation of variables ξ and t and the so-called Runge phenomenon (function waving), that considerably affects the process of calculation of derivatives.

Moreover, the $\{v_n(\xi,t)\}$ functions not always give a complete system. In order to avoid these disadvantageous phenomena a local approximation of temperature within an element (Fig. 2) is proposed according to (8), using finite number of polynomials $v_i(\xi,t)$. The whole area of the solution is to be discretized that is shown in Figure 1.

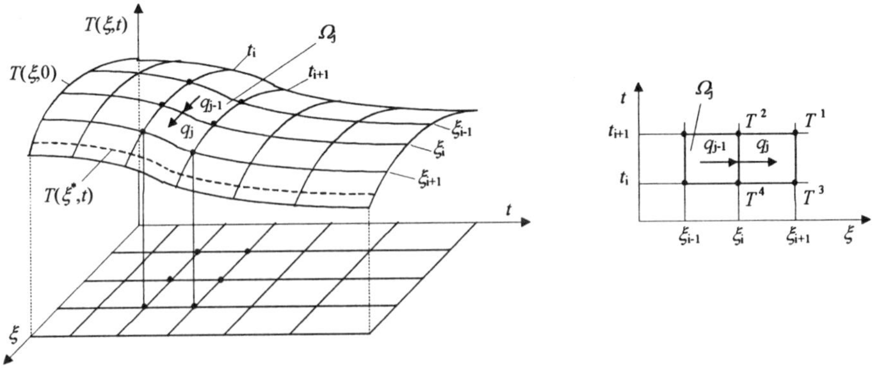

Figure 1: Solution surface in space-time domain.

Let us consider the problem of interpolation of the solution $T(\xi,t)$ within the element (Fig. 2).

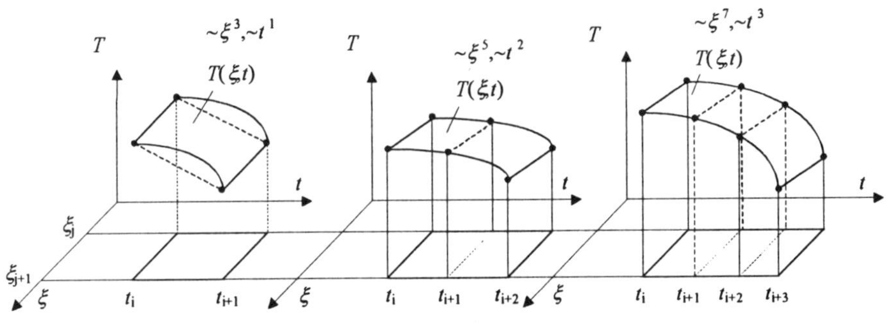

Figure 2. Approximation of solutions performed with the use of finite elements of various node numbers.

As the step of dimensionless time Δt is small as against the interval of $\xi \in <0,1>$ variable, within an element a new variable expressed with relationship $\overline{\xi} = \xi - \xi_j$, $\xi_j \le \xi \le \xi_{j+1}$ appears more convenient. Thus, temperature field within the element is expressed with a formula

$$T_j(\overline{\xi},t) = \sum_{n=1}^{N} A_n^j \cdot v_n(\overline{\xi},t), \quad (\overline{\xi},t) \in \Omega_j \tag{11}$$

As the function $v_n(\xi,t)$ satisfies the formula (6), $v_n(\overline{\xi},t)$ satisfies also the following equation

$$\frac{\partial v_n}{\partial t} = \frac{\partial^2 v_n}{\partial \overline{\xi}^2} \tag{12}$$

The method of solving the equation (1) consists in sequential solutions of the equation (1) within time intervals $<t_i, t_{i+k}>$, k = N/2-1, where N is the node number of the element (N=4,6,8).

For even node number N the ξ power degree amounts to N-1, while power degree of t is equal N/2-1. Each element node is characterized by two coordinates (ξ_k,t_k). Hence, in order to determine the A_n^l constants in (11), assuming that temperatures of element nodes are known, one gets (below l indices will be omitted for better clarity)

$$T(\overline{\xi}_k,t_k) = T_k = \sum_{n=1}^{N} A_n \cdot v_n(\overline{\xi}_k,t_k), \quad k = 1,2,\ldots,N \tag{13}$$

that provides an equation system permitting to define the $\{A_n\}$ coefficients of all elements.

$$\begin{bmatrix} v_1(\overline{\xi}_1,t_1) & v_2(\overline{\xi}_1,t_1) & v_3(\overline{\xi}_1,t_1) & \cdots & v_N(\overline{\xi}_1,t_1) \\ v_1(\overline{\xi}_2,t_2) & v_2(\overline{\xi}_2,t_2) & v_3(\overline{\xi}_2,t_2) & \cdots & v_N(\overline{\xi}_2,t_2) \\ v_1(\overline{\xi}_3,t_3) & v_2(\overline{\xi}_3,t_3) & v_3(\overline{\xi}_3,t_3) & \cdots & v_N(\overline{\xi}_3,t_3) \\ \ldots & \ldots & \ldots & \ldots & \ldots \\ v_1(\xi_N,t_N) & v_2(\xi_N,t_N) & v_3(\xi_N,t_N) & \cdots & v_N(\xi_N,t_N) \end{bmatrix} \begin{Bmatrix} A_1 \\ A_2 \\ A_3 \\ \ldots \\ A_N \end{Bmatrix} = \begin{Bmatrix} T_1 \\ T_2 \\ T_3 \\ \ldots \\ T_N \end{Bmatrix} \tag{14}$$

or

$$[v]\{A\}=\{T\} \tag{15}$$

and having inverted the matrix

$$\{A\}=[v]^{-1}\{T\}=[U]\{T\}, \quad A_n=\sum_{k=1}^{N} U_{nj}\cdot T^k \tag{16}$$

Substitution of (14) into (9) allows obtaining the base functions $\varphi(\xi,t)$ characterizing the element, i.e.

$$T(\xi,t)=\sum_{n=1}^{N}\left(\sum_{k=1}^{N} U_{nk}\cdot T^k\right)\cdot v_n(\xi,t)=\sum_{k=1}^{N}\left(\sum_{n=1}^{N} U_{nk}\cdot v_n(\xi,t)\right)\cdot T^k=$$

$$=\sum_{k=1}^{N}\varphi_k(\xi,t)\cdot T^k, \quad \varphi_k(\xi,t)=\sum_{n=1}^{N} U_{nk}\cdot v_n(\xi,t), \ t\in\langle t_i,t_{i+k}\rangle, \ \text{k=N/2-1} \tag{17}$$

Hence, the base function $\varphi(\xi,t)$, i = 1,...,N are characterized by the following properties

1. $$\left(\frac{\partial}{\partial t}-\frac{\partial^2}{\partial \xi^2}\right)\varphi_k(\xi,t)=0 \ \ as \ \left(\frac{\partial}{\partial t}-\frac{\partial^2}{\partial \xi^2}\right)v_n(\xi,t)=0, \ \ n=1,\dots,N \tag{18}$$

2. $$\varphi_k(\xi_j,t_j)=\begin{cases}1 & k=j\\ 0 & k\neq j \quad (\xi_j,t_j)\text{ - element nodes}\end{cases} \tag{19}$$

k = 1,...,N, j = 1,...,N

3 $$\sum_{k=1}^{N}\varphi_k(\xi,t)=1 \tag{20}$$

4. $$\frac{\partial\varphi_k(\xi,t)}{\partial\xi}=\sum_{n=1}^{N} U_{nk}\cdot\frac{\partial v_n(\xi,t)}{\partial\xi}=\sum_{n=1}^{N} U_{nk}\cdot v_{n-1}(\xi,t) \tag{21}$$

$$\sum_{k=1}^{N}\frac{\partial\varphi_k(\xi,t)}{\partial\xi}=0$$

5. $$\frac{\partial\varphi_k(\xi,t)}{\partial t}=\sum_{n=1}^{N} U_{nk}\,\frac{\partial v_n(\xi,t)}{\partial t}=\sum_{n=1}^{N} U_{nk}\cdot v_{n-2}(\xi,t) \tag{22}$$

In order to solve the equation (1) with the conditions (2)-(4) the space-time domain is divided into space-time finite elements (Fig. 3). The smaller is the element diameter, the better the solution (17) approximates an accurate solution.

In the time $t=t_i$ the nodes 3 and 4 (a 4-node element), the nodes 5 and 6 (a 6-node element), and the nodes 7 and 8 (an 8-node element) are located at the line, the initial temperature of which is known. Table 1 specifies known and unknown parameters of the element, with regard to its node number.

Table 1: Characteristics of space-time elements.

No	Element	The order of T(ξ,t) function with regard to ξ and t	Unknown parameters
1	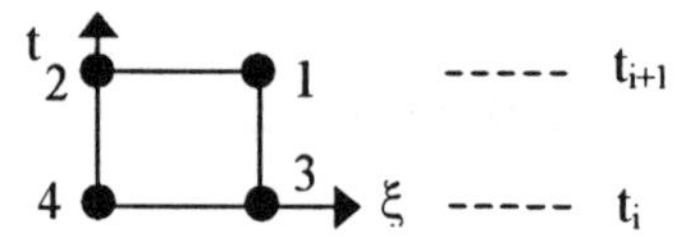	$\sim t, \sim \xi^3$	$T_1, T_2,\ t = t_{i+1}$
2	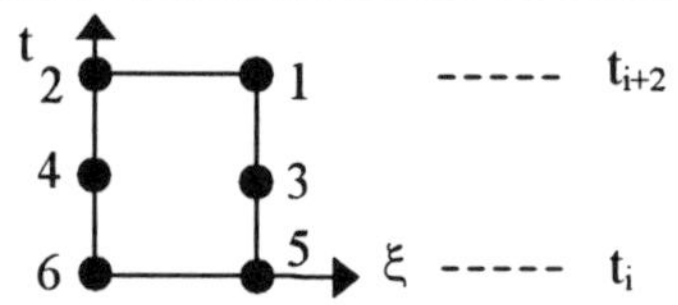	$\sim t^2, \sim \xi^5$	$T_1, T_2,\ t = t_{i+2}$ $T_3, T_4,\ t = t_{i+1}$
3	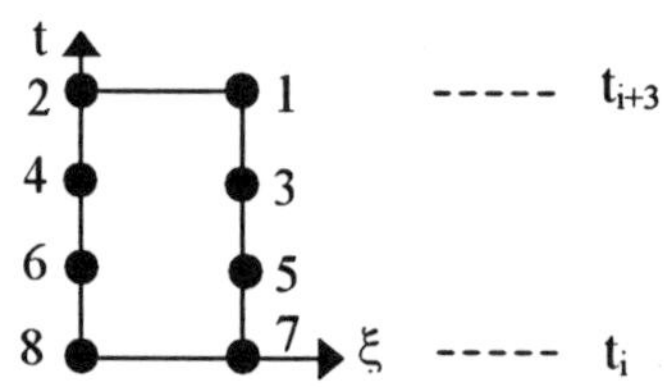	$\sim t^3, \sim \xi^7$	$T_1, T_2,\ t = t_{i+3}$ $T_3, T_4,\ t = t_{i+2}$ $T_5, T_6,\ t = t_{i+1}$

The Table 1 shows that crossing from j-th to j+1-th element temperature distribution remains continuous, that means that $T_j(\xi=1,t)=T_{j+1}(\xi=0,t)$, as these distributions are expressed by the polynomials of the same degree with regard to t and are determined for the same temperature values in the nodes between adjacent elements.

The boundary-initial condition may be observed assuming given temperature values in the nodes of space-time mesh or only forcing observance of the boundary-initial condition in the sense of minimization of mean square deviation - the functional (10).

Observance of energy balance in the whole area filled with finite elements requires the heat leaving a j-th element to be equal to the heat crossing its common boundary with a j+1-th element. Taking into account that for approximation of temperature distribution within an element finite number of polynomials are used, the heat flow through the $\xi = \xi_{j+1}$ boundary between the elements is not a continuous function.

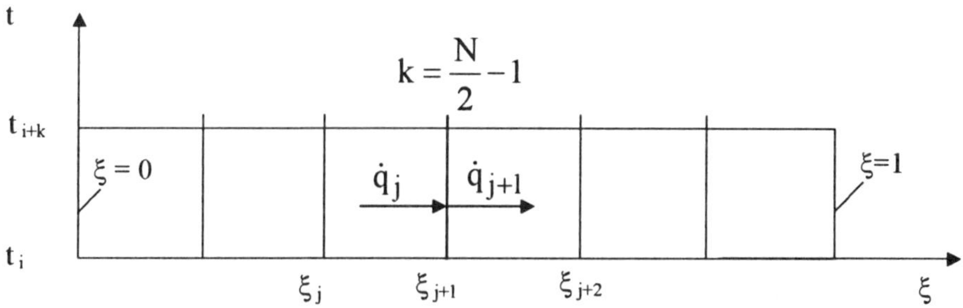

Figure 3. Heat balance at the boundaries of elements.

Deficit of the heat flowing between the elements caused by the approximation process amounts to

$$\delta q_j(t) = -\frac{\partial T_j(\xi_{j+1}, t)}{\partial \xi} + \frac{\partial T_{j+1}(\xi_{j+1}, t)}{\partial \xi} = q_j - q_{j+1} \tag{23}$$

It provides a basis for formulation of a functional. The functional is a sum of heat deficits at the element boundaries, and takes a form

$$I = \sum_{j=1}^{L} \int_{t_i}^{t_{i+k}} [\delta q_j(\tau)]^2 d\tau = \sum_{j=1}^{L} \int_{t_i}^{t_{i+k}} [q_j(\tau) - q_{j+1}(\tau)]^2 d\tau, \quad k = N/2-1 \tag{24}$$

Derivatives occurring in (25) are calculated on the grounds of equation (17), i.e.

$$\frac{\partial T_j(\xi, t)}{\partial \xi} = \frac{\partial}{\partial \xi} \sum_{i=1}^{N} \left(\sum_{k=1}^{N} u_{k_i} \cdot v_k(\xi, t) \right) \cdot T^i = \sum_{i=1}^{N} \left(\sum_{k=1}^{N} u_{k_i} \cdot v_{k-1}(\xi, t) \right) T^i = \sum_{i=1}^{N} \varphi_{k-1}(\xi, t) \cdot T^i \tag{25}$$

Unknown temperatures in the nodes of finite element mesh are searched minimizing the functional

$$J_i = \sum_{l=1}^{L} \int_{x_2}^{x_{l+1}} \left[T_l(\xi,t_i) - T_l^o(\xi,t_i)\right]^2 d\xi + \int_{t_i}^{t_{i+k}} \left[T_1(0,\tau) - g(\tau)\right]^2 d\tau +$$

$$\int_{t_i}^{t_{i+k}} \left[T_L(1,\tau) - h(\tau)\right] d\tau + \sum_{j=1}^{L} \int_{t_i}^{t_{i+k}} \left[q_j(\tau) - q_{j+1}(\tau)\right]^2 d\tau, \quad i = 0,k,2k,\ldots;\ k = N/2-1 \quad (26)$$

where $T_l^o(\xi,t_i)$ is the initial temperature t_i time. Unknowns are the node temperatures.

3. Inverse problem

The inverse problem considered in the paper consists in searching unknown pattern of 1st-, 2nd- or 3-rd-type boundary condition at the x=1 surface, Fig. 4, assuming that temperatures at one (or more) lines located within the area are known. In Figure 4 known temperature patterns within the area are marked with dashed lines. As the unknown boundary condition is sought numerically, it should be defined in boundary nodes of the mesh. Unknown values of boundary condition in the mesh nodes located on the boundary $\xi = 1$ and in time interval

$<t_i, t_{i+k}>$, $k = N/2$-1 will be sought by minimization of the functional

$$J_E = \left\| T\left(\xi^*, t^*, \{h\}\right) - T^* \right\|^2 \quad (27)$$

where $T(\xi, t, \{h\})$ is the solution of the simple problem expressed in parametric form by the boundary condition (3) at the $\xi = 1$ line. On the other hand, (ξ^*, t^*) are coordinates of the point, where the temperature T^* is imposed (measured). Minimization of the functional (27) leads to a solution of the inverse problem in a matrix form

$$\{T\} = [STAB]\{T_o\} + \sum_{m=1}^{M} Z_m \cdot T_m^* + \{WA\} \cdot g_o + \{WB\} \cdot g_1 + WC\} \cdot g_2 + \{WD\} \cdot g_3 \quad (28)$$

where g_o, g_1, g_2, g_3 are the values of g(t) function (3) in element nodes (in this case N=8). Vector $\{T_o\}$ gives the temperature distribution in previous time layer.

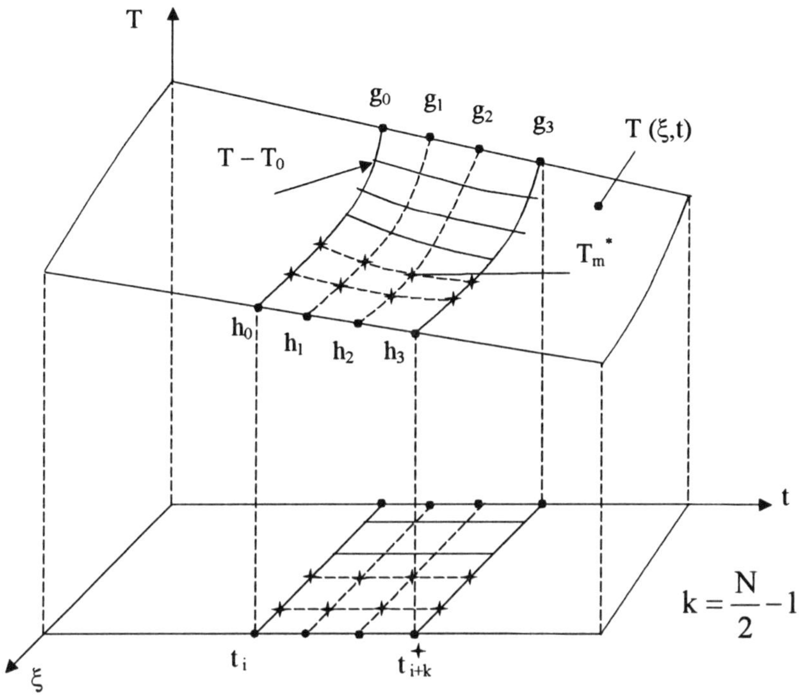

Figure 4. Solution in time layer <t_i, t_{i+k}>.

Temperature disturbance T_m^* in the point (ξ_m, t_m) equal to δT_m^* results in disturbance of temperature T_o amounting to δT_o (beside the first time-layer), so, the disturbed temperature distribution may be expressed by the relationship

$$\{T + \delta T\} = [STAB] \cdot \{T_o + \delta T_o\} + \sum_{m=1}^{M} Z_m \cdot \left(T_m^* + \delta T_m^*\right) +$$

$$+ \{WA\} \cdot g_o + \{WB\} \cdot g_1 + \{WC\} \cdot g_2 + \{WD\} \cdot g_3 \tag{29}$$

Subtracting (28) from (29) gives

$$\{\delta T\} = [STAB] \cdot \{\delta T_o\} + \sum_{m=1}^{M} Z_m \cdot \delta T_m^* \tag{30}$$

Hence, if spectral radius of the matrix [STAB], r([STAB])<1, the inverse problem is stable.

4. Numerical calculations

For purposes of numerical calculations the temperature distribution in the points (ξ^*, t^*) has been assumed as equal to the solution of a direct problem under the conditions

$$T(\xi,0) = T_o = 1, \quad \left.\frac{\partial T}{\partial \xi}\right|_{\xi=0} = 0, \quad -\left.\frac{\partial T}{\partial \xi}\right|_{\xi=1} = Bi\left(T - T_f\right), \quad Bi = \frac{\alpha \cdot l}{\lambda} \tag{31}$$

where Bi is Biot's number (3rd boundary condition) and T_f is the temperature of liquid that surrounds the layer. If $Bi \to \infty$ the surface $\xi = 1$ is subject to the 1st type boundary condition. The following values have been assumed for the calculations: the mesh in ξ-direction is divided into 30 parts, time step $\Delta\tau$ = 0.5s, the layer thickness l = 58 mm, thermal conductivity λ = 27,2 W/m K, heat transfer coefficientα = 20 000 W/m · K, location of temperature measurement points - 2 mm from the edge (ξ^* = 0,9655).

Figure 5 shows the pattern of heat-transfer coefficient determined on the grounds of solution of the inverse problem using the 1st type boundary condition (3) for linear (N=4), square (N=6), and cubic (N=8) approximation of $T(\xi,t)$ function with regard to time. Stability of the method is maintained. Values of spectral radii are marked in Figure 5.

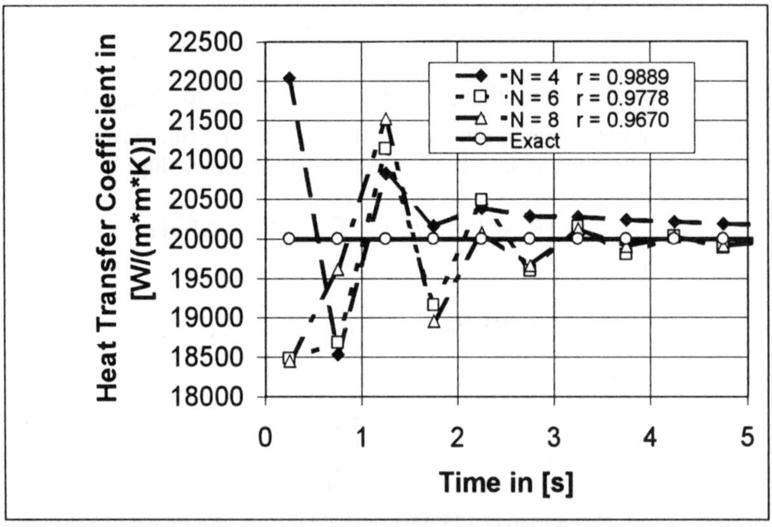

Figure 5. Comparison of accurate and approximate solutions of the inverse problem.

5. Comparison of fem method with classical and new basic functions

Comparison of FEM method with classical and new basic functions is shown in Tables 2, 3, 4. Table 5 presents generating functions for selected partial differential equations.

Table 2: Comparison of FEM approaches with classical and improved basic functions.

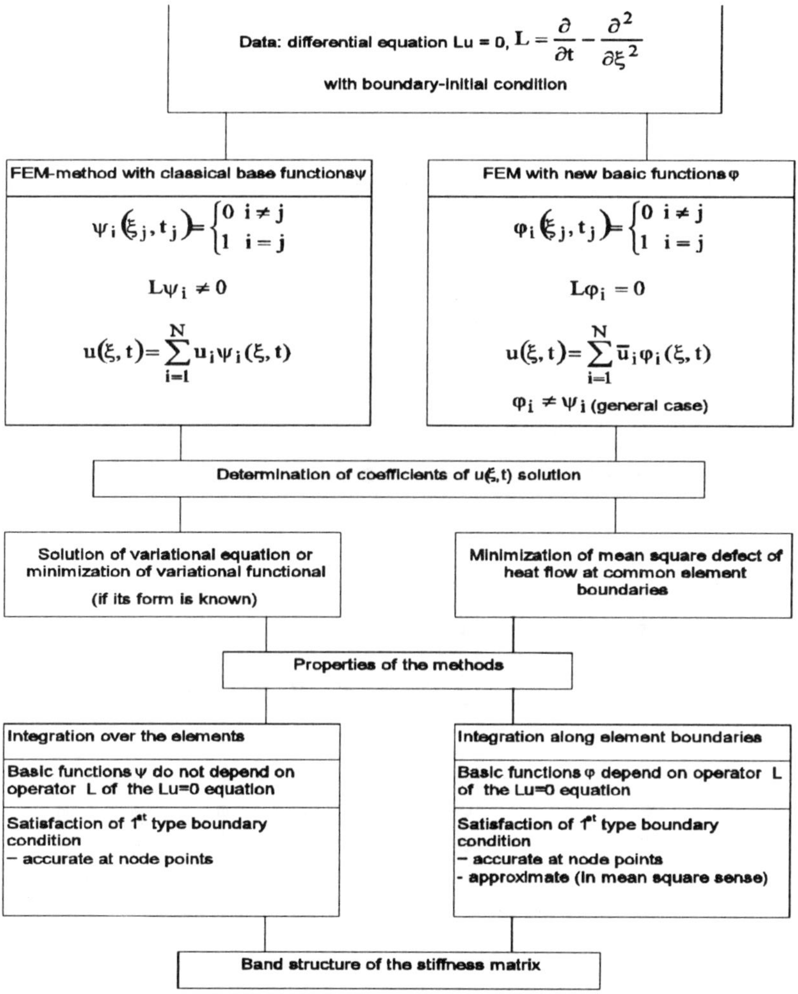

Table 3: Stability of the inverse problem

Approximation of T(ξ,t) Function with regard to t	FEM: classic	FEM: improved
linear	+	+
square	-	+
cubic	-	+

Table 4: Approximation order of T(ξ,t) solution in a finite element - ξ-direction, for the same approximation order in t-direction

	FEM: improved $T(\xi,t)=\sum\varphi_i(\xi,t)\cdot T^i$	FEM: classic $T(\xi,t)=\sum\Psi_i(\xi,t)\cdot T^i$
t	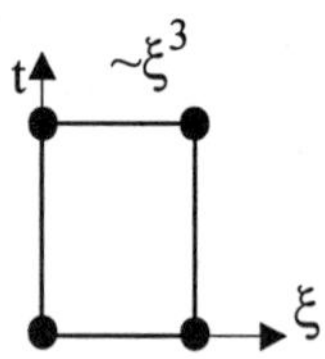	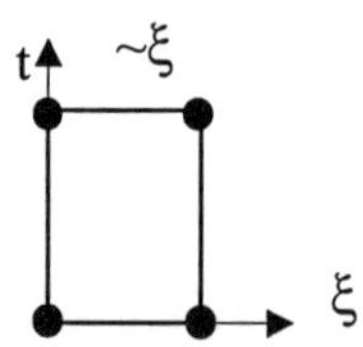
t^2	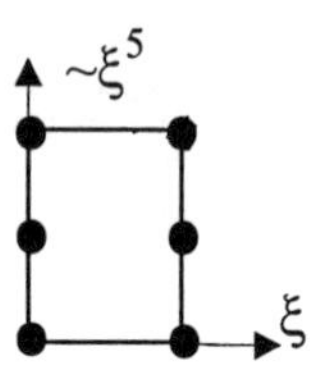	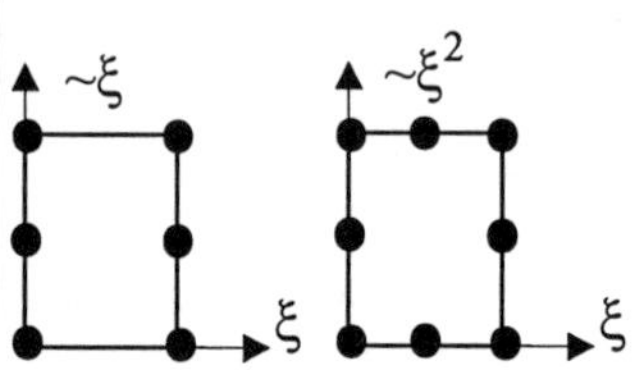
t^3	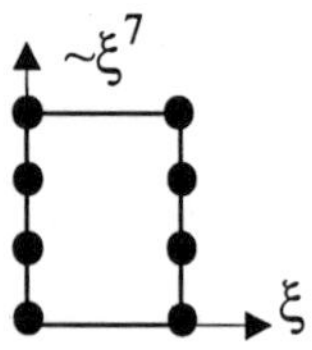	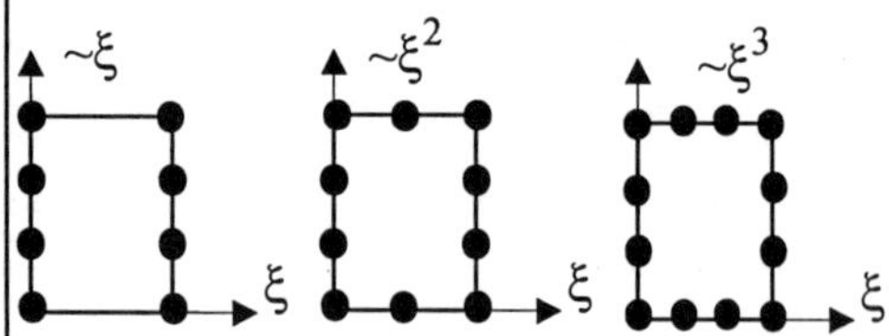

Table 5: Examples of generating functions for selected differential equations [2]

Differential equation	Generating function
$\frac{\partial^2 u}{\partial x^2}+\frac{\partial^2 u}{\partial y^2}=0$	$exp[p(x+i\cdot y)]$
$\frac{\partial^2 u}{\partial x^2}+\frac{\partial^2 u}{\partial y^2}+\frac{\partial^2 u}{\partial z^2}=0$	$exp\left[p\cdot x+q\cdot y+i\cdot\sqrt{p^2+q^2}\cdot z\right]$
$\frac{\partial^2 u}{\partial x^2}+\frac{\partial^2 u}{\partial y^2}+\frac{\partial^2 u}{\partial z^2}-\alpha^2\cdot u=0$	$exp\left[\alpha\frac{p\cdot x+q\cdot y+s\cdot z}{\sqrt{p^2+q^2+s^2}}\right]$, $p^2+q^2+s^2>0$
$\frac{\partial^4 u}{\partial x^4}+2\frac{\partial^4 u}{\partial x^2\partial y^2}+\frac{\partial^4 u}{\partial y^4}=0$	$exp[p(x+i\cdot y)]$
$\frac{\partial^4 u}{\partial x^4}+2\frac{\partial^4 u}{\partial x^2\partial y^2}+\frac{\partial^4 u}{\partial y^4}-\alpha^4\cdot u=0$	$exp\left[\alpha\frac{p\cdot x+q\cdot y}{\sqrt{p^2+q^2}}\right]$, $p^2+q^2>0$
$\frac{\partial u}{\partial t}=\frac{\partial^2 u}{\partial x^2}+\frac{\partial^2 u}{\partial y^2}+\frac{\partial^2 u}{\partial z^2}$	$exp\left[p\cdot x+q\cdot y+s\cdot z+\left(p^2+q^2+s^2\right)\cdot t\right]$
$\frac{\partial^2 u}{\partial t^2}=\frac{\partial^2 u}{\partial x^2}+\frac{\partial^2 u}{\partial y^2}+\frac{\partial^2 u}{\partial z^2}$	$exp\left[p\cdot x+q\cdot y+s\cdot z+\sqrt{p^2+q^2+s^2}\cdot t\right]$
$\frac{\partial^2 u}{\partial t^2}=\frac{\partial^4 u}{\partial x^4}+2\frac{\partial^4 u}{\partial x^2\partial y^2}+\frac{\partial^4 u}{\partial y^4}$	$exp\left[p\cdot x+q\cdot y+\left(p^2+q^2\right)\cdot t\right]$

6. Summary

The use of heat polynomials for determining the base FEM functions permitted to obtain improved accuracy of the approximation for the same element node number as compared to classical finite element method.

On the other hand, introduction of functional (24) to the process of solution of the problem of heat flow gave stable solution of the inverse problem, also in the case of square and cubic approximation of the solution T(ξ, t) with regard to the variable t. The order of approximation of the solution T(ξ, t) with regard to ξ is associated with approximation order in t-direction.

The paper provides generating functions for selected partial differential equations. Results of the calculations prove high effectiveness of the use of heat polynomials for purposes of solving the inverse problems of heat flow.

Acknowledgements

The present work has been effected in the framework of KBN Grant 8T 10B01913 and author's stay as holder of Humboldt Fellowship at the Chair of Steam and Gas Turbines TU Dresden - Germany.

References

1. P.C. Rosenbloom, D.V. Widder , *Trans.Am.Math.Soc.*, **92**, 220-266, (1959).
2. M.J. Ciałkowski, Selected methods and algorithms for solution of inverse problem of heat conduction equation *WNPP* 1-189, (1996) (in Polish).
3. M.J. Ciałkowski, *Report of Thermal Engineering Chair of TU Poznań*, No56-220/1998, 1-30 (in Polish).
4. M.J. Ciałkowski, *X Sympozjum Wymiany Ciepła i Masy, Świeradów Zdrój*, 1-6, (September 1998).

HEAT POLYNOMIALS APPLIED TO DIRECT AND INVERSE HEAT CONDUCTION PROBLEMS

MICHAŁ CIAŁKOWSKI

Poznan University of Technology, ul. Piotrowo 3, 60-965 Poznan, Poland

E-mail: cialkows@sol.put.poznan.pl

SYLWIA FUTAKIEWICZ AND LESZEK HOŻEJOWSKI

Kielce University of Technology, Al. 1000-lecia P.P. 7, Kielce, Poland

E-mail: grysa@eden.tu.kielce.pl

The paper gives theoretical principles of the heat polynomials method. Heat polynomials for multidimensional heat conduction problems in the Cartesian coordinate system as well as for time dependent axisymmetrical problems in the cylindrical coordinate system are generated. The description of the method in reference to direct and inverse problems is given.

1 Introduction

Though computer methods and techniques instantly develop, analytical and semi-analytical methods have not lost their importance. The method described in this paper belongs to the latter category.

In genaral, all methods of solving the initial-boundary problems can be divided into three groups, depending on whether the obtained solution satisfies the differential equation and the boundary conditions accurately or approximately:

Group I

- differential equation solution: analytical,
- initial and boundary conditions: fulfilled accurately

The group consists of methods based on spectral and integral analysis and others.

Group II

- differential equation solution: approximate,
- initial and boundary conditions: fulfilled approximately.

The group consists of (among others) FEM, BEM, finite differences and their combinations.

Group III

- differential equation solutions: exact,
- initial and boundary condition: fulfilled approximately.

The method employing heat polynomials belongs to this group.

2 Heat polynomials determining and properties.

The heat polynomials have been introduced by Rosenbloom and Widder [1] who considered expanding the so-called temperature functions in series of heat polynomials and associated functions. Also, properties of the heat polynomials as well as convergence criterion were discussed in the paper.

In our paper we show how heat polynomials (for various problems) can be generated. The approach allows to generalise the method for two- and three-dimensional problems.

It is not difficult to note that the function [1]

$$w(x,t,p) = e^{px+p^2 t} \tag{1}$$

with p being a parameter (real or complex), satisfies the heat conduction equation:

$$\frac{\partial^2 w}{\partial x^2} = \frac{\partial w}{\partial t} \qquad x \in R, \quad t > 0 \tag{2}$$

Let us examine its basic properties. When expanding the function $w(x,t,p)$ in a power series with respect to the parameter p we arrive to the formula

$$w(x,t,p) = e^{px+p^2 t} = \sum_{n=0}^{\infty} v_n(x,t) \cdot p^n \qquad t \geq 0, \quad x \in R \tag{3}$$

The coefficients of the expansion (3) are the polynomial solutions of (2) and for that reason they have been called „heat polynomials". In [1] $v_n(x,t)$ are defined as coefficients of $p^n/n!$; the present approach gives better numerical properties of $v_n(x,t)$.

Differentiating the relation (3) with respect to x leads to the following formula:

$$\frac{\partial}{\partial x} e^{px+p^2 t} = \frac{\partial}{\partial x} \sum_{n=0}^{\infty} v_n(x,t) \cdot p^n = \sum_{n=0}^{\infty} \frac{\partial v_n}{\partial x} \cdot p^n \tag{4}$$

On the other hand, by virtue of (3), we have

$$\frac{\partial}{\partial x} e^{px+p^2 t} = p e^{px+p^2 t} = p \sum_{n=0}^{\infty} v_n(x,t) \cdot p^n = \sum_{n=1}^{\infty} v_{n-1}(x,t) p^n \tag{5}$$

Comparing the right sides of eqs. (4) and (5) we arrive to

$$\frac{\partial v_0}{\partial x} + \sum_{n=1}^{\infty} \frac{\partial v_n}{\partial x} \cdot p^n = \sum_{n=1}^{\infty} v_{n-1}(x,t) p^n \tag{6}$$

The respective coefficients of p^n on both sides of the eq. (6) have to be the same. Hence

$$\frac{\partial v_0(x,t)}{\partial x} = 0, \quad \frac{\partial v_n(x,t)}{\partial x} = v_{n-1}(x,t), \quad n = 1,2,\ldots \tag{7}$$

Differentiating the relation (7) with respect to x we obtain

$$\frac{\partial^2 v_n(x,t)}{\partial x^2} = \frac{\partial v_{n-1}(x,t)}{\partial x} = v_{n-2}(x,t) \quad , \quad n = 2,3,... \tag{8}$$

and generally

$$\frac{\partial^k v_n(x,t)}{\partial x^k} = v_{n-k}(x,t) \quad , \quad n = k, k+1,... \tag{9}$$

Because the polynomials $v_n(x,t)$ satisfy the heat conduction equation, hence

$$\frac{\partial v_0(x,t)}{\partial t} = 0 \ , \ \frac{\partial v_1(x,t)}{\partial t} = 0 \ , \ \frac{\partial v_n(x,t)}{\partial t} = v_{n-2}(x,t) \ , \ n = 2,3,... \tag{10}$$

Now an explicit form of the heat polynomial of degree n will be shown. When we put $p=0$ in (3), we get $v_0(x,t)=1$. Next, differentiating the formula (3) with respect to p and then putting $p=0$ leads to the explicit form of successive polynomials. Finally, the n-th heat polynomial reads [1]

$$v_n(x,t) = \sum_{k=0}^{[n/2]} \frac{x^{n-2k} t^k}{(n-2k)!\,k!} \tag{11}$$

where $[\cdot]$ denotes the greatest integer of „$\cdot$".

The same result may be obtained by expanding the functions $exp(px)$ and $exp(p^2t)$ into Taylor series and then taking their product.

The recurrence formula is more advisable as far as numerical applications are concerned. As shown in [2], we integrate $w(x,t,p)$ with a weight function $(x+2pt)$:

$$\int_0^p e^{px+p^2t}(x+2pt)dp = \int_0^p \frac{d}{dp}\left(e^{px+p^2t}\right)dp = e^{px+p^2t} - 1 = \sum_{n=0}^{\infty} v_n(x,t)p^n - 1 \tag{12}$$

Using the formula (3) we have

$$\int_0^p \sum_{n=0}^{\infty} v_n(x,t)p^n(x+2pt)dp = x\sum_{n=0}^{\infty} v_n(x,t)\frac{p^{n+1}}{n+1} + 2t\sum_{n=0}^{\infty} v_n(x,t)\frac{p^{n+2}}{n+2} \tag{13}$$

Comparing right sides of eqs. (12) and (13) we find

$$\sum_{n=0}^{\infty} v_n(x,t)p^n - 1 = x\sum_{n=1}^{\infty} v_{n-1}(x,t)\frac{p^n}{n} + 2t\sum_{n=2}^{\infty} v_{n-2}(x,t)\frac{p^n}{n} \tag{14}$$

Hence, we get the following recurrence formula:

$$
\begin{aligned}
&v_0(x,t) = 1 \\
&v_1(x,t) = x \\
&v_n(x,t) = \frac{x}{n} v_{n-1}(x,t) + \frac{2t}{n} v_{n-2}(x,t) \quad , \quad n \geq 2
\end{aligned}
\tag{15}
$$

It is worth to note that the coefficients in the formula (15) are decreasing and for any fixed *(x,t)* we have, [2]:

$$
\lim_{n\to\infty} \left| \frac{\partial^k v_n(x,t)}{\partial x^k} \right| = \lim_{n\to\infty} \left| v_{n-k}(x,t) \right| < \infty \quad , \quad n=k,k+1,... \tag{16}
$$

and

$$
\lim_{n\to\infty} \left| \frac{\partial v_n(x,t)}{\partial t} \right| = \lim_{n\to\infty} \left| v_{n-2}(x,t) \right| < \infty \quad , \quad n=2,3,... \tag{17}
$$

3 Heat polynomials method applied to transient direct and inverse heat conduction problems.

In the paper [3] H. Yano, S. Fukutami and A. Kieda proposed a new technique for solving a transient heat conduction problem. In their approach the temperature of a solid was approximated with a linear combination of the heat polynomials. Now we shall present generalisation of their approach and show the way of applying this approach to inverse heat conduction problems.

Consider a transient heat conduction problem described in dimensionless coordinates with the equation

$$
\frac{\partial^2 T}{\partial x^2} = \frac{\partial T}{\partial t}, \qquad 0 < x < 1, \ t > 0 \tag{18}
$$

and the conditions

$$
k_1 \frac{\partial T}{\partial x} + h_1 T = \varphi(t) \qquad \text{at } x=0, \ t > 0 \tag{19}
$$

$$
k_2 \frac{\partial T}{\partial x} + h_2 T = \psi(t) \qquad \text{at } x=1, \ t > 0 \tag{20}
$$

$$
T = F(x) \qquad \text{at } 0 \leq x \leq 1, \ t = 0 \tag{21}
$$

In the k-th time interval chosen as follows: $k\Delta t \leq t \leq (k+1)\Delta t$, $\Delta t > 0$ (t - dimensionless time, $k = 0,1,2,...$) the approximate solution of the problem (18)-(21) will be found as a linear combination of the heat polynomials $v_n(x,t)$:

ERRATA

These two pages were inadvertently omitted from the volume.

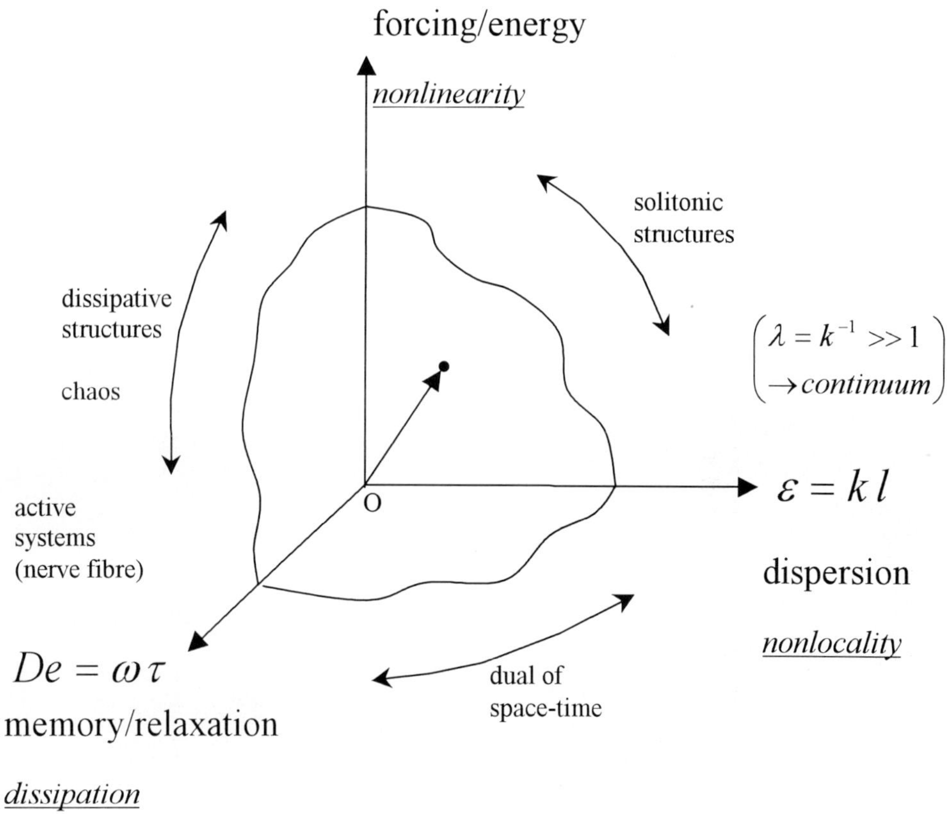

The author of this diagram is Prof. Gerard A. Maugin from the Université Pierre et Marie Curie (Paris VI), Modelisation en Mécanique, Paris, France.

Mit finanzieller Unterstützung der Stiftung für deutsch-polnische Zusammenarbeit aus Mitteln der Bundesrepublik Deutschland
(Z finansowym wsparciem Fundacji Współpracy Polsko-Niemieckiej ze środków Republiki Federalnej Niemiec
With the financial support of the Polish-German Collaboration Foundation based on funds of the Federal Republic of Germany)

and also

The Batory Foundation, Warsaw, Poland
The State Committee for Scientific Research, Warsaw, Poland
The Poznań Society of Friends of Arts and Sciences, Poznań, Poland

Participants of TRECOP '98

$$\theta_k(x,t)=\sum_{n=0}^{N} c_n^{(k)} v_n(x,t) \tag{22}$$

The coefficients $c_n^{(k)}$ are selected to minimise the functional which is the sum of three components corresponding to
- initial condition

$$\left\|\theta_k(x,k\Delta t)-\theta_{k-1}(x,k\Delta t)\right\|^2=\int_0^1\left(\theta_k(x,k\Delta t)-\theta_{k-1}(x,k\Delta t)\right)^2 dx \tag{23}$$

where $\theta_{-1}(x,0)=F(x)$.
- condition on the boundary $x=0$

$$\left\|\theta_k(0,t)-\varphi(t)\right\|^2=\int_{k\Delta t}^{(k+1)\Delta t}\left[k_1\frac{\partial\theta_k(0,t)}{\partial x}+h_1\theta_k(0,t)-\varphi(t)\right]^2 dt \tag{24}$$

- condition on the boundary $x=1$

$$\left\|\theta_k(1,t)-\psi(t)\right\|^2=\int_{k\Delta t}^{(k+1)\Delta t}\left[k_2\frac{\partial\theta_k(1,t)}{\partial x}+h_2\theta_k(1,t)-\psi(t)\right]^2 dt \tag{25}$$

Finally, we minimise the functional

$$\begin{aligned} I_k=&\int_0^1\left(\theta_k(x,k\Delta t)-\theta_{k-1}(x,k\Delta t)\right)^2 dx+\\ &+\int_{k\Delta t}^{(k+1)\Delta t}\left[\left(k_1\frac{\partial\theta_k(0,t)}{\partial x}+h_1\theta_k(0,t)-\varphi(t)\right)^2+\left(k_2\frac{\partial\theta_k(1,t)}{\partial x}+h_2\theta_k(1,t)-\psi(t)\right)^2\right]dt \end{aligned} \tag{26}$$

$k=0,1,\ldots$. Instead of $\theta_k(x,t)$ with $t\in\langle k\Delta t,(k+1)\Delta t\rangle$ we can use $\theta_k(x,\tau)$ with $\tau=t-k\Delta t$. The unknown coefficients $c_n^{(k)}$ $(n=0,1,\ldots,N)$ are solution of the system of linear equations

$$\frac{\partial I_k}{\partial c_n^{(k)}}=0 \qquad (n=0,1,\ldots,N) \tag{27}$$

The method can be applied to inverse heat conduction problems as shown in [4]. If the temperature at the inner points, say x_1 and x_2, $0< x_1<x_2<1$, is known, we find the solution of the problem for $x\in[x_1,x_2]$ and extrapolate it on $[0,1]$.

4 Generalisation on 2- and 3-dimensional problems.

The idea of heat polynomials is easy to be generalised for 2D and 3D. The key problem is to find proper generating functions. They are given in the table 1.

Table 1. Generating functions

Space dimension	*2D*	*3D*
Differential equation	$\frac{\partial^2 T}{\partial x^2}+\frac{\partial^2 T}{\partial y^2}=\frac{\partial T}{\partial t}$	$\frac{\partial^2 T}{\partial x^2}+\frac{\partial^2 T}{\partial y^2}+\frac{\partial^2 T}{\partial z^2}=\frac{\partial T}{\partial t}$
Generating function	$w=e^{px+qy+t(p^2+q^2)}$	$w=e^{px+qy+sz+t(p^2+q^2+s^2)}$
Heat polynomials	$V_{mn}(x,y,t)=v_m(x,t)v_n(y,t)$	$U_{lmn}(x,y,t)=v_l(x,t)v_m(y,t)v_n(z,t)$

p,q,s - parameters

Application of V_{mn} and U_{lmn} to direct or inverse problems is the same as described in the previous paragraph.

5 Heat polynomials in steady state problems

In steady state problems the harmonic polynomials will be used. The generating function in this case is

$$w=e^{pz}=\sum_{n=0}^{\infty}(u_n(x,y)+\mathrm{i}v_n(x,y))p^n \tag{28}$$

where $z=x+\boldsymbol{i}y$ ($\boldsymbol{i}$ - square root of -1). Alternatively we define the harmonic polynomials as the real and imaginary part, respectively, of the holomorphic function $z^n/n!$. Hence we have

$$\begin{aligned} u_n(x,y)&=\mathrm{Re}\left(z^n/n!\right)\\ v_n(x,y)&=\mathrm{Im}\left(z^n/n!\right)\end{aligned} \qquad n=0,1,... \tag{29}$$

The recurrence formulas for u_n and v_n are the following:

$$\begin{aligned} &u_0=1, \quad v_0=0\\ &u_{n+1}=\frac{x}{n+1}u_n-\frac{y}{n+1}v_n\\ &v_{n+1}=\frac{x}{n+1}v_n+\frac{y}{n+1}u_n \quad , \qquad n=1,2,... \end{aligned} \tag{30}$$

It is worth to mention that the coefficients on the right side of the derived formulas are decreasing for each fixed value of x and y. The technique of using such polynomials for solving steady state problems does not differ from the one described for transient problems.

6 Heat polynomials in the polar and cylindrical coordinate system.

As in the previous paragraphs, we want to define the polynomials satisfying the heat conduction equation in polar coordinates:

$$\frac{\partial^2 T}{\partial r^2}+\frac{1}{r}\frac{\partial T}{\partial r}=\frac{\partial T}{\partial t} \tag{31}$$

The transformation from (x,y) to (r,φ) applied to $e^{\frac{p}{\sqrt{2}}(x+y)+p^2 t}$ gives $e^{\frac{p}{\sqrt{2}}r(\cos\varphi+\sin\varphi)+p^2 t}$ as the generating function. In case of temperature not depending on φ, we remove the variable φ by integrating with respect to φ which yields

$$\frac{1}{2\pi r}\int_0^{2\pi} e^{\frac{p}{\sqrt{2}}r(\cos\varphi+\sin\varphi)+p^2 t}\, r d\varphi = I_0(pr)e^{p^2 t} \tag{32}$$

where $I_0(.)$ denotes the modified Bessel functions of the first of the order zero.

Expanding both functions, $e^{p^2 t}$ and $I_o(pr)$, in a power series, we arrive to the formula

$$I_0(pr)e^{p^2 t}=\sum_{n=0}^{\infty} q_n(r,t)p^{2n} \tag{33}$$

where the coefficients $q_n(r,t)$, called heat polynomials, given with the formula

$$q_n(r,t)=\sum_{k=0}^{n}\frac{2^{2k-2n}}{[(n-k)!]^2\,k!}r^{2n-2k}t^k \tag{34}$$

are polynomials of degree $2n$.

The heat polynomials, Q_{mn}, for axisymmetric problems in the cylindrical coordinates are the products of $v_n(z,t)$ and $q_m(r,t)$, i.e. $Q_{mn}(r,z,t)=v_n(z,t)q_m(r,t)$.

7 Acknowledgements

The first author did his research in part as Humboldt-Fellowship at The Chair of Steam and Gasturbines of TU Dresden and in part under grant KBN 8T10B01913.

References

1. P.C.Rosenbloom and D.V.Widder, Expansions in terms of heat polynomials and associated functions, *Trans. Am. Math. Soc.* **92** (1959) pp. 220-266.
2. M.Ciałkowski, Solutions of heat conduction problems with the use of heat polynomials (in Polish), *Raport Katedry Techniki Cieplnej Politechniki Poznanskiej* BW56-220/98 (1998) pp. 1-30.
3. H.Yano, S.Fukutani and A.Kieda, A boundary residual method with heat polynomials for solving unsteady heat conduction problems, *Journal of the Franklin Institute* **316**,.4 (1983) pp. 291-298.
4. Futakiewicz S., Hożejowski L., Heat polynomials method in the n-dimensional direct and inverse heat conduction problems, in: *Advanced Computational Methods in Heat Transfer V*, ed. by Nowak A.J., Brebbia C.A, Białecki R., Zerroukat M. (Computational Mechanics Publications, 1998) pp 103-112.

MINKOWSKI SPACE AS A MODEL OF ANISOTROPIC PORE SPACE OF PERMEABLE MATERIALS: MODELLING OF THE SKELETON PORE STRUCTURE

M. CIESZKO
Department of Environmental Mechanics
Pedagogical University of Bydgoszcz
85-064 Bydgoszcz, Chodkiewicza 30
E-mail: mkk @ rose.man.poznan.pl

In the paper the anisotropic pore space structure of a rigid permeable porous materials is described. A new macroscopic model of saturated porous medium is proposed in which a fluid flow through porous skeleton of anisotropic pore structure is considered as a motion of the material continuum in the plane anisotropic metric space (Minkowski space) immersed in Euclidean one. The metrics of both spaces are used to determine the double measures of any line, surface and volume elements. Within such approach the pure geometrical macroparameters characterising anisotropic pore structure of porous materials, namly: the tortuosity, the volume and surface porosity. It has been shown that the isotropic pore structure is described only by two independent scalar parameters.

1 Introduction

Most of the existing macrocontinuum models of mechanical behaviour of fluid-saturated porous materials (e.g. [1]-[3]) are based on the fundamental principle of the mixture theory ([4], [5]) in which the physical components of the medium, i.e. porous skeleton and fluid filling its pores are considered as superimposed continua of equivalent kinematics. In the mixture theory the kinematical privilege of porous skeleton when compared with a fluid motion is not taken into account explicitly. This leads to considerable difficulties in the complete description of the skeleton pore structure and the related problem of the dynamical coupling between fluid and skeleton. These difficulties appear in description of the isotropic structure of porous skeleton and in the anisotropic case remain still unsolved.

The subject of this paper is to model the anisotropic pore space structure in rigid porous materials capable of fluid flow through pores. In the paper the new macroscopic model of a saturated porous medium is proposed in which a fluid flow through porous skeleton of an anisotropic pore structure is considered as a motion of the material continuum in the plane anisotropic metric space. This model takes into consideration the fundamental fact of kinematics of a fluid-saturated porous solid that permeable skeleton forms the real space for a fluid motion. In such approach restrictions imposed by the structure of porous skeleton on a fluid motion correspond to those imposed by the metric structure of an anisotropic space on a motion of the material continuum (a fluid). Therefore quantities describing anisotropic properties of the metric space can be interpreted as parameters characterising the pore space structure of the skeleton.

The concept of the plane anisotropic space is used in the modern differential geometry as Minkowski space ([6], [7]). This space and its generalisation -Finsler space (the curved one) ([6], [8]) becomes useful in the modelling of various physical, biological and mechanical phenomena ([9]-[11]).

In the paper the notion of Minkowski space immersed in Euclidean one is used as a model of the pore space of porous materials immersed in the physical space. The new metric tensors of surface and volume elements in these spaces have been proposed that are directly related to the metric tensors of distance. This enables one to define the geometrical parameters characterising pore structure of such materials: the tortuosity, the volume and surface porosity.

2 Definitions and notations

Vectors and tensors. We denote by $\mathcal{V}$ a three-dimensional real vector space and by $\mathcal{V}^*$ the dual vector space of $\mathcal{V}$. If $\mathbf{u} \in \mathcal{V}$ and $\mathbf{v}^* \in \mathcal{V}^*$ then the scalar $\mathbf{u}\cdot\mathbf{v}^* = \mathbf{v}^*\cdot\mathbf{u} \in R$ we shall call the dual product of a vector $\mathbf{u}$ and covector $\mathbf{v}^*$ and by "dot" $((\;)\cdot(\;))$ we shall denote the bilinear operation (the dual multiplication) defined on $\mathcal{V}$ and $\mathcal{V}^*$.

The multilinear transformations of vector spaces are called tensors. They are elements of linear spaces formed by the tensor products of vector spaces. For example, a tensor $\mathbf{A} \in \mathcal{V}\otimes\mathcal{V}^*$ is an endomorphism of $\mathcal{V}$ and $\mathcal{V}^*$. For $\mathbf{u} \in \mathcal{V}$ and $\mathbf{v}^* \in \mathcal{V}^*$ we have

$$\mathbf{A}\cdot\mathbf{u} \in \mathcal{V}, \qquad \mathbf{v}^*\cdot\mathbf{A} \in \mathcal{V}^*.$$

The tensor product of vectors $\mathbf{u}\otimes\mathbf{v}^*$ is the simplest form of a tensor in the space $\mathcal{V}\otimes\mathcal{V}^*$.

Since, in the further considerations we shall have to do with tensors of order three and four, to simplify operations on such objects we introduce new, alternative notations for tensor products. For $\mathbf{t}, \mathbf{u}, \mathbf{v}, \mathbf{w} \in \mathcal{V}$ we take

$$\mathbf{t}\otimes\mathbf{u} \equiv \begin{pmatrix}\mathbf{t}\\ \mathbf{u}\end{pmatrix} \in \otimes\mathcal{V}^2 = \mathcal{V}\otimes\mathcal{V}, \qquad \mathbf{t}\otimes\mathbf{u}\otimes\mathbf{v} \equiv \begin{pmatrix}\mathbf{t}\\ \mathbf{u}\\ \mathbf{v}\end{pmatrix} \in \otimes\mathcal{V}^3,$$

$$\mathbf{t}\otimes\mathbf{u}\otimes\mathbf{v}\otimes\mathbf{w} \equiv \begin{pmatrix}\mathbf{t}\;\mathbf{v}\\ \mathbf{u}\;\mathbf{w}\end{pmatrix} \equiv \begin{pmatrix}\mathbf{t}\otimes\mathbf{v}\\ \mathbf{u}\otimes\mathbf{w}\end{pmatrix} \equiv \begin{pmatrix}\mathbf{t}\\ \mathbf{u}\end{pmatrix}\otimes\begin{pmatrix}\mathbf{v}\\ \mathbf{w}\end{pmatrix} \in \otimes\mathcal{V}^4. \tag{1}$$

Such representations of the tensor products renders their factors accessible for various operations, e.g. the dual multiplication or transposition. Taking into account the fact that each tensor can be represented by a linear combination of tensor products of basis vectors of $\mathcal{V}$ (and of $\mathcal{V}^*$), the notations (1) will have also consequences for notations of operations performed on a general form of tensors.

Representations of surface and volume elements. Similarly as a vector is the algebraic model of a directed segment, the skew-symmetric tensors of order two and three are natural models of surface and volume elements [12], respectively. These elements spanned by two $(\mathbf{u}, \mathbf{v})$ and three $(\mathbf{u}, \mathbf{v}, \mathbf{w})$ linearly independent vectors can be represented by their exterior products $\mathbf{u}\wedge\mathbf{v}$ and $\mathbf{u}\wedge\mathbf{v}\wedge\mathbf{w}$. We have

$$\sigma \equiv \mathbf{u}\wedge\mathbf{v} \equiv \mathbf{u}\otimes\mathbf{v} - \mathbf{v}\otimes\mathbf{u} \equiv \begin{bmatrix}\mathbf{u}\\ \mathbf{v}\end{bmatrix} = \begin{pmatrix}\mathbf{u}\\ \mathbf{v}\end{pmatrix} - \begin{pmatrix}\mathbf{v}\\ \mathbf{u}\end{pmatrix},$$

$$\vartheta \equiv \mathbf{u}\wedge\mathbf{v}\wedge\mathbf{w} \equiv \begin{bmatrix}\mathbf{u}\\ \mathbf{v}\\ \mathbf{w}\end{bmatrix} \equiv \begin{pmatrix}\mathbf{u}\\ \mathbf{v}\\ \mathbf{w}\end{pmatrix} + \begin{pmatrix}\mathbf{v}\\ \mathbf{w}\\ \mathbf{u}\end{pmatrix} + \begin{pmatrix}\mathbf{w}\\ \mathbf{u}\\ \mathbf{v}\end{pmatrix} - \begin{pmatrix}\mathbf{u}\\ \mathbf{w}\\ \mathbf{v}\end{pmatrix} - \begin{pmatrix}\mathbf{v}\\ \mathbf{u}\\ \mathbf{w}\end{pmatrix} - \begin{pmatrix}\mathbf{w}\\ \mathbf{v}\\ \mathbf{u}\end{pmatrix}.$$

The surface elements σ form the three-dimensional space $\wedge\mathcal{V}^2$ of the skew-symmetric tensors of order two,

$$\sigma \in \wedge\mathcal{V}^2 \subset \otimes\mathcal{V}^2 ,$$

and the volume elements ϑ form the one-dimensional space $\wedge\mathcal{V}^3$ of the skew-symmetric tensors of order three,

$$\vartheta \in \wedge\mathcal{V}^3 \subset \otimes\mathcal{V}^3 .$$

Any two elements of the space $\wedge\mathcal{V}^3$ are related by a scalar multiplier. For example, two skew-symmetric tensors of the form

$$\begin{bmatrix}\mathbf{u}\\ \mathbf{v}\\ \mathbf{w}\end{bmatrix}, \quad \begin{bmatrix}\mathbf{A}\cdot\mathbf{u}\\ \mathbf{A}\cdot\mathbf{v}\\ \mathbf{A}\cdot\mathbf{w}\end{bmatrix} \in \wedge\mathcal{V}^3$$

are related by the scalar that is the determinant of an automorphism $\mathbf{A} \in \mathcal{V}\otimes\mathcal{V}^*$. We have

$$\begin{bmatrix}\mathbf{A}\cdot\mathbf{u}\\ \mathbf{A}\cdot\mathbf{v}\\ \mathbf{A}\cdot\mathbf{w}\end{bmatrix} = \begin{pmatrix}\mathbf{A}\\ \mathbf{A}\\ \mathbf{A}\end{pmatrix} \begin{matrix}\cdot\\ \cdot\\ \cdot\end{matrix} \begin{bmatrix}\mathbf{u}\\ \mathbf{v}\\ \mathbf{w}\end{bmatrix} = \det(\mathbf{A})\begin{bmatrix}\mathbf{u}\\ \mathbf{v}\\ \mathbf{w}\end{bmatrix} . \tag{2}$$

This identity may be treated as the definition of determinant of a tensor $\mathbf{A}$.

The normed vector spaces. The vector space $\mathcal{V}$ is the normed space if there is a real-valued function (a norm) defined on $\mathcal{V}$

$$\mathrm{L} : \mathcal{V} \to \mathrm{R} , \qquad \mathrm{L}(\mathbf{u}) = \mathrm{u} ,$$

which satisfies the following axioms:

$$\mathrm{L}(\mathbf{u}) > 0 \qquad \text{for } \mathbf{u} \neq \mathbf{0} \text{ and } \mathrm{L}(\mathbf{0}) = 0 \tag{3}$$

$$\mathrm{L}(\mathrm{k}\,\mathbf{u}) = \mathrm{k}\,\mathrm{L}(\mathbf{u}) \qquad \text{for } \mathrm{k} > 0 , \tag{4}$$

$$\mathrm{L}(\mathbf{u} + \mathbf{v}) \leq \mathrm{L}(\mathbf{u}) + \mathrm{L}(\mathbf{v}) \tag{5}$$

for all $\mathbf{u}, \mathbf{v} \in \mathcal{V}$. We assume that $\mathrm{L}(\mathbf{u})$ is of the class C^{∞}.
The norm $\mathrm{L}(\mathbf{u})$ enables one to define a distance between elements of $\mathcal{V}$ satisfying all axioms of the generalised metric. We have

$$\mathrm{d}(\mathbf{u}, \mathbf{v}) \equiv \mathrm{L}(\mathbf{v} - \mathbf{u}) . \tag{6}$$

The metric (6) is invariant under translation and, in general, does not have to be symmetric

$$\mathrm{d}(\mathbf{u}, \mathbf{v}) \neq \mathrm{d}(\mathbf{v}, \mathbf{u}) .$$

Similarly, the norm for any other linear space can be defined.

Affine spaces. The pair $(P, \mathcal{V})$ composed of point P and of vector space $\mathcal{V}$ we shall identify with the affine point space. It is possible because structures of both objects are isomorphic. The point P is called the reference point and $\mathcal{V}$ is the space of placement vectors of points. The affine space $(P, \mathcal{V})$ is normed if the space $\mathcal{V}$ of the placement vectors is normed.

3 Geometry of Euclidean point space

The Euclidean point space is the simplest special case of the normed affine point space. The affine space is Euclidean when its norm is given by expression

$$\mathrm{L}^2(\mathbf{u}) = \mathbf{u}\cdot\mathbf{M}\cdot\mathbf{u} \equiv \mathbf{M} : \begin{pmatrix} \mathbf{u} \\ \mathbf{u} \end{pmatrix} \tag{7}$$

where $\mathbf{M} \in \mathcal{V}^* \otimes \mathcal{V}^*$ is independent of $\mathbf{u}$ and is called the metric tensor of Euclidean space.
The tensor $\mathbf{M}$ is non-singular, symmetric and positive definite and therefore expression (7) satisfies the axioms of the symmetric norm.

The pair $(\mathcal{V}, \mathbf{M})$ form Euclidean vector space and triple $(P, (\mathcal{V}, \mathbf{M}))$ Euclidean point space. This space because of affine structure is plane and due to independence of the metric tensor $\mathbf{M}$ of the measured vector $\mathbf{u}$ is homogeneous and isotropic. Euclidean point space is a model of the physical space used in the classical mechanics.

The existence of the metric of distance in Euclidean space $(P, (\mathcal{V}, \mathbf{M}))$ enables one to define the metric for elements of surface $\sigma \in \wedge\mathcal{V}^2$ and volume $\vartheta \in \wedge\mathcal{V}^3$. We take

$$\mathrm{S}^2(\sigma) = \sigma : \mathbf{S} : \sigma \ , \qquad \mathbf{V}^2(\vartheta) = \vartheta : \mathbf{V} : \vartheta \tag{8}$$

where

$$\mathbf{S} = \frac{1}{2!}\begin{pmatrix}\mathbf{M}\\ \mathbf{M}\end{pmatrix}, \qquad \mathbf{V} = \frac{1}{3!}\begin{pmatrix}\mathbf{M}\\ \mathbf{M}\\ \mathbf{M}\end{pmatrix} \tag{9}$$

are the metric tensors in the spaces of surface elements $\wedge\mathcal{V}^2$ and volume elements $\wedge\mathcal{V}^3$, respectively. These tensors are of order four and six.

The metrics defined by expressions (8) satisfy all axioms of the norm. They are positive definite, homogeneous, symmetric and strictly convex.

4 Geometry of Minkowski point space

The normed affine point space the metrics of which is defined by expression (6) has the general properties (3)-(5) is called Minkowski space [6]. This space, similarly as the Euclidean one, is plane and the anisotropic properties and luck of symmetry of distance determine the peculiar features of its internal, geometrical structure. The metric tensor of distance in Minkowski space takes the form

$$\mathbf{M}_A(\mathbf{u}) = \frac{1}{2}\frac{\partial^2 \mathrm{L}^2(\mathbf{u})}{\partial\mathbf{u}\partial\mathbf{u}} \qquad \in \mathcal{V}^*\otimes\mathcal{V}^* \tag{10}$$

and due to the homogeneity of $\mathrm{L}(\mathbf{u})$ given by (4) has the following properties:

$$\mathbf{u}\cdot\mathbf{M}_A(\mathbf{u})\cdot\mathbf{u} \equiv \mathrm{L}^2(\mathbf{u}) , \tag{11}$$

$$\mathbf{M}_A(k\,\mathbf{u}) \equiv \mathbf{M}_A(\mathbf{u}) \qquad \text{for } k > 0 , \tag{12}$$

$$2\mathbf{M}_A(\mathbf{u})\cdot\mathbf{u} \equiv \frac{\partial \mathrm{L}^2(\mathbf{u})}{\partial\mathbf{u}} \qquad \in \mathcal{V}^* . \tag{13}$$

In expressions (10) and (13) the gradient of a function $L^2(\mathbf{u})$ is denoted by

$$\frac{\partial}{\partial \mathbf{u}} L^2(\mathbf{u}) \ .$$

It is defined by the identity

$$\mathbf{v}\cdot\frac{\partial}{\partial \mathbf{u}} L^2(\mathbf{u}) \equiv \frac{\partial}{\partial h} L^2(\mathbf{u}+h\mathbf{v})\Big|_{h=0}$$

for all $\mathbf{v} \in \mathcal{V}$.

From (12) results that the metric tensor $\mathbf{M}_A(\mathbf{u})$, in general, depends on direction of the measured vector $\mathbf{u}$ and is independent of its length. This property of tensor $\mathbf{M}_A(\mathbf{u})$ defines the anisotropic properties of Minkowski space.

We use the metric tensor $\mathbf{M}_A(\mathbf{u})$ determining distance of points in Minkowski space to define metrics of surface and volume elements in this space. Similarly as in Euclidean space, these metrics we take in the form

$$S_A^2(\sigma) = \sigma : \mathbf{S}_A(\mathbf{r}) : \sigma \ , \qquad V_A^2(\vartheta) = \vartheta \vdots \mathbf{V}_A(\mathbf{r}) \vdots \vartheta \tag{14}$$

where

$$\mathbf{S}_A(\mathbf{r}) = \alpha(\mathbf{r})\frac{1}{2!}\begin{pmatrix}\mathbf{M}_A(\mathbf{r})\\ \mathbf{M}_A(\mathbf{r})\end{pmatrix} , \qquad \mathbf{V}_A(\mathbf{r}) = \alpha(\mathbf{r})\frac{1}{3!}\begin{pmatrix}\mathbf{M}_A(\mathbf{r})\\ \mathbf{M}_A(\mathbf{r})\\ \mathbf{M}_A(\mathbf{r})\end{pmatrix} \tag{15}$$

are the metric tensors in the spaces of surface elements $\wedge\mathcal{V}^2$ and volume elements $\wedge\mathcal{V}^3$, respectively. These tensors are of order four and six. The vector $\mathbf{r}$ in (15) is orthogonal to the surface element σ with respect to the metric $\mathbf{M}_A(\mathbf{r})$. It satisfies condition

$$\sigma \cdot \mathbf{M}_A(\mathbf{r}) \cdot \mathbf{r} = \mathbf{0} \ .$$

Moreover, in expressions (15) the scalar-valued vector function $\alpha(\mathbf{r})$ has been introduced to ensure the measure of volume elements $(14)_2$ to be independent of the vector $\mathbf{r}$.

Applying the definition of determinant (2) and introducing, additionally, the Euclidean metrization to Minkowski space we obtain

$$\vartheta \vdots \begin{pmatrix}\mathbf{M}_A\\ \mathbf{M}_A\\ \mathbf{M}_A\end{pmatrix} = \vartheta \vdots \begin{pmatrix}\mathbf{M}\\ \mathbf{M}\\ \mathbf{M}\end{pmatrix} \vdots \begin{pmatrix}\mathbf{M}^{-1}\cdot\mathbf{M}_A\\ \mathbf{M}^{-1}\cdot\mathbf{M}_A\\ \mathbf{M}^{-1}\cdot\mathbf{M}_A\end{pmatrix} = \det(\mathbf{M}^{-1}\cdot\mathbf{M}_A)\, \vartheta \vdots \begin{pmatrix}\mathbf{M}\\ \mathbf{M}\\ \mathbf{M}\end{pmatrix} \ .$$

Then, the measure $(14)_2$ is independent of $\mathbf{r}$ if the coefficient $\alpha(\mathbf{r})$ is taken in the form

$$\alpha(\mathbf{r}) = \frac{\beta}{\det(\mathbf{M}^{-1}\cdot\mathbf{M}_A)} \tag{16}$$

where β is a positive constant parameter.
In that case the Minkowski measure $(14)_2$ of a volume element ϑ becomes proportional to the Euclidean measure $(8)_2$. We have

$$\beta\, V_A^2(\vartheta) = V^2(\vartheta) . \tag{17}$$

The parameter β is independent of the Minkowski metric $\mathbf{M}_A(\mathbf{r})$. It allows the possibility that the metrics of surface and volume elements in Minkowski space are not fully determined by the metric of distance.
From equality (17) results that Minkowski measure of volume $V_A(\vartheta)$ satisfy all axioms of the norm because they are satisfied by the Euclidean measure $V(\vartheta)$.

It can be shown that the metric of surface elements $(14)_1$ satisfy all axioms of the norm. It is positive definite, homogeneous and strictly convex.

5. Minkowski space as a model of anisotropic pore space of permeable porous materials

We apply the notions of Euclidean and Minkowski spaces for modelling of a pore space in permeable porous materials and to define parameters characterising its structure.

The mutually connected pores in permeable porous materials form a very complicated net of channels that determine the real space for a fluid motion. Therefore fluid particles moving in pores between any two points of the medium have much longer way to travel than distance between these points is in the physical space. In the case when the structure of the pore space is anisotropic length of particles way depend on the direction determined by these points in the porous medium and may take various values for the opposite direction of a fluid flow. Such geometry of a fluid flow in porous materials justifies the use of Minkowski metric space as a model of the pore space in porous medium.

Taking into account that porous skeleton is immersed in the physical space modelled within the classical mechanics by Euclidean point space, the anisotropic Minkowski space as a model of pore space shall be considered as immersed in Euclidean space. At the same time we shall identify vectors of points placement in both spaces. Such immersion imposes restrictions on relations of measures of the line, surface and volume elements in both spaces. It results from the physically justified requirement that immersion should not disturb the geometrical structure of the base (Euclidean) space. It means that the minimum length of a line element and the maximum volume of a volume element measured with respect to Minkowski metric can not exceed the values of measures, of these elements, with respect to Euclidean metric.

For any vector $\mathbf{u} \in \mathcal{V}$ and any volume element $\boldsymbol{\vartheta} \in \wedge\mathcal{V}^3$ we obtain the following inequalities

$$\mathbf{u}\cdot\mathbf{M}_A(\mathbf{u})\cdot\mathbf{u} \quad \geq \quad \mathbf{u}\cdot\mathbf{M}\cdot\mathbf{u}\,, \tag{18}$$

$$\boldsymbol{\vartheta} : \mathbf{V}_A(\mathbf{r}) :\boldsymbol{\vartheta} \quad \leq \quad \boldsymbol{\vartheta}:\mathbf{V}:\boldsymbol{\vartheta}\,. \tag{19}$$

From (18) and (19) results that measures of surface elements are also restricted. For any $\boldsymbol{\sigma} \in \wedge\mathcal{V}^2$ we have

$$\boldsymbol{\sigma}: \mathbf{S}_A(\mathbf{r}) : \boldsymbol{\sigma} \;\leq\; \boldsymbol{\sigma}: \mathbf{S} : \boldsymbol{\sigma}\;. \tag{20}$$

6 Pore structure parameters of permeable materials

Introduction of the measures of line, surface and volume elements in a pore space modelled by the anisotropic metric space enables one to define their relations to the measures of these elements in the physical (Euclidean) space. These relations are: the tortuosity of pores, the volume and surface porosity. They characterise pore space structure of a porous medium and play important role in the mechanics of fluid-saturated porous materials.

6.1 Tortuosity of pores

The tortuosity parameter of pores in direction determined by any vector $\mathbf{u} \in \mathcal{V}$ in the porous medium we define as the ratio of length of $\mathbf{u}$ with respect to Minkowski metric to the length of this vector with respect to Euclidean metric. Denoting the tortuosity by $\delta(\mathbf{u})$ we obtain

$$\delta^2(\mathbf{u}) = \left(\frac{\mathrm{u_A}}{\mathrm{u}}\right)^2 = \frac{\mathbf{u}\cdot\mathbf{M}_A(\mathbf{u})\cdot\mathbf{u}}{\mathbf{u}\cdot\mathbf{M}\cdot\mathbf{u}} \tag{21}$$

or

$$\delta^2(\mathbf{n}) \;=\; \mathbf{n}\cdot\mathbf{M}_A(\mathbf{n})\cdot\mathbf{n} \tag{22}$$

where $\mathbf{n} = \mathbf{u}/\mathrm{u}$ is the unit Euclidean vector (versor; $\mathbf{n}\cdot\mathbf{M}\cdot\mathbf{n} = 1$).

The parameter $\delta(\mathbf{u})$ is a homogeneous scalar valued function of order zero

$$\delta(k\mathbf{u}) = \delta(\mathbf{u})$$

for any $k > 0$.

From (22) results that the metric tensor $\mathbf{M}_A(\mathbf{n})$ fully determine the value of tortuosity $\delta(\mathbf{n})$ and therefore can be interpreted as the tortuosity tensor of pores. This tensor characterises the anisotropic properties of the pore space structure of permeable porous materials.

The tensor $\mathbf{M}_A(\mathbf{n})$ will describe the isotropic pore structure if it is proportional to the Euclidean metric tensor $\mathbf{M}$. According to definition (22) we obtain

$$\mathbf{M}_A(\mathbf{n}) = \delta^2 \mathbf{M} . \tag{23}$$

Taking into account the restriction (18) we obtain that tortuosity parameter $\delta(\mathbf{n})$ takes values no less than one,

$$\delta(\mathbf{n}) \geq 1 . \tag{24}$$

6.2 Volume porosity

Similarly as the tortuosity we can define the volume porosity f_v of porous materials as the ratio of measures of any volume element $\vartheta \in \wedge\mathcal{V}^3$ in Minkowski and Euclidean spaces. We have

$$f_v^2 = \left(\frac{\mathrm{V_A}}{\mathrm{V}}\right)^2 = \frac{\vartheta \vdots \mathbf{V}_A \vdots \vartheta}{\vartheta \vdots \mathbf{V} \vdots \vartheta} = \mathbf{E}_3 \vdots \mathbf{V}_A \vdots \mathbf{E}_3 \tag{25}$$

where $\mathbf{E}_3 = \vartheta / V$ is the Euclidean unit volume element ($\mathbf{E}_3 \vdots \mathbf{V} \vdots \mathbf{E}_3 = 1$).

The volume porosity parameter f_v due to the restriction (19) has to satisfy the condition

$$f_v \leq 1 . \tag{26}$$

The definition (25) and relation (17) enables one to determine the constant parameter β in the coefficient $\alpha(\mathbf{r})$ of metric tensors (15). We obtain

$$\beta = f_v^2 . \tag{27}$$

6.3 Surface porosity

The third parameter characterising pore space structure of porous materials is the surface porosity $\lambda(\sigma)$. This parameter is defined as the ratio of measures of any surface element $\sigma \in \wedge\mathcal{V}^2$ with respect to Minkowski and Euclidean metrics. Taking into account definitions $(8)_1$ and $(14)_1$ we have

$$\lambda^2(\boldsymbol{\sigma})=\left(\frac{S_A}{S}\right)^2 = \frac{\boldsymbol{\sigma}:\mathbf{S}_A(\mathbf{N}):\boldsymbol{\sigma}}{\boldsymbol{\sigma}:\mathbf{S}:\boldsymbol{\sigma}} \tag{28}$$

or

$$\lambda^2(\sigma) = \mathbf{E}_2:\mathbf{S}_A(\mathbf{N}):\mathbf{E}_2 \tag{29}$$

where $\mathbf{E}_2 = \boldsymbol{\sigma} / S$ is the Euclidean unit surface element,

$$\mathbf{E}_2:\mathbf{S}:\mathbf{E}_2 = 1 \ ,$$

and $\mathbf{N}$ is the Minkowski versor orthogonal to the surface element $\boldsymbol{\sigma}$ (and also to $\mathbf{E}_2$) with respect to the metric $\mathbf{M}_A(\mathbf{N})$.

From the definition (29) results that the metric tensor of surface element $\mathbf{S}_A(\mathbf{r})$ can be interpreted as the tensor of surface porosity.

The surface porosity parameter λ due to the restriction (20) has to satisfy the condition

$$\lambda \leq 1 \ . \tag{30}$$

Representing $\mathbf{E}_2$ by its Euclidean normal versor $\mathbf{n}$ as follows

$$\mathbf{E}_2 = \mathbf{E}_3^{\cdot\mathbf{M}\cdot\mathbf{n}} \ ,$$

the definition (29) takes form

$$\lambda^2(\sigma) = \mathbf{n}\cdot\mathbf{M}_S(\mathbf{N})\cdot\mathbf{n} \tag{31}$$

where

$$\mathbf{M}_S(\mathbf{N}) = {}^{\mathbf{M}\cdot}\mathbf{E}_3:\mathbf{M}_A(\mathbf{N}):\mathbf{E}_3{}^{\cdot\mathbf{M}} \tag{32}$$

is the second order tensor of surface porosity.

Applying the definition of determinant (2) and the identity

$$\mathbf{E}_3:\begin{pmatrix}\mathbf{M}\\ \mathbf{M}\end{pmatrix}:\mathbf{E}_3{}^{\cdot\mathbf{M}} = 2\mathbf{I} \ ,$$

from (32) we obtain

$$\mathbf{M}_S(\mathbf{N}) = f_v^2\ \mathbf{M}\cdot\mathbf{M}_A^{-1}(\mathbf{N})\cdot\mathbf{M} \tag{33}$$

where $\mathbf{I} \in \mathcal{V}\otimes\mathcal{V}^*$ is the identity automorphism of spaces $\mathcal{V}$ and $\mathcal{V}^*$.

It is seen that the surface porosity tensor $\mathbf{M}_S(\mathbf{N})$ is directly related to the inverse tensor of tortuosity of pores.

In the case of isotropic pore space structure, due to the relation (23), tensor (33) reduces to the form

$$\mathbf{M}_S(\mathbf{N}) = \left(\frac{f_v}{\delta}\right)^2 \mathbf{M} \ ,$$

and from definition (31) we obtain

$$f_v = \lambda\delta \ . \tag{34}$$

The relation (34) shows that only two independent parameters describe the isotropic pore space structure of porous materials.

7 Final remarks

In the paper the new physical model of a fluid motion in a rigid porous material of anisotropic pore space structure has been proposed. In this model the permeable pore space of porous skeleton is considered as a real space of fluid motion and is described by Minkowski metric space. Such approach in a natural way introduces into description parameters characterising pore structure of porous skeleton that play the fundamental role in the dynamic behaviour of fluid filling porous medium.

The concept of the anisotropic space as a model of pore space in which the fluid flows axceedes the range of the fundamental notions of the mixtures theory usually applied in description of such a medium and creates new possibilities for this description

It should be also underlined that such approach is the convenient starting point for the modelling of deformable fluid-saturated porous materials of anisotropic pore structure where the concept of deforming anisotropic space (Finsler space) would have to be used.

Acknowledgements

This work was supported by the grant of Polish Scientific Research Council (Project No 7T 07A 01309).

References

1. A. Bedford, D.S. Drumheller, *Int. J. Engng. Sci.* **21**, 863-960, (1983);
2. M.A. Biot, *J. Acoust. Soc. Am.* **28**, 161-191, (1956);
3. J.Kubik, *TiPM*, **9**, 15-24, (1992);

4. C. Truesdell R.A. Toupin, *Handbuch der Physik*, Vol. III/1, (Springer-Verlag 1960);
5. R.M. Bowen, *Theory of Mixtures* in *Continuum Physics*, ed. A.C. Eringen, (Academic Press, New York, 1960);
6. H. Rund, *The Differential Geometry of Finsler Spaces*, (Springer-Verlag, Berlin-Gottingen-Heidelberg, 1959);
7. A.C. Thompson, *Minkowski Geometry*, Encyclopedia of Mathematics and its Applications, (Cambridge University Press, 1996);
8. M. Matsumoto, *Foundations of Finsler Geometry and Spatial Finsler Spaces*, (Kaiseisha Press, Shigaken, 1986);
9. A. Bejancu, *Finsler Geometry and Applications*, (Ellis Horwood, New York, 1990);
10. P.L. Antonelli, R.S. Ingarden, M. Matsumoto, *Theory of Spays and Finsler Spaces with Application in Physics and Biology*, (Kluver Academic Publishers, 1993);
11. J. Saczuk, *Finslerian Foundations of Solid Mechanics*, Reports of Institute of Fluid-Flow Machinery, 472/1427, (1996);
12. M. Bowen, C.C. Wang, *Introduction to Vectors and Tensors*, (Plenum Press, New York, London, 1976).

An application of a geometrical framework to anelastic systems

M.Dolfin[1], M.Francaviglia[2], P.Rogolino[1]

[1]Dept. of Mathematics, University of Messina
[2]Dept. of Mathematics, University of Torino

Abstract

In this paper we use a geometrical framework developed by the authors in previous papers to analyse the thermodynamics of simple materials far from equilibrium.

We apply the model to the case of large strain in anelastic systems in which irreversible phenomena are described by an internal variable. Two different cases, of spatially uniform and of spatially non-uniform internal variable, are considered. Two different decompositions for the tensor of total strain are examinated and the thermodynamic conseguences discussed. In both applications the entropy form is explicitly given.

1 The thermodynamical bundle

In this section we briefly recall the main framework we already introduced in previous papers [1],[2]. We consider a material element ([3]) and define the state space at time t as the set B_t of the state variables which "fit" the configuration of the element at time t and we assume that B_t has the structure of a finite-dimensional manifold. The "total state space" is then given by the disjoint union

$$\mathcal{B} = \bigcup_t \{t\} \times B_t \tag{1}$$

with a given natural structure of a fibre bundle over the real line $\mathbb{R}$ where time flows [4, 5]. If the istantaneous state space B_t does not vary in time (i.e., there is an abstract "universal state space" B such that $B_t \simeq B$ for all instants of time t), then $\mathcal{B}$ is trivial, i.e. $\mathcal{B}$ has the topology of the Cartesian product

$$\mathcal{B} \simeq \mathbb{R} \times B. \tag{2}$$

In the sequel we shall assume for the sake of simplicity that this simplifying topological assumption holds. Now we consider the abstract *space of processes* which, according to [6], consists of a set Π of functions

$$P_t^i \ : \ [0,t] \to \mathcal{G} \tag{3}$$

where $[0,t]$ is any time interval, the space $\mathcal{G}$ is a suitable target space suggested by the model (usually a vector space), i is a label ranging in an unspecified index set for all allowed processes and $t \in \mathbb{R}$ is called the *duration* of the process P_t^i. A continuous function is defined

$$\rho : \Pi \to C^0(B_0, B_t) \tag{4}$$

in such a way that for any instant of time t and for any process $P_t^i \in \Pi$ it generates a continuous mapping called the *transformation induced* by the process. We define now a further function of time by the following

$$\lambda_b^i(\tau) = \begin{cases} b & \text{if } \tau = 0 \quad \text{with } b \in D^i \\ \rho_t^i(b) & \text{if } \tau \in]0,t] \end{cases} \tag{5}$$

where D^i is the domain of the process, so that setting

$$\lambda_b^i(t) = \rho_t^i(b) = \Phi^i(t,b) \tag{6}$$

we obtain a continuous mapping

$$\Phi^i(t,b) \ : \ \mathbb{R} \times B \to B \tag{7}$$

which defines the flows in the thermodynamic bundle generated by the process considered. The transformation for the system is thence a function which for every local trivialization of the thermodynamic bundle gives

$$\sigma \ : \ t \to [t, \lambda_b(t)]. \tag{8}$$

With these positions the transformation is interpreted as a curve in $\mathcal{B}$, the union of all the state spaces, such that it intersects the instantaneous state space just once, i.e. the transformation σ is a *section* of the thermodynamic bundle ([4],[5]).

2 Applications

We consider a simple material element which undergoes irreversible processes due to changes in a tensorial internal variable $\mathbf{F}^p$. To clarify ideas such a system is of an anelastic type, e.g. elasto-plastic or elasto-viscoplastic.
For the deformation gradient $\mathbf{F}$ we take the Lee decomposition

$$\mathbf{F} = \mathbf{F}^e \mathbf{F}^p \tag{9}$$

being $\mathbf{F}^e$ the elastic part of the *deformation gradient* $\mathbf{F}$.
We want to consider, from a thermodynamical point of view, the two different situations of spatially uniform and non-uniform internal variable. For this last case the results for the different decomposition $\mathbf{F} = \mathbf{F}^p\mathbf{F}^e$ [7] are obtained and the differences discussed.

2.1 Spatially uniform internal variable

The state space can be given by the elastic and anelastic part of the deformation gradient $\mathbf{F}$, given by $\mathbf{F}^e$ and $\mathbf{F}^p$ respectively, the internal energy e and the vector $\beta = \frac{1}{\mu}\nabla\theta$, where the constant μ is the mass density and θ is the temperature. So, the state space is topologically given by:

$$B = Lin(\mathcal{V}) \oplus \mathbb{R} \oplus \mathcal{V} \oplus W, \tag{10}$$

where $\mathcal{V}$ is the translation space of $\mathbb{R}$ and W is a finite-dimensional vector space.
The general process P_t is a piecewise continuous function whose values are

$$P_t(\tau) = [\mathbf{L}(\tau), h(\tau), \gamma(\tau)] \tag{11}$$

where $\mathbf{L}$ is the instantaneous value of the *velocity gradient*, $\gamma = \dot{\beta}$ and $h = -\frac{1}{\mu}\nabla \cdot \mathbf{q}$, $\mathbf{q}$ being the *heat flux vector* given as a stationary vertical field on the bundle:

$$\mathbf{q} : \mathbb{R} \times B \to \mathcal{V} \tag{12}$$

With these positions the target space turns out to be

$$\mathcal{G} = Lin(\mathcal{V}) \oplus \mathbb{R} \oplus \mathcal{V}. \tag{13}$$

The process $P : [0, t] \to \mathcal{G}$ maps an interval of the real line into the fibre $\mathcal{G}$ of the bundle.

Another stationary field on the thermodynamic bundle $\mathbb{R} \times B$ is the *stress field*:

$$\mathbf{T} : \mathbb{R} \times B \to Sym(\mathcal{V}). \tag{14}$$

A further stationary field is the *temperature* field

$$\theta : \mathbb{R} \times B \to \mathbb{R}^{++}, \tag{15}$$

where $\mathbb{R}^{++}$ is the set of real positive numbers.

The system of dynamical equations is

$$\begin{cases} \dot{\mathbf{F}}^e(\tau) = \mathbf{L}^e(\tau)\mathbf{F}^e(\tau) \\ \dot{\mathbf{F}}^p(\tau) = \mathbf{L}^p(\tau)\mathbf{F}^p(\tau) \\ \dot{e}(\tau) = \mathbf{T}(\mathbf{b}) \cdot \mathbf{L}(\tau) + h(\tau) \\ \dot{\beta}(\tau) = \gamma(\tau) \end{cases} \tag{16}$$

being here and in the sequel $\cdot$ the total contraction over all possible indices.

According to [8] the entropy action is defined on $\mathbb{R} \times TB$ by setting

$$s(t) = \int_0^t \frac{h(\tau)}{\theta[b(\tau)]} d\tau + \int_0^t \frac{1}{\theta^2} \{\mathbf{q}[b(\tau)] \cdot \beta(\tau)\} d\tau, \tag{17}$$

which in turn defines a 1-form Ω_ω in the thermodynamic bundle $\mathbb{R} \times B$ called the "entropy 1-form". The integral of Ω_ω along the solution curve σ gives s. Using the system (16) we get a modified entropy form $\tilde{\Omega}_\omega$ given by:

$$\begin{aligned} s(\rho_t, b, t) = \quad & \int_\sigma \tilde{\Omega}_\omega = \int_\sigma -\frac{\mathbf{T}[(\mathbf{F}^e)^{-1}]^T}{\theta} \cdot d\mathbf{F}^e - \frac{\mathbf{T}(\mathbf{F}^{-1})^T \mathbf{F}^e}{\theta} \cdot d(\mathbf{F}^p)^T + \\ & + \frac{1}{\theta} de + \frac{1}{\theta^2} \mathbf{q} \cdot \beta dt \end{aligned} \tag{18}$$

where $\mathbf{X}^T$ denotes the transpose of the matrix $\mathbf{X}$. Following [8], the function s defined by (18) is just an "action", which in order to let the system satisfy the second law of thermodynamics should admit an "upper potential" S (see [8], page 50) such that the following inequality holds:

$$S(b_t) - S(b) \geq s(\rho_t, b, t). \tag{19}$$

From the properties of the exterior differential, necessary conditions for the existence of the entropy function during the analysed process have been obtained in [1] (with minor modifications on the choice of the state variables).

2.2 Spatially non-uniform internal variable

We consider now the simple material just treated which undergoes irreversible processes far from equilibrium, characterized by spatially non-uniform internal variables. We refer in particular to a model investigated by Maugin in [9]. The decomposition for the deformation gradient is again the Lee decomposition (9). In this case the state space is given by $(\mathbf{F}^e, \mathbf{F}^p, e, \beta, \nabla \mathbf{F}^p)$. Topologically the state space B is thence given by

$$B = Lin(\mathcal{V}) \oplus \mathbb{R} \oplus \mathcal{V} \oplus W \oplus Lin(W). \tag{20}$$

Processes are assumed to be the same as in the case of the general theory of simple materials, i.e. they are piecewise continuous functions

$$P_t = [L(t), h(t), \gamma(t)] \tag{21}$$

As in the preceding case, three stationary fields $\mathbf{q}$, $\mathbf{T}$, θ are introduced on $\mathbb{R} \times B$.

When the gradients of the internal variables are involved, a complete set of balance laws for them is needed. Therefore, evolution-diffusion equations, rather then evolution equations only, have to be given. Under these assumptions, when the outcome of dissipative structures is involved, a different relation between the heat flux $\mathbf{q}$ and the entropy flux $\mathbf{J}_s$ has to be postulated. This is due to the presence of an additional term, called *extra entropy flux* in the literature, which is assumed to include contributions from the flux of internal variables. More precisely we set:

$$\mathbf{J}_s = \frac{1}{\theta}\mathbf{q} + \mathbf{k} \tag{22}$$

The entropy function given by (19) satisfies the Clausius-Duhem inequality

$$\dot{S} + \nabla \cdot \mathbf{J}_s \geq 0 \tag{23}$$

which taking $\theta > 0$ into account can be written as

$$\theta \dot{S} + \nabla \cdot (\theta \mathbf{J}_s) - (\mathbf{J}_s \cdot \nabla)\theta \geq 0 \tag{24}$$

We introduce now the free energy per unit mass Ψ

$$\Psi = \Psi[\mathbf{F}^e, \beta, (\mathbf{F}^p)^T, \nabla(\mathbf{F}^p)^T] \equiv e - \theta S \tag{25}$$

Using equation (25) together with (24) and (16) we obtain the following relation:

$$-(\dot{\Psi} + S\dot{\theta}) + \mathbf{T} \cdot \mathbf{L} - (\mathbf{J}_s \cdot \nabla)\theta + \nabla \cdot (\theta \mathbf{J}_s) + h \geq 0 \tag{26}$$

Introducing the relation for the entropy flux and after some manipulations one obtains:

$$-(\dot{\Psi} + S\dot{\theta}) + \mathbf{T} \cdot \mathbf{L} - (\mathbf{J}_s \cdot \nabla)\theta + \nabla \cdot (\theta \mathbf{k}) \geq 0 \tag{27}$$

Using the expression for the free energy (25) we obtain the following equation:

$$\dot{\Psi} = \hat{\mathbf{T}} \cdot \dot{\mathbf{F}}^e - \hat{S}\dot{\beta} + \mathcal{A} \cdot \dot{(\mathbf{F}^p)}^T + \nabla \cdot [\mathbf{B} \cdot \dot{(\mathbf{F}^p)}^T] \tag{28}$$

with the positions

$$\hat{\mathbf{T}} = \frac{\partial \Psi}{\partial \mathbf{F}^e}, \ \mathbf{B} = \frac{\partial \Psi}{\partial \nabla (\mathbf{F}^p)^T}, \ \hat{S} = -\frac{\partial \Psi}{\partial \beta}, \ A = \frac{\partial \Psi}{\partial (\mathbf{F}^p)^T} \tag{29}$$

and

$$\mathcal{A} = A - \nabla \cdot \mathbf{B} = \frac{\delta \Psi}{\delta (\mathbf{F}^p)^T} \tag{30}$$

where $\frac{\delta}{\delta \mathbf{F}^p}$ is the variational derivative, i.e. $\frac{\delta}{\delta (\mathbf{F}^p)^T} = \frac{\partial}{\partial (\mathbf{F}^p)^T} - \nabla \left[\frac{\delta}{\delta \nabla (\mathbf{F}^p)^T} \right]$. By substituting equation (28) into (27) we have now the inequality

$$\begin{aligned}[-\hat{\mathbf{T}} + \mathbf{T}(\mathbf{F}^e)^{-1})^T] \cdot \dot{\mathbf{F}}^e + (\hat{S} \cdot \nabla - S)\dot{\theta} + [\mathbf{T}(\mathbf{F}^{-1})^T \mathbf{F}^e - \mathcal{A}] \cdot \dot{(\mathbf{F}^p)}^T + \\ + \nabla \cdot [\theta \mathbf{k} - \mathbf{B} \cdot \dot{(\mathbf{F}^p)}^T] - (\mathbf{J}_s \cdot \nabla)\theta \geq 0 \end{aligned} \tag{31}$$

where $\hat{S} \cdot \nabla - S$ is a linear scalar-valued operator in the argument $\dot{\theta}$.
We remark that S and $\mathbf{T}[(\mathbf{F}^e)^{-1}]^T$ are assumed to be both independent on $\dot{\theta}$ and $\dot{\mathbf{F}}^e$, while the remaining coefficients in (37) can in general depend on their respective factors; since (37) cannot change sign, [10], following an earlier argument of [9] we obtain

$$\mathbf{T}[(\mathbf{F}^e)^{-1}]^T = \hat{\mathbf{T}} = \frac{\partial \Psi}{\partial \mathbf{F}^e}, \ \hat{S} \cdot \nabla \dot{\theta} = S\dot{\theta} = -\frac{\partial \Psi}{\partial \beta}\dot{\theta} \tag{32}$$

The extra entropy flux at this point can be defined as:

$$\mathbf{k} = \frac{1}{\theta} \mathbf{B} \cdot \dot{(\mathbf{F}^p)}^T \tag{33}$$

and (37) reduces to the following residual dissipation inequality [9]

$$\Phi \equiv [\mathbf{T}(\mathbf{F}^{-1})^T\mathbf{F}^e - \mathcal{A}] \cdot \dot{(\mathbf{F}^p)}^T - (\mathbf{J}_s \cdot \nabla)\theta \geq 0 \tag{34}$$

which can be split into the following two parts:

$$\begin{aligned} \Phi_{intr} &\equiv [\mathbf{T}(\mathbf{F}^{-1})^T\mathbf{F}^e - \mathcal{A}] \cdot \dot{(\mathbf{F}^p)}^T \geq 0 \\ \Phi_{th} &\equiv -(\mathbf{J}^s \cdot \nabla)\theta \geq 0 \end{aligned} \tag{35}$$

where it is possible to recognize the two different dissipative processes as the intrinsic and thermal dissipations respectively. As already noticed by Maugin in [9], from equations (34)-(35) one can notice that the forms expressing the entropy production closely resemble those obtainable in the case of internal variables without dissipative structure, in the sense that the standard bilinear behaviour is preserved with the addition of extra addenda.

We are now able to calculate the integral of the entropy 1-form Ω_ω along the solution curve σ in the thermodynamical bundle. Using the system (16) we get a modified entropy form $\tilde{\Omega}_\omega$ given by:

$$\begin{aligned} s = \int_\sigma \tilde{\Omega}_\omega = \int_\sigma -\frac{\mathbf{T}[(\mathbf{F}^e)^{-1}]^T}{\theta} \cdot d\mathbf{F}^e + \frac{1}{\theta}de + \frac{1}{\theta^2}\mathbf{q} \cdot \beta dt - \\ -\left[\frac{\mathbf{T}[(\mathbf{F})^{-1}]^T\mathbf{F}^e}{\theta} + \nabla \cdot \left(\frac{\mathbf{B}}{\theta}\right)\right] \cdot d(\mathbf{F}^p)^T - \frac{\mathbf{B}}{\theta}d(\nabla\mathbf{F}^p)^T \end{aligned} \tag{36}$$

2.3 Decomposition $\mathbf{F} = \mathbf{F}^p\mathbf{F}^e$

Using the decomposition $\mathbf{F} = \mathbf{F}^p\mathbf{F}^e$ results are slightly different. The independent variables are the same as before $[(\mathbf{F}^e)^T, \beta, \mathbf{F}^p, \nabla\mathbf{F}^p]$ and the final inequality reads

$$\begin{aligned} [-\hat{\mathbf{T}} + \mathbf{T}[(\mathbf{F}^{-1})]^T\mathbf{F}^p] \cdot \dot{(\mathbf{F}^e)}^T + (\hat{S} \cdot \nabla - S)\dot{\theta} + [\mathbf{T}(\mathbf{F}^p)^{-1} - \mathcal{A}] \cdot \dot{\mathbf{F}}^p + \\ +\nabla \cdot [\theta\mathbf{k} - \mathbf{B} \cdot \dot{\mathbf{F}}^p] - (\mathbf{J}_s \cdot \nabla)\theta \geq 0 \end{aligned} \tag{37}$$

In the hypothesis that $\mathbf{T}(\mathbf{F}^{-1})^T\mathbf{F}^p$ does not depend on $\dot{\mathbf{F}}^e$ then

$$\hat{\mathbf{T}} = \mathbf{T}(\mathbf{F}^{-1})^T\mathbf{F}^p \tag{38}$$

Choosing

$$\mathbf{k} = \frac{1}{\theta}\mathbf{B} \cdot \dot{\mathbf{F}}^p \tag{39}$$

the dissipation inequality reduces to

$$\begin{aligned} &\Phi_{intr} \equiv \{\mathbf{T}[(\mathbf{F}^p)^{-1}]^T - \mathcal{A}\} \cdot \dot{\mathbf{F}}^p \geq 0 \\ &\Phi_{th} \equiv -(\mathbf{J}^s \cdot \nabla)\theta \geq 0 \end{aligned} \tag{40}$$

One notices that, different to the previous case in which the total deformation gradient influences the intrinsic dissipation, here only the plastic part $\mathbf{F}^p$ appears into equation $(40)_1$. The integral of the entropy 1-form Ω_ω along the solution curve σ now gives

$$\begin{aligned} s(\rho_t, b, t) = \int_\sigma \tilde{\Omega}_\omega = \int_\sigma -\frac{\mathbf{T}(\mathbf{F}^{-1})^T\mathbf{F}^p}{\theta} \cdot d(\mathbf{F}^e)^T + \frac{1}{\theta} de + \frac{1}{\theta^2}\mathbf{q} \cdot \beta dt - \\ -\left[\frac{\mathbf{T}[(\mathbf{F}^p)^{-1}]^T}{\theta} + \nabla \cdot \left(\frac{\mathbf{B}}{\theta}\right)\right] \cdot d\mathbf{F}^p - \frac{\mathbf{B}}{\theta} \cdot d(\nabla \mathbf{F}^p) \end{aligned} \tag{41}$$

References

[1] Dolfin M. Francaviglia M., Rogolino P. A geometric perspective on irreversible thermodynamics with internal variables. Journal Non-Equilibrium Thermodynamics (in printing).

[2] Dolfin M. Francaviglia M., Rogolino P. A geometric model on the thermodynamics of simple materials. Periodica Polytechnica - Budapest (in printing).

[3] Wang C.C. On the geometric structures of simple bodies, a mathematical foundation for the theory of continuous didtributions of dislocations. Arch. Rat. Mech. Anal. 27, (1967),33–94.

[4] Francaviglia M. Elements of differential and Riemannian geometry. Bibliopolis, Napoli, (1988).

[5] Marsden J., Abraham R. Foundations of mechanics. Addison-Wesley Publ. Co., (1978).

[6] Noll W. A new mathematical theory of simple materials. Arch. Rat. Mech. Anal. 48, (1972),1-50.

[7] Clifton R.J. On the equivalence of $\mathbf{f}^e\mathbf{f}^p$ and $\bar{\mathbf{f}}^p\bar{\mathbf{f}}^e$. Journal of Applied Mechanics.

[8] Owen D.R., Coleman B.D. A mathematical foundation for thermodynamics. Arch. Rat. Mech. Anal. 54, (1974),1-104.

[9] Maugin G.A. Internal Variables and Dissipative Structures J. Non-Equilib. Thermodyn. 15, (1990),173-192.

[10] Maugin G. A. Continuum Mechanics of Electromagnetic Solids. North-Holland, Amsterdam, (1988).

APPLYING SPECIAL ATEB-FUNCTIONS IN CONSTRUCTION BOUNDARY PROBLEM SOLUTIONS FOR SOME NONLINEAR WAVE EQUATIONS

CH.T.DROGOMIRETSKA
Department of Applied Mathematics,
Lviv State Polytechnical University,
12, S. Bandera Str., Lviv, 290646, Ukraine,
E-mail christdr @ polynet.lviv.ua

Development of Krylov-Bogolubov-Mitropolskiy's asymptotic methods based on the theory of special Ateb-functions to construct asymptotic solutions of boundary problems described by general nonlinear wave equation and different types of linear and nonlinear nonperturbed and perturbed boundary conditions is proposed. The worked up methods are used for the investigation of some dynamic systems including systems of pivots which material obeyes nonlinear elasticity law.

1 Formulation of the problem

Many problems of mechanics, physics and vibration theory require nonlinear differential equations with power nonlinearity in comparison with linear which more exactly describe real processes.

Many difficuilties arise when studing nonlinear systems because there are no universal methods of construction solutions of nonlinear boundary problems.

Asymptotic methods worked up by Krylov M.M., Bogolubov M.M., Mytropolskiy Yu. A. [1,2] are very effective in researching nonlinear vibrations. However, classic asymptotic methods of solving nonlinear boundary problems are not effective enough because of the limited possibilities of zero approximation presentation in closed form.

The aim of this work is to combine the theory of special Ateb-functions [3,4] and the ideas of classical asymptotic methods [1,2] to build solutions of boudary problems described by autonomous

$$\left(\frac{\partial u}{\partial t}\right)^m \left(\frac{\partial^2 u}{\partial t^2}\right)^p - \alpha \left(\frac{\partial u}{\partial x}\right)^n \left(\frac{\partial^2 u}{\partial x^2}\right)^q = \varepsilon F\left(x, u, \frac{\partial u}{\partial t}, \frac{\partial u}{\partial x}, \frac{\partial^2 u}{\partial x^2}\right) \quad (1)$$

and nonautonomous

$$\left(\frac{\partial u}{\partial t}\right)^m \left(\frac{\partial^2 u}{\partial t^2}\right)^p - \alpha(\tau) \left(\frac{\partial u}{\partial x}\right)^n \left(\frac{\partial^2 u}{\partial x^2}\right)^q = \varepsilon F\left(\tau, x, u, \frac{\partial u}{\partial t}, \frac{\partial u}{\partial x}, \frac{\partial^2 u}{\partial t^2}\right) + \varepsilon_1 Q(x,\tau) Y(\Theta) \quad (2)$$

general nonlinear wave equations, where $u = u(x,t)$; $\alpha > 0$; $\tau = \varepsilon t$ — "slow" time; $\alpha(\tau)$ — slowly changable coefficient; p, q and m, n are numbers of the type $\frac{2k_1+1}{2k_2+1}$ and $\frac{2k_1}{2k_2+1}$ (k_1, $k_2 = 0, 1, 2, \ldots$) accordingly; $\varepsilon F\left(x, u, \frac{\partial u}{\partial t}, \frac{\partial u}{\partial x}, \frac{\partial^2 u}{\partial x^2}\right)$, $\varepsilon F\left(\tau, x, u, \frac{\partial u}{\partial t}, \frac{\partial u}{\partial x}, \frac{\partial^2 u}{\partial x^2}\right)$ — known generaly speaking nonlinear functions of their arguments, $\varepsilon_1 Q(x,\tau) Y(\Theta)$ — small $2\Pi_t$-periodic exitation force, $\Pi_t = \left(\frac{p}{n+p+q}, \frac{m+p}{m+2p}\right)$.

In the case $m = n = 0$, $p = q = 1$ equations (1), (2) reduce to linear autonomous and nonautonomous wave equations.

2 Spesial Ateb-functions and their properties

Special Ateb-functions $sa(\bar{n}, \bar{m}, w)$ and $ca(\bar{m}, \bar{n}, w)$ were introducet [4] as an inversion of uncomplete Beta-function

$$w = \frac{1}{2} B_y\left(\frac{1}{\bar{n}+1}, \frac{1}{\bar{m}+1}\right) = \frac{1}{2} \int_0^{0 \le y \le 1} t^{-\frac{\bar{n}}{\bar{n}+1}} (1-t)^{-\frac{\bar{m}}{\bar{m}+1}} \, dt,$$

were $\bar{n}$, $\bar{m}$ are numbers of the type $\frac{2k_1+1}{2k_2+1}$ $(k_1, k_2 = 0, 1, 2, \ldots)$. They are 2Π-periodical $\left(\Pi = \Pi(\bar{m}, \bar{n}) = B\left(\frac{1}{\bar{n}+1}, \frac{1}{\bar{m}+1}\right)\right)$, odd or even acordingly and when $\bar{m} = \bar{n} = 1$ reduce to ordinary trigonometric functions $\sin w$ and $\cos w$.

3 Boundary problems for general nonlinear wave equations

For boundary problems described by equations (1), (2) the Fourier method of variables separation is applicible.

3.1 Nonperturbed general nonlinear wave equation (1) and nonperturbed boundary conditions

For the boundary problems described by nonperturbed ($\varepsilon = 0$) equation (1) and nonperturbed boundary conditions

$$\left[\frac{\partial u}{\partial x}\right]_{x=0} = 0, \quad \left[\frac{\partial u}{\partial x}\right]_{x=\ell} = 0; \tag{3}$$

$$[u]_{x=0} = 0, \quad \left[\frac{\partial u}{\partial x}\right]_{x=\ell} = 0; \tag{4}$$

$$[u]_{x=0} = 0, \quad [u]_{x=\ell} = 0; \tag{5}$$

$$[u]_{x=0}=0, \quad \left[\left(\frac{\partial u}{\partial x}\right)^{\mathcal{K}}+\gamma u^{\mathcal{K}}\right]_{x=\ell}=0; \tag{6}$$

$$\left[\frac{\partial u}{\partial x}\right]_{x=0}=0, \quad \left[\left(\frac{\partial u}{\partial x}\right)^{n+q}+\beta\left(\frac{\partial u}{\partial t}\right)^{m}\left(\frac{\partial^2 u}{\partial t^2}\right)^{p}\right]_{x=\ell}=0 \tag{7}$$

the solutions can be given in a closed form using special Ateb-functions

$$u_{0,k}(x,t)=aX_k(x)T_k(\varphi_k)=$$

$$=a\ sa\left(\frac{m+p}{q},\frac{q}{n+q},\nu_k(a)+\psi\right)sa\left(\frac{n+q}{p},\frac{p}{m+p},\omega_k(a)t\right), \tag{8}$$

where $\mathcal{K}$, γ, β — constants; $\omega_k(a)=\partial\varphi_k/\partial t$; $\nu_k(a)$, $\omega_k(a)$ — values taken from boundary and initial conditions.

3.2 Autonomous general nonlinear wave equation and nonperturbed boundary conditions

For the boundary problems described by autonomous equation (1) and nonperturbed boundary conditions (3)-(7) the single frequency solution is built in the form of asymptotic decomposition

$$u(x,t)=u_0(x,t)+\varepsilon u_1(x,a,\varphi)+\varepsilon^2\cdots,$$

where $u_0(x,t)$ — the solution (8) of corresponding nonperturbed ($\varepsilon=0$) boundary problem; a and φ as time functions are connected with differential correlations

$$\begin{aligned}\tfrac{dq}{dt}&=\varepsilon A_1(a)+\varepsilon^2\cdots\\ \tfrac{d\varphi}{dt}&=\omega(a)+\varepsilon B_1(a)+\varepsilon^2\cdots\end{aligned} \tag{9}$$

Values $A_1(a)$, $B_1(a)$ are found basing on energetic methods [2] and on

Theorem 1. Let $u_1(x,a,\varphi)$ — a twice continuously differential in the variable x and φ, 2Π-periodocal in the variable φ function; $aX(x)T(\varphi)$ — the solution of nonperturbed ($\varepsilon=0$) equation (1). Then

$$\int\limits_0^{2\Pi}\int\limits_0^{\ell}\left(a^{m+p-1}\omega^{m+2p}(a)\left(m\left(\frac{\partial u_0}{\partial\varphi}\right)^{m-1}\left(\frac{\partial^2 u_0}{\partial\varphi^2}\right)^{p}\frac{\partial u_1}{\partial\varphi}+\right.\right.$$

$$\left.\left.+p\left(\frac{\partial u_0}{\partial\varphi}\right)^{m}\left(\frac{\partial^2 u_0}{\partial\varphi^2}\right)^{p-1}\frac{\partial^2 u_1}{\partial\varphi^2}\right)\right)-$$

$$-\alpha_0 a^{n+q-1}\left(n\left(\frac{\partial u_0}{\partial x}\right)^{n-1}\left(\frac{\partial^2 u_0}{\partial x^2}\right)^q\frac{\partial u_1}{\partial x}++q\left(\frac{\partial u_0}{\partial x}\right)^n\left(\frac{\partial^2 u_0}{\partial x^2}\right)^{q-1}\frac{\partial^2 u_1}{\partial x^2}\right)\times$$

$$\times X(x)\left\{\begin{array}{l} sa\left(\frac{n+q}{p},\frac{p}{m+p},\varphi\right) \\ ca\left(\frac{p}{m+p},\frac{n+q}{p},\varphi\right)\end{array}\right\}dxd\varphi=0.$$

3.3 Autonomous general nonlinear wave equation and perturbed boundary conditions

The boundary problem described by equation (1) and perturbed boundary conditions

$$[u]_{x=0}=\varepsilon\eta_1\left(u,\frac{\partial u}{\partial x},\frac{\partial u}{\partial t}\right),\quad\left[\frac{\partial u}{\partial x}\right]_{x=\ell}=\varepsilon\omega_1\left(u,\frac{\partial u}{\partial x},\frac{\partial u}{\partial t}\right);\qquad(10)$$

$$[u]_{x=0}=\varepsilon\eta_2\left(u,\frac{\partial u}{\partial x},\frac{\partial u}{\partial t}\right),\left[\left(\frac{\partial u}{\partial x}\right)^{\mathcal{K}}+\gamma u^{\mathcal{K}}\right]_{x=\ell}=\varepsilon\omega_2\left(u,\frac{\partial u}{\partial x},\frac{\partial u}{\partial t}\right);\quad(11)$$

$$\left[\frac{\partial u}{\partial x}\right]_{x=0}=\varepsilon\eta_3\left(u,\frac{\partial u}{\partial x},\frac{\partial u}{\partial t}\right),$$

$$\left[\left(\frac{\partial u}{\partial x}\right)^{n+q}+\beta\left(\frac{\partial u}{\partial t}\right)^m\left(\frac{\partial^2 u}{\partial t^2}\right)^p\right]_{x=\ell}=\varepsilon\omega_3\left(u,\frac{\partial u}{\partial x},\frac{\partial u}{\partial t}\right),\qquad(12)$$

where $\omega_i\left(u,\frac{\partial u}{\partial x},\frac{\partial u}{\partial t}\right)$, $\eta_i\left(u,\frac{\partial u}{\partial x},\frac{\partial u}{\partial t}\right)$ $(i=1,2,3)$ — known analytical functions, is solved with solution presentation in the form

$$u_1(x,a,\varphi)=V_1(x,a,\varphi)+W_1(x,a,\varphi),$$

where $W_1(x,a,\varphi)$ — auxiliary functions which are the solutions of the boundary problem for the equation

$$\frac{\partial^2 W_1(x,a,\varphi)}{\partial x^2}\equiv 0$$

and boundary conditions (10)-(12). Then boundary problem (1), (10)-(12) comes to the problem with nonperturbed boundary conditions in $V_1(x,a,\varphi)$ function wich is solved by analogy with the problem (1), (3)-(7) with nonperturbed boundary conditions.

3.4 Nonautonomous general nonlinear wave equation and nonperturbed boundary conditions

For the boundary problem described by equation (2) and boundary conditions (5) the solution is presented in the form

$$u(x,t) = a\ sa\left(\frac{m+p}{q}, \frac{q}{n+q}, \frac{\Pi_x x}{\ell}\right) sa\left(\frac{n+q}{p}, \frac{p}{m+p}, \omega(a,\tau)t\right),$$

where $\omega(a,\tau) = \frac{\partial\varphi}{\partial t}$; $\Pi_x = B\left(\frac{q}{m+p+q}, \frac{n+q}{n+2q}\right)$. Amplitude a and phase difference $\psi = \varphi - \Theta$ are defined from the system of differential equations

$$\frac{da}{dt} = \varepsilon A_1(a,\tau) + \varepsilon^2 \cdots$$

$$\frac{d\psi}{dt} = \omega(a,\tau) - \nu(a) + \varepsilon B_1(a,\tau) + \varepsilon^2 \cdots ,$$

where $\nu(a) = \frac{d\Theta}{dt}$.

Values $A_1(a,\tau)$, $B_1(a,\tau)$ are found on the basis of energetic method [2] and theorem 1 by means of representing them as well as external exitation force $\varepsilon_1 Q(x,\tau)Y(\Theta)$ in the form of Fourier series and compareing coefficients under the same sine and cosine in the first approximation equation.

In case $m = 0$, $p = q = 1$ equations (1), (2) describe free and forced longitudinal vibration of pivot which material obeyes nonlinear elasticity law

$$\sigma = E_n(\varepsilon_x)^{n+1},$$

where σ — stress; ε_x — relative deformation; E_n — proportion coefficient; n — power exponent. The worked up methods allow to build the solutions of boundary problems which describe free longitudinal vibration of pivot with constant length and cross section for different conditions of pivot fixing, pivot systems, pivots with concentrated masses and elastic elements.

References

1. Bogolubov N.N., Mitropolskiy Yu. A. Assymptotic methods in nonlinear vibration theory, Moscow (1958) [In Russian].

2. Mitropolskiy Yu. A., Moseenkov B.I. Assymptotic solutions of equations in partial derivatives, Kiev (1976) [In Russian].

3. Rozenberg R.M. The Ateb (H)-functions and thear properties.-Quart. Appl. Math. (1963).

4. Senyk P.M. Uncomplete Beta-function inversion, Ukr. Math. Journ **21**, 3 (1969) [In Ukrainian].

CARDIAC DYNAMICS INVOLVING ARRHYTHMIAS AND CONTINUUM MECHANICS

J. ENGELBRECHT, M. VENDELIN, O. KONGAS
Institute of Cybernetics at Tallinn Technical University
Akadeemia tee 21, 12617 Tallinn, Estonia

R.v. HERTZEN
Helsinki University of Technology
Otakaari 1, FIN-02150 Espoo, Finland

The modelling of cardiac dynamics must include suitable description of activation, cell energetics and muscle contraction. Using the concepts of continuum mechanics, it is possible to model explicitly elastic and dissipative potentials, interaction of various fields, observable and internal variables, all framed by thermodynamics. Here the activation is modelled by a periodically driven nerve pulse equation that is able to describe various arrhythmias and coexiting attractors. The existence of coexisting attractors suggests that sudden changes in cardiac functioning might be due to coexisting attractors rather than the changes in system parameters. The contraction includes active and passive stresses. The active stress is described by the distribution - moment approximation of the Huxley-type equations where cell energetics is modelled by internal variables. The passive stress is calculated using a potential of the hardly compressible material.

1 Introduction

Continuum mechanics has been and is an excellent starting point in many interdisciplinary studies. It offers thermodynamical background, clear notions of conservation laws, elastic and dissipative potentials, interaction of various fields, observable and internal variables, etc. Biological tissues including myocardium display complicated mechanical behaviour due to their numerous constituents and electrical activity. Thus the modelling of cardiac dynamics must include the proper understanding of the activation, cell energetics and muscle contraction. Although the heart muscle is a mixture of many components like various fibres, vessels, and fluids, the general view to describe its behaviour is based on continuum mechanics of anisotropic materials [1,2]. Here we demostrate how to model first the activation and second, the stress state of the myocardium muscle. The muscle includes several types of inhomogeneities: like the inhomogeneity of the muscle and the inhomogeneities of the energy supply and activation. After such a stress state model is validated on the basis of experimental data involving stable excitation, the activation is added that may be irregular, i.e. arrhythmias are also included [1,3].

2 Activation

The rhythm of the heart is set by a sinoatrial (SA) node, a small region of a specialized tissue in the right atrium, which generates a spontaneous electrical rhythm. The activation then spreads over the atria causing their contraction. In normal individuals the atria and ventricles are electrically coupled only by way of a small strip of special tissue, an atrioventricular (AV) node. The activation then propagates further through this bundle. The fractal-like structure of the latter quarantees nearly synchronized contraction of the ventricular muscle cells.

Early models derived in 1920's to explain the rhythms and arrhythmias of the heart were based on difference equations or on coupled nonlinear electric oscillators. Later studies, accounting also for spacial aspects, were performed utilizing the concept of an excitable medium. Recently, periodically driven and coupled oscillators have been subject to considerable interest due to the success of nonlinear dynamics. The overview on the modelling can be found in [1, 4, 5]. Our results are based on the analysis of a periodically driven nerve pulse equation (NPE), a particular form of an asymmetric van der Pol equation, derived from basic cable equations [6]. The NPE models the driven action potential oscillations in the His-Purkinje system.

The time-dependence of the action potential w is described by

$$\ddot{w} + \epsilon(w - w_1)(w - w_2)\dot{w} + w = Ig(\omega t) + C \tag{1}$$

where $\epsilon, w_1, w_2, I\omega, C$ are constants and g represents the drive. This equation is solved numerically with various drive types and the bifurcation structures computed over the wide range of parameters [7]. The overall bifurcation structure of Eq. (1) displaying the bifurcation lines of period one solutions in the (I, ω) - control plane is shown in Fig. 1 for the delta drive. Here dotted and solid lines denote period doublings and Neimark-Sacker bifurcations, respectively. A detailed study of this study is demonstrated in Fig. 2, where coexisting attractors (CA) are displayed. At $I = 1.146, \omega = 2.8$ an extra stimulus (ES) was applied that kicked the 3:1 regime to a quasiperiodic mode. Another ES brought the system back to the original regime. This result suggests that a sudden change in cardiac functioning indicates a possibility of existing CAs rather than the changes in system parameters.

3 Contraction

The stress developed in the myocardium is divided into active stress σ_a and passive stress σ_p generated by muscle fibres and determined by the elastic response of the tissue to deformation, respectively.

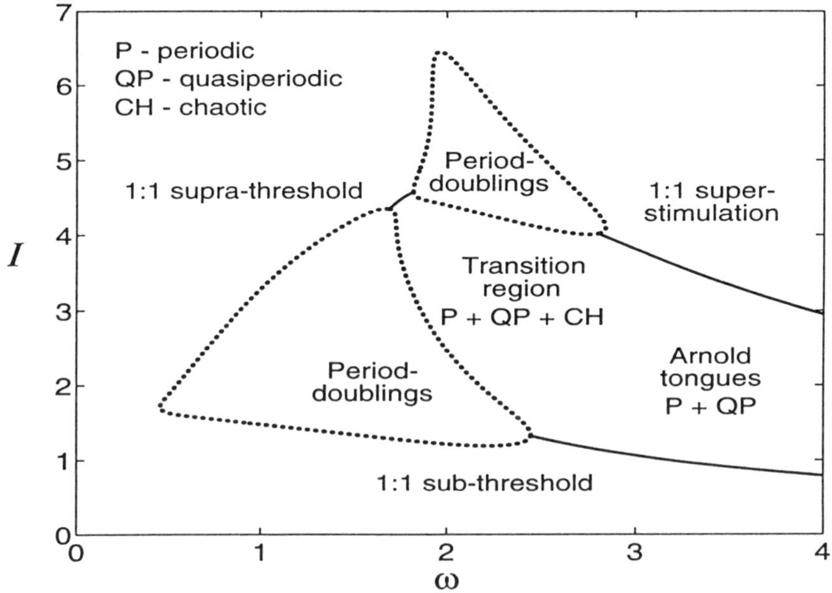

Figure 1: The skeleton of the bifurcation structure of Eq. (1).

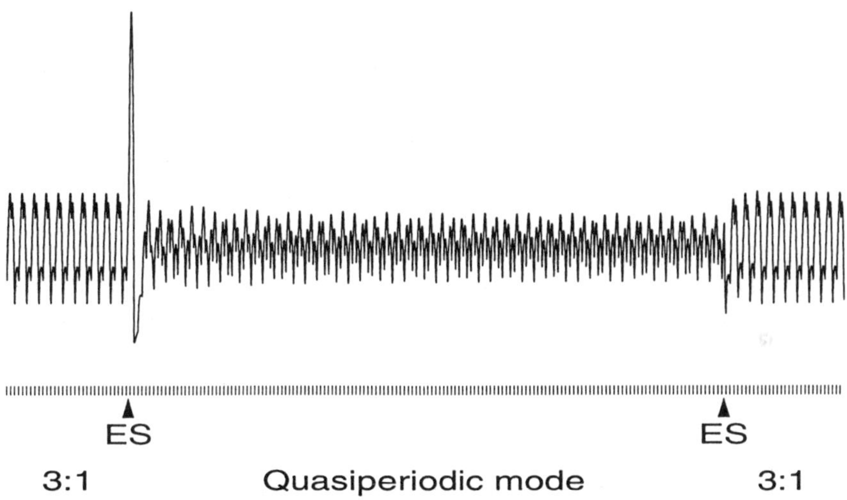

Figure 2: Co-existing attractors.

The active stress of the myocardium is described by the distribution-moment (DM) approximation of the Huxley-type equations [8] based on three moments: M_0, M_1, and M_2. The developed active stress is determined by M_1:

$$\sigma_a = \gamma l_s M_1, \tag{2}$$

where σ_a is Cauchy stress, γ is constant, and l_s is sarcomere length and active stress tensor σ_a is $\sigma_a \vec{e}_f \vec{e}_f$. Vector $\vec{e}_f$ is a unit vector along the fibre. The distribution moments obey the following equations:

$$\frac{dM_0}{dt} = A(t)b_0 - G_0(M_0, M_1, M_2), \tag{3}$$

$$\frac{dM_1}{dt} = A(t)b_1 - G_1(M_0, M_1, M_2) + vM_0, \tag{4}$$

$$\frac{dM_2}{dt} = A(t)b_2 - G_2(M_0, M_1, M_2) + 2vM_1, \tag{5}$$

where

$$b_i = \int_{-\infty}^{+\infty} z^i f(z) dz, \tag{6}$$

$$G_i = \int_{-\infty}^{+\infty} z^i [f(z) + g(z)] n(z,t) dz, \tag{7}$$

$$n(z,t) = n(z, M_0(t), M_1(t), M_2(t)) \tag{8}$$

with $v = dl_s/dt$. The distribution $n(z,t)$ is approximated by the Gaussian distribution $n(z, M_0(t), M_1(t), M_2(t))$ [8]. Functions $f(z)$ and $g(z)$ are bounding and unbounding functions [9]. Actually, in terms of continuum physics, M_0, M_1 and M_2 are the internal variables, and their governing equations (3) - (5) are of the usual kinetic type. These equation, however, involve one additional internal variable $A(t)$ which represents the relative amount of activated actomyosin sites. It is regulated by the intracellular Ca^{++} concentration and governed by its own kinetic equation.

The DM equations (2) - (4) are connected with deformation of the tissue through l_s and its first derivative v. To find the sarcomere length in deformed state, conservation of the number of sarcomeres in the fibre could be used:

$$\frac{df}{l_s} = \frac{df_0}{l_{s0}}, \tag{9}$$

where df_0 is the lenght of vector $\vec{df_0}$ along the fibre in undeformed state and df is the length of the same vector in deformed state. The length in deformed state obeys

$$df^2 = df_0^2 \vec{e_f}^c \cdot \mathbf{C} \cdot \vec{e_f}, \tag{10}$$

where $\mathbf{C} = \mathbf{F}^\mathbf{C} \cdot \mathbf{F}$ and $\mathbf{F}$ is deformation gradient tensor. Thus, for l_s we have

$$l_s = l_{s0}\frac{df}{df_0} = l_{s0}\sqrt{\vec{e_f}^{\,c} \cdot \mathbf{C} \cdot \vec{e_f}} \tag{11}$$

and for active stress

$$\sigma_a = \gamma l_{s0}\sqrt{\vec{e_f}^{\,c} \cdot \mathbf{C} \cdot \vec{e_f} M_1}. \tag{12}$$

To solve the system of ODE ((2) - (4) and (8)) the deformation gradient tensor $\mathbf{F}$ is required as a function of time. $\mathbf{F}$ is found from the conservation of momentum:

$$\nabla \cdot (\sigma_a \vec{e_f}\vec{e_f} + \sigma_\mathbf{P}) = 0, \tag{13}$$

where $\sigma_\mathbf{P}$ is an elastic response of the tissue to deformation

$$\sigma_\mathbf{P} = \frac{1}{J}\mathbf{F} \cdot \mathbf{S_P} \cdot \mathbf{F}^\mathbf{C}, \tag{14}$$

$$S_{ij} = \frac{1}{2}\left(\frac{\partial W}{\partial E_{ij}} + \frac{\partial W}{\partial E_{ji}}\right), \tag{15}$$

$$\mathbf{E} = \frac{1}{2}(\mathbf{F}^c \cdot \mathbf{F} - \mathbf{I}) \tag{16}$$

and W is a strain-energy function while S_{ij} and E_{ij} denote the components of $\mathbf{S_P}$ and $\mathbf{E}$, respectively. The myocardium is modelled as hardly compressible material

$$W = W_e + a(det(2\mathbf{E} + \mathbf{I}) - 1)^2 \tag{17}$$

and the compressibility is determined by constant a.

Summary

The analysis of the activation has shown the possibility to model several arrhythmias by a rather simple mathematical model - the NPE. It is shown that
- The NPE reproduces closely the physiological quantities and the bifurcation structure of the Purkinije fibre preparations and can be used as a model for this type of a tissue.
- Using the NPE as a model, the non-reentrant bistability phenomenon of purely nonlinear nature is demonstrated.

The activation plays a regulatory role in myocardium contraction. Our model is expected to describe mechanical contraction in response to complicated activation patterns and is characterized by several internal variables.

To sum up, the behaviour of such a complicated structure as the human heart can be modelled by using the continuum mechanics with internal variables. The internal variables describe the cell energetics, mostly by certain kinetic equations. However, the number of parameters to be determined experimentally is high and needs a special attention.

The results are obtained in the close cooperation of authors with Helsinki University of Technology, Eindhoven University of Technology and Maastricht University. The Nuffic Fund (M.V.), CIMO (O.K.) and the Estonian Science Foundation (J.E., O.K., M.V.) have supported the research.

References

1. L.Glass, P.Hunter, A.McCulloch (eds.) Theory of Heart, Springer, New York et al. (1991).
2. D.H. van Campen, J.M.Huyghe, P.H.M. Bovendeerd and T.Arts, Eur. J. Mech. A/Solids, **13**, Suppl., 19 (1994).
3. O.Kongas, R.von Hertzen and J.Engelbrecht. Chaos, Solitons & Fractals (to be published in 1998).
4. B.J.West. Fractal Physiology and Chaos in Medicine. World Scientific, Singapore *et al.* (1990).
5. J.Jalife (ed.). Mathematical Approaches to Cardiac Arrhythmias. Annals N. Y. Acad. Sci., **591** (1990).
6. J.Engelbrecht. An Introduction to Asymmetric Solitary Waves. Longman, Harlow (1991).
7. O.Kongas. Nonlinear Dynamics in Modelling Cardiac Arrhytmias. PhD thesis, Tallinn Techn. Univ. (1998).
8. G.Zahalak. Mathematical Biosciences, **55**, 89 (1981).
9. R.B.Panerai. Biomech **13**, 929 (1980).

ON A PROBLEM OF BOUNDARY TEMPERATURE IDENTIFICATION IN A CYLINDRICAL LAYER

SYLWIA FUTAKIEWICZ, KRZYSZTOF GRYSA, LESZEK HOŻEJOWSKI
Kielce University of Technology, Al. 1000-lecia PP 7, 25-314 Kielce, Poland
E-mail: grysa@eden.tu.kielce.pl

In this paper the boundary temperature history along the surface of a cylindrical layer is inferred from the remotely sensed data. The desired temperature is estimated with the use of heat polynomials (HPs). This solution is compared with the other one obtained from the approach in which a set of reccurential Helmholtz equations is substituted for the heat conduction equation and potentials for Helmholtz equation are exploited to obtain the boundary temperature.

1 Introduction

In many technical applications some parts of machines or technical devices may practically be considered as a part of a cylindrical layer. Thermal conditions in which the machine works, frequently exclude a chance of direct measurements of the boundary temperature. In this case temperature sensors have to be placed within the solid. In order to obtain the searched boundary temperature an inverse heat conduction problem (IHCP) has to be solved. The IHCPs belong to the class of ill-posed problems in the Hadamard sense. Therefore, the fundamental question is the problem of stability of the solution to measurement inaccuracies.

2 Heat polynomials

The heat polynomials (HPs) introduced by Rosenbloom and Widder [1] for 1D problems in Cartesian coordinates can be defined as coefficients of p^n in a power series expansion

$$e^{px+p^2t} = \sum_{n=0}^{\infty} w_n(x,t)p^n \tag{1}$$

and have the following explicit form [1]:

$$w_n(x,t) = \sum_{k=0}^{[n/2]} \frac{x^{n-2k}t^k}{(n-2k)!k!}, \tag{2}$$

where [.] denotes the greatest integer of "." The HPs satisfy the dimensionless heat conduction equation

$$\frac{\partial^2 T}{\partial x^2} = \frac{\partial T}{\partial t} \tag{3}$$

To obtain HPs in polar coordinates (r,φ) we consider the generating function $e^{p(x+y)+2p^2t}$. After transformation from (x,y) to (r,φ) we obtain the generating function $u_{pol} = e^{rp(\cos\varphi+\sin\varphi)+2tp^2}$ and in case of axisymmetry we remove the variable φ by integrating with respect to φ which yields:

$$\frac{1}{2\pi}\int_0^{2\pi} e^{rp(\cos\varphi+\sin\varphi)+2tp^2} r d\varphi = e^{2tp^2} I_0\left(\sqrt{2}pr\right) \tag{4}$$

Expanding both functions, e^{2tp^2} and $I_0(\sqrt{2}pr)$, in a power series we arrive to the formula

$$e^{2tp^2} I_0\left(\sqrt{2}pr\right) = \sum_{n=0}^{\infty} v_n(r,t)p^{2n} \tag{5}$$

in which the heat polynomials $v_n(r,t)$ read

$$v_n(r,t) = \sum_{n=0}^{n} \frac{2^{-n+2k}}{k![(n-k)!]^2} r^{2n-2k} t^k \tag{6}$$

They are polynomials of degree 2n and they satisfy the equation

$$\frac{\partial^2 T}{\partial r^2} + \frac{1}{r}\frac{\partial T}{\partial r} = \frac{\partial T}{\partial t} \tag{7}$$

3 Heat polynomials applied to heat conduction problems in an infinite hollow cylinder

Consider the following problem of heat conduction in an infinite homogeneous and isotropic hollow cylinder ($a<r<b$), described by eq. (7) and conditions as follows:

$$T(r,0) = T_0(r) \quad \text{for} \quad a < r < b \tag{8}$$

$$\left.\begin{aligned} k_1 T(a,t) + h_1 \frac{\partial T(r,t)}{\partial r}\Big|_{r=a} &= \phi(t) \\ k_2 T(a,t) + h_2 \frac{\partial T(r,t)}{\partial r}\Big|_{r=b} &= \psi(t) \end{aligned}\right. \quad \text{for } t > 0 \tag{9}$$

The solution $T(r,t)$ of the problem (7)-(9) will be approximated with a linear combination of the polynomials $v_n(r,t)$:

$$T(r,t) \cong \Theta^{(i)}(r,t) = \sum_{n=0}^{N} c_n^{(i)} v_n(r,t) \tag{10}$$

The superscript „i" corresponds to the i-th time interval in which the approximation (10) is valid, i.e. $0 = t_0 < t_1 < t_2 < \ldots$, and $t \in (t_{i-1}, t_i\rangle$, $i=1,2,\ldots$. The function $\Theta^{(i)}(r,t)$ is a polynomial of degree $2N$ and satisfies the equation (8). The coefficients $c_n^{(i)}$ are chosen to minimize the functional $J^{(i)}$

$$\begin{aligned} J^{(i)} = & \int_a^b \left(\Theta^{(i)}(r,t_{i-1}) - \Theta^{(i-1)}(r,t_{i-1})\right)^2 dr + \\ & + \int_{t_{i-1}}^{t_i} \left(k_1 \Theta^{(i)}(a,t) + h_1 \left.\frac{\partial \Theta^{(i)}(r,t)}{\partial r}\right|_{r=a} - \phi(t) \right)^2 + \\ & + \int_{t_{i-1}}^{t_i} \left(k_2 \Theta^{(i)}(a,t) + h_2 \left.\frac{\partial \Theta^{(i)}(r,t)}{\partial r}\right|_{r=b} - \psi(t) \right)^2 \end{aligned} \tag{11}$$

where $\Theta^{(0)}(r,0)=T_0(t)$ and $i=1,2,\ldots$. The necessary condition for minimum leads to the system of linear algebraic equations in which the coefficients $c_n^{(i)}$ are unknowns.

$$\frac{\partial J^{(i)}}{\partial c_{\mathrm{n}}^{(\mathrm{i})}} = 0 \qquad \text{for} \quad n = 1,2,\ldots,N \tag{12}$$

The heat polynomials method can be applied also to the IHCPs, as shown in [2]. If the conditions are known at the inner points, say r_L^* and r_U^*, $a<r_L^*<r_U^*<b$, then validity of the solution in the region $<r_L^*, r_U^*>$ can be extended for all $r \in <a,b>$.

4 Potentials for Helmholtz equation applied to heat conduction problems

Consider the heat conduction equation

$$\left(\nabla^2 - \frac{\partial}{\partial t}\right) T(x,t) = 0 \quad , \quad x \in \Omega \subset R^m \quad , \quad t \in (0,\infty) \tag{13}$$

When the first derivative of temperature with respect to time is replaced by the backward finite difference, the equation (13) changes into a set of reccurential Helmholtz equations of the form

$$\left(\nabla^2 - p^2\right)T_k(x) = f_k(\mathrm{x}) \quad , \quad x \in \Omega \subset R^m \quad , \quad k = 1,2,... \tag{14}$$

where $T_k(x)=T(x,k\Delta t)$, $f_k(x)= - p^2 T_{k-1}(x)$, $k=1,2,...$, Δt stands for a time step, $p^2=1/\Delta t$, $T_0(x) = 0$. Let $S_m(x,p|h)$ denote the simple layer potential and $V_m(x,p|t)$ - the volume potential for the Helmholtz equation (14), with m being space dimension, where

$$S_m(x,p|h) = \int_{\partial\Omega} G_m(x-\xi,p)h(\xi)dS(\xi) \quad , \quad \xi \in \partial\Omega \quad , \quad \Omega \subset R^m \tag{15}$$

$$V_m(x,p|f) = -\int_{\Omega} G_m(x-y,p)f(y)dV(y) \quad , \quad x,y \in \Omega \subset R^m \tag{16}$$

Here h is a density of the simple layer potential and $G_m(x\text{-}y,p)$ stands for the fundamental solution of the equation (14), [3]. Using the potentials, one may express the approximate solution of the equation (13) in the form

$$T_k(x) = S_m(x,p|h_k) + V_m(x,p|f_k) \tag{17}$$

where the density h is the solution of a proper integral equation obtained from the boundary condition in which $T(x,t)$ is replaced by the right side of (17).

The same replacement takes place also in the case of an IHCP in the condition(s) describing an internal temperature (heat flux) response(s).

For axisymmetrical problem in a cylindrical layer, when two internal temperature responses are given at r_L^* and r_U^* where $a< r_L^* < r_U^* <b$, approximate solution obtained from the procedure described above has the form

$$\begin{aligned} T_k(r) = &\left[T_{Lk}^* + p^2 V_T\left(r_L^*,p|T_{k-1}\right)\right]\frac{M(r,r_U^*)}{M(r_L^*,r_U^*)} + \\ &+\left[T_{Uk}^* + p^2 V_T\left(r_U^*,p|T_{k-1}\right)\right]\frac{M(r_L^*,r)}{M(r_L^*,r_U^*)} - p^2 V_T\left(r,p|T_{k-1}\right) \end{aligned} \tag{18}$$

where $I_0(.)$ and $K_0(.)$ denote the modified Bessel functions of the first and second kind, respectively, of the order zero, $a<r<b$ and

$$M(r,y) = I_0(pr)K_0(py) - I_0(py)K_0(pr) \tag{19}$$

$$V_T(r,p|f) = -\left\{K_0(pr)\int_a^r zI_0(pz)f(z)dz + I_0(pr)\int_a^r zK_0(pz)f(z)dz\right\} \tag{20}$$

where T_{Lk}^* and T_{Uk}^* are internal temperature responses at r_L^* and r_U^*, respectively.

5 Numerical example

Consider a problem of the inner surface temperature identification for a turbine casing. The geometrical model for the turbine casing is a long hollow cylinder.The following data are used for calculations: thermal diffusivity $\alpha = 6.626 \times 10^{-6}\ m^2/s$, inner radius $a = 0.70\ m$, outer radius $b = 0.78\ m$, initial temperature of the turbine casing $T_0 = 250\ ^oC$. The outer surface of the casing is insulated. Additionally, a set of discrete data concerning temperature at the inner point of the casing, $r^* = 0.74\ m$, is given. Note that any measurement of temperature contains errors. The measurements were taken with a time step Δt from *45 s* at the begining up to *130* s at the end of the process (after nondimensionalisation the time step is from 0.0466 up to 0.186). The measured temperatures are given in the table 1.

Table 1. Temperature at $r = r^*$.

Time [s]	Temperature measured at $r=r^*$ [^{0}C]	Smoothed temp. [^{0}C]	Time [s]	Temperature measured at $r=r^*$ [^{0}C]	Smoothed temp. [^{0}C]
15	-	250.71	330	-	305.08
30	-	252.52	390	-	313.92
45	250.0	255.00	450	322.27	322.27
60	-	257.74	540	333.94	333.94
75	-	260.57	630	344.69	344.69
90	265.97	263.37	720	354.60	354.60
135	271.46	271.46	780	-	360.77
180	281.33	280.33	870	-	369.41
225	288.13	288.13	900	372.14	372.14
270	295.96	295.56			

Identification of the inner surface temperature with the use of formula (18) does not give acceptable results when only experimental data are taken into account. The reason is that the procedure requires more measurements. The missing data are obtained from numerical simulation. Finally, all the measurements (real and simulated) have been smoothed with the splines..They are presented in the table 1 as „smoothed temperature". The inner surface temperature calculated then with the formula (18) is shown in Fig.1 and in Fig.2 (dots).

Next, the HPs were used to calculate the inner surface temperature. In the functional (11) the second integral becomes a finite sum because $\phi(t)$ is a set of discrete data in this case. The results obtained for the measured temperatures are shown in Fig. 1 (solid line). For the "improved" data the solution is presented in Fig.2 (solid line). The results obtained by heat polynomials method from measured temperatures and those obtained from smoothed data do not differ much.

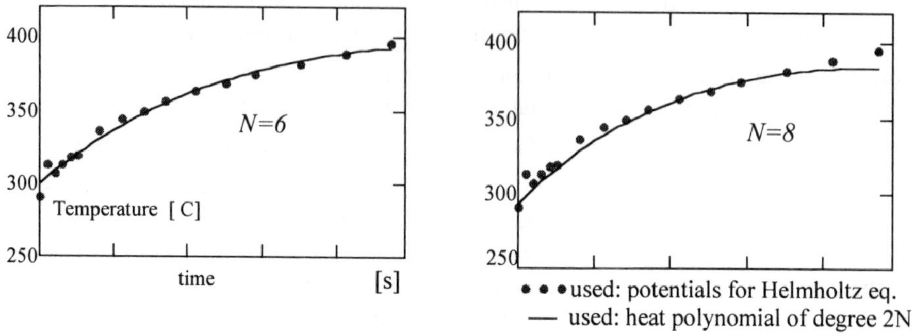

Figure 1. Results for inaccurate and incomplete input data (Hps). Dots describe solution for improved temperatures obtained with the use of potentials for Helmholtz eq.

Since both considered methods give similar results, the computed inner surface temperature is supposed to be close to its true value. However, the solution obtained by heat polynomials method practically does not depend on which input data have been taken: measured or "smoothed".

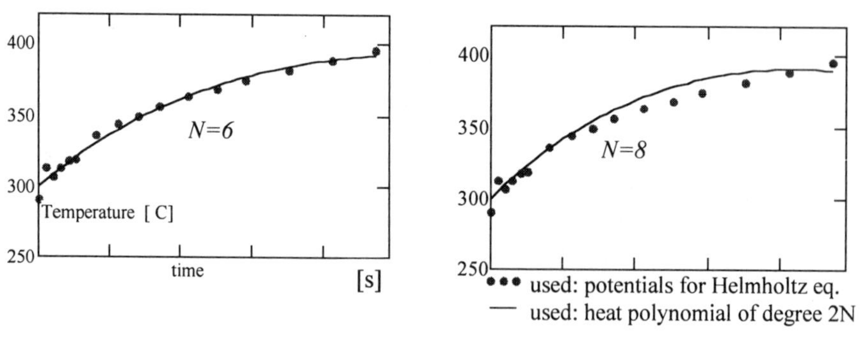

Figure 2. Results for improved input data (both methods)

6 Final remarks

The present methods of solving the considered IHCP show that even in case of ill-posed problems stable approximate solution may be obtained in a relatively simple way. In the Helmholtz potentials approach the boundary conditions are satisfied at any chosen moment of time and the heat conduction equation is satisfied approximately. In the heat polynomial method the heat conduction equation is satisfied strictly in all the moments of time and the boundary and initial conditions are satisfied approximately. The HPs method seems to be more effective. In the Helmholtz potentials approach, too small time steps may result in unstability, especially in case of inaccurate internal response, [4].

Anyway, both methods lead to almost the same results. It means that for IHCPs we have two relatively simple methods that can verify each other.

References

1. P.C.Rosenbloom and D.V.Widder, Expansions in terms of heat polynomials and associated functions, *Trans. Am. Math. Soc.* **92** (1959) pp. 220-266.
2. Futakiewicz S., Hożejowski L, Heat polynomials method in solving the direct and inverse heat conduction problems in a cylindrical system of coordinates, in: *Advanced Computational Methods in Heat Transfer V*, ed. by Nowak A.J., Brebbia C.A, Białecki R., Zerroukat M. (Computational Mechanics Publications, 1998) pp 71-79.
3. Grysa K., On the exact and approximate methods of solving the inverse problems of temperature fields (Politechnika Poznańska , Rozprawy, nr 204, Poznań, 1989) (in Polish).
4. Grysa K., Maciąg A., Stability investigation for the inverse heat conduction problems, in: *Advanced Computational Methods in Heat Transfer V*, ed. by Nowak A.J., Brebbia C.A, Białecki R., Zerroukat M. (Computational Mechanics Publications, 1998) pp 93-102.

PROPAGATION OF THE PERIODIC WAVES IN AN ELASTIC LAYER COVERED BY SPRING MASSES

IRENA GOŁĘBIOWSKA, MACIEJ DUTKIEWICZ
Faculty of Civil and Environmental Engineering, University of Technology and Agriculture
Kaliskiego 7, 85-796 Bydgoszcz,
Poland,
E-mail: macdut@mail.atr.bydgoszcz.pl

The plane problem of the propagation of periodic waves in an elastic isotropic layer of depth 2 h with high number of small spring masses applied to the bottom and top surface of the layer was studied. A dispersive equation for a symmetric vibration case was received.
The phase velocity was calculated in the relation to the nondimensional frequency parameter for the first and the second mode of the vibration. Numerical analysis of the equation shows that for small frequencies the phase velocity can be reduced in the case of suitable choice of parameters of the system, applied to the layer.

1 Introduction

In engineering practice the propagation of elastic waves in an elastic medium is an important problem, particularly in the seismology.
The problems of propagation of waves in the elastic medium and in the layer were considered in [4], [5], [6], [8], [9].
Furthermore propagation of waves in the half-space and in the elastic layer covered by different resonating elements is solved in [2], [3].
In above-mentioned papers general characteristic equation dealing with propagation of surface waves in the elastic medium with resonating elements were achieved. It is shown that suitable selection of system parameters causes the considerable reducing of the velocity of the surface wave.
In this paper the propagation of periodic waves in the elastic layer with the system of spring - masses, added to the medium, is analysed.

2 Formulation and solution of the problem

We consider the plane problem of the propagation of periodic waves in the elastic isotropic layer of depth 2 h with high number of small spring masses applied to the bottom and top surface of the layer (Fig. 1).
The reaction of spring masses is assumed as being uniformly distributed. Additionally, the asumption has been made, that elastic spring elements react to vertical motion of elastic layer surface only.
The elastic wave is transfered in the direction of the x axis with the velocity c.

The aim of the paper is to determine the influence of changing of parameters of spring - mass system on the velocity of the wave.

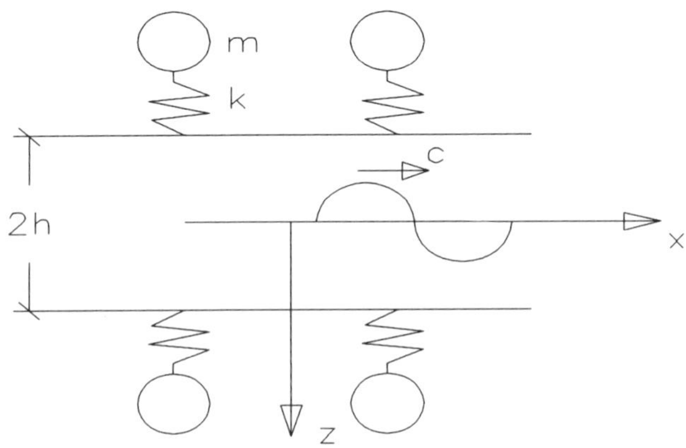

Figure 1. Coordinate system and propagating wave

The equations of motion of elastic medium may be written in the following form :

$$\mu\nabla^2 u + (\lambda + \mu)\text{grad div } u - \rho\ddot{u} = 0 \tag{1}$$

where ∇^2 = Laplacian operator, u = vector of displacement, ρ = mass density, and λ, μ are Lame's constants for a linear elastic material.
Equation (1) is supplemented by the stress-strain relations :

$$\sigma_{ij} = \lambda\delta_{ij}\varepsilon_{kk} + 2\mu\varepsilon_{ij} \quad (i,j,k = 1,2,3) \tag{2}$$

the strain tensor components :

$$\varepsilon_{ij} = \frac{1}{2}(u_{i,j} + u_{j,i}) \qquad (i,j = 1,2,3) \tag{3}$$

and the following boundary conditions :

$$\sigma_{zz}\left(x,\pm h,0\right) = -\gamma w(x,\pm h)$$
$$\sigma_{xz}\left(x,\pm h,0\right) = 0 \tag{4}$$

where δ_{ij} = Kronecker symbol, $\gamma = \dfrac{k_e\omega^2}{\omega_e^2 - \omega^2}$, k_e , ω_e are parameters of spring masses.

When we uncouple equation (1), we can receive the following form :

$$\nabla_1^2\Phi - \frac{1}{c_1^2}\ddot{\Phi} = 0$$
$$\nabla_1^2\Psi - \frac{1}{c_2^2}\ddot{\Psi} = 0, \tag{5}$$

where

$$c_1 = \sqrt{\frac{\lambda + 2\mu}{\rho}} \quad \text{- velocity of the longitudinal wave,}$$
$$c_2 = \sqrt{\frac{\mu}{\rho}} \quad \text{- velocity of the transverse wave} \tag{6}$$

and the functions Φ and Ψ are elastokinetic potentials.

We search the solution of wave equations (5) with the asumption :

$$\Phi(x,z,t) = \hat{\Phi}(x,z)e^{-i(\alpha x - \omega t)}$$
$$\Psi(x,z,t) = \hat{\Psi}(x,z)e^{-i(\alpha x - \omega t)} \tag{7}$$

Substituting (7) into (5) leads to differential equations :

$$[\frac{d^2}{dz^2} - (\alpha^2 - k_1^2)]\hat{\Phi}(z) = 0$$
$$[\frac{d^2}{dz^2} - (\alpha^2 - k_2^2)]\hat{\Psi}(z) = 0 \tag{8}$$

with solutions in the form :

$$\hat{\Phi}(z) = A\sinh\nu_1 z + B\cosh\nu_1 z$$
$$\hat{\Psi}(z) = C\sinh\nu_2 z + D\sinh\nu_2 z \tag{9}$$

where $k1 = \frac{\omega}{c_1}, k_2 = \frac{\omega}{c_2}, \alpha = \frac{\omega}{c}, \nu_1 = \sqrt{\alpha^2 - k_1^2}, \nu_2 = \sqrt{\alpha^2 - k_2^2}$.

Substituting (9) into (7) leads to :

$$\Phi(x,z,t) = (A\sinh\nu_1 z + B\cosh\nu_1 z)e^{-i(\omega t - \alpha x)}$$
$$\Psi(x,z,t) = (C\sinh\nu_2 z + D\sinh\nu_2 z)e^{-i(\omega t - \alpha x)} \tag{10}$$

The components of the displacement and the stress tensor are related to the potential by means of :

$$u = \frac{\partial\Phi}{\partial x} - \frac{\partial\Psi}{\partial z}, \qquad w = \frac{\partial\Phi}{\partial x} + \frac{\partial\Psi}{\partial x}$$
$$\sigma_{zz} = 2\mu\left(\frac{\partial^2\Phi}{\partial z^2} + \frac{\partial^2\Psi}{\partial x\partial z}\right) + \lambda\nabla_1^2\Phi \tag{11}$$
$$\sigma_{xz} = \mu\left(2\frac{\partial^2\Phi}{\partial x\partial z} + \frac{\partial^2\Psi}{\partial x^2} - \frac{\partial^2\Psi}{\partial z^2}\right)$$

Substituting functions Φ and Ψ from (10) to equations (11), then to the suitable boundary conditions (4), we can receive four equations with unknown constants A, B, C, D.

Because displacement u(x) has the symmetric solution, w(z) - antisymmetric solution, we can state that stresses : σ_{zz} are symmetric, σ_{xz} are antisymmetric in relation to plane $z = 0$.

Using above mentioned attributes of the symmetry we can receive the characteristic equation :

$$2i\alpha\nu_1\sinh\nu_1 h(2i\mu\alpha\nu_2\cosh\nu_2 h +$$
$$-\gamma i\alpha\sinh\nu_2 h) + [(2\mu\nu_1^2 - \lambda\alpha^2 + \lambda\nu_1^2)\cos\nu_1 h + \tag{12}$$
$$-\gamma\nu_1\sinh\nu_1 h][(\alpha^2 + \nu_2^2)\sin\nu_2 h] = 0$$

For $\mu = \lambda$ ($\nu = 0.25$) we can simplify the equation (12) to the form :

$$\begin{aligned}&(2s^2-1)^2 \operatorname{ctgh}(h_0\sqrt{s^2-n^2})+\\&-4s^2\sqrt{s^2-1}\sqrt{s^2-n^2}\operatorname{ctgh}(h_0\sqrt{s^2-1})+\\&+\frac{m^* h_0\sqrt{s^2-n^2}}{(1-1^2)}=0\end{aligned} \tag{13}$$

where $s=\frac{c_2}{c}, h_0=\frac{\omega h}{c_2}, n=\frac{k_1}{k_2}, m^*=\frac{m_e}{\rho h}, 1=\frac{\omega}{\omega_e}$

The first and the second mode of the vibration with suitable phase velocity in the relation to nondimensional frequency parameter αh, for fixed $\nu = 0.25$, $1 = 0.995$ and for different value of the m^* are shown in Figure 2 and 3.

The broken line shows the values of $\frac{c}{c_2}$ obtained by Tolstoy and Usdin [10]

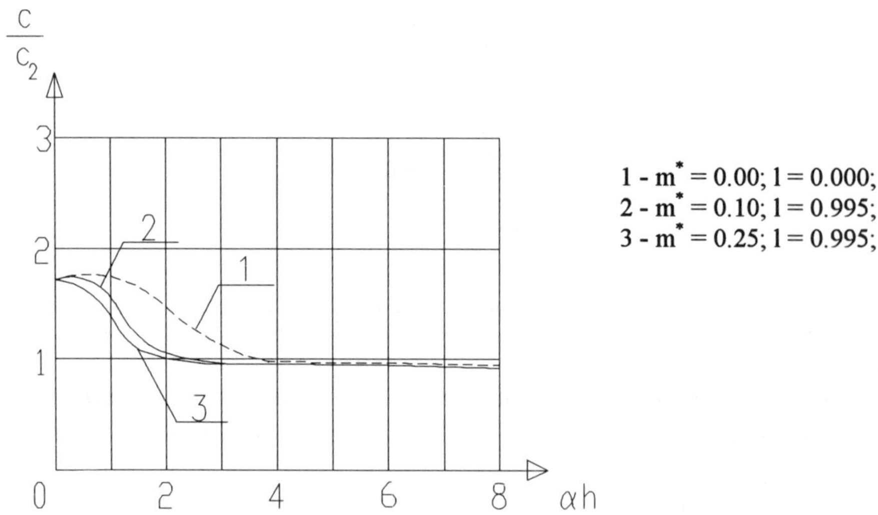

Figure 2. Phase velocity c for various masses m^*. Mode 1.

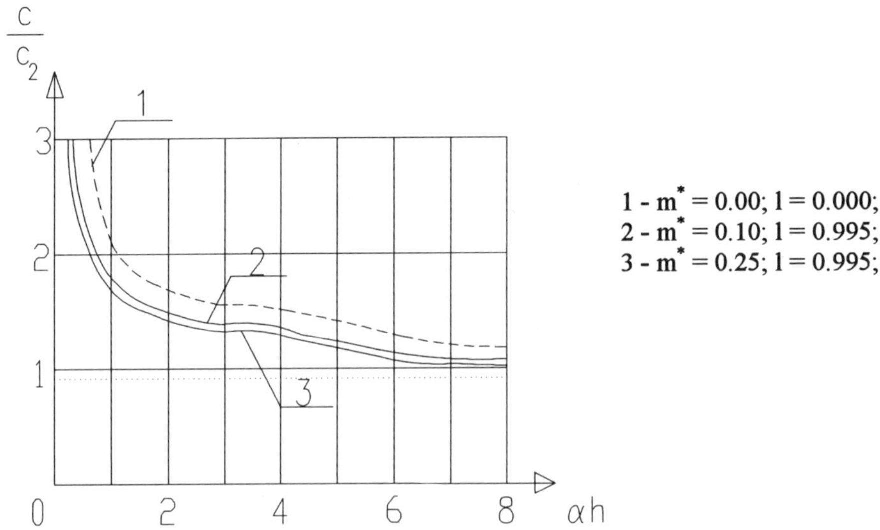

Figure 3. Phase velocity c for various masses m^*. Mode 2.

3 Discussion

It is shown that phase velocity of the propagating wave depends on the parameters of the spring - mass system (Fig. 2, 3).The required velocities of the surface wave can be obtained by appropriate selection of these parameters.
Figure 2 and 3 show the influence of the different value of masses used in the model for known value $l = 0.995$ and $\nu = 0.25$ on the velocity of the propagation of the surface wave.
The numerical example shows that for assumed quantities phase velocity c decreases from the value $1.633c_2$ for $\alpha h \to 0$ to the asymptotic value $0.919c_2$ for $\alpha h \to \infty$ for the first mode of the vibration, and c decreases from the infinity for $\alpha h \to 0$ to the asymptotic value c_2 for $\alpha h \to \infty$ for the second mode of the vibration.

References

1. Gogoladze W. G., Dispersion of the Rayleigh wave in the layer, Publ. Sejsm. Inst. A.N. USSR (1947), v.119.

2. Gołębiowska I., Surface waves propagation in a half-space with a surface layer of vibration dampers, XVIth Symposium Vibrations and Waves' 94, Poznań - Błażejewko, Poland (1994) pp. 141-142.
3. Gołębiowska I., Propagation of surface wave in a half-space with resonating elements, Proc. of the Third International Conference on Vibration Problem (ICOVP-96), University of North Bengal, India (1996) pp. 4-5.
4. Knopoff L., On Rayleigh wave velocities, Bull.Seism. Soc. Amer. (1952) Vol. 42.
5. Lamb H., On waves in an elastic plate, Proc. Roy. Soc. (1910) vol. 93.
6. Love A. E. H., Some problems in geodynamics, Cambridge (1911).
7. Nowacki W., Structural dynamic, Arkady, Warsaw (1972).
8. Rayleigh J. W., On waves propagated along the plane surface on an elastic solid, Proc. Lond. Math. Soc. (1885) 17, 4.
9. Rayleigh J. W., On the free vibration of an infinite plate of homogeneous isotropic elastic matter, Proc. Math. Soc. (1889) vol. 20.
10. Tolstoy J., Usdin E., Dispersive properties of stratified elasic and liquid media. Bull. Seism. Soc. Amer. (1954) vol. 44.

SOME REMARKS ON THE CONDUCTIVITY COEFFICIENT IDENTIFICATION

KRZYSZTOF GRYSA, ARTUR MACIAG
Kielce University Of Technology, Al. 1000-lecia PP 7, 25-314 Kielce, Poland

E-mail: grysa@eden.tu.kielce.pl

A new metod of the conductivity coefficient identification is presented. Three functions of time are used. Two of them describe the displacement and temperature of the top surface of a flat slab. The third one describes internal temperature response at a point close to the top surface of the slab. The method is verified by a numerical experiment.

1 Introduction

To consider any heat transfer problem in a solid body one has to know its thermophysical properties. They are determined for variuos materials with the use of many methods. In this paper a new metod of the conductivity coefficient identification is presented and verified by a numerical experiment.

Temperature field and displacements in a homogeneous and isotropic flat slab are considered. Its bottom surface is immobilized and thermally insulated. The top surface is kept in a constant temperature (i.e. is cooled or heated). The temperature and displacement histories at inner points change along thickness of the slab. Then, we consider a one-dimensional temperature field and uniaxial strain state.

In order to find the conductivity coefficient three functions are used. Two of them describe the displacement and temperature of the top surface of the slab. The third one describes internal temperature response at a point close to the top surface. They may be also given in the form of sets of discrete data (from an experiment). Then, using a simple formula, the values of the searched coefficient are calculated.

2 Basic equations and conditions

In the theory of thermal stresses the following displacement equation holds for the strain state described above [1,2]:

$$\frac{\partial^2 u}{\partial x^2} - \frac{1}{c^2}\frac{\partial^2 u}{\partial t^2} = k\frac{\partial T}{\partial x} \tag{1}$$

In the considered case the initial and boudary conditions for displacement read

$$u(x,0)=0 \quad , \quad \left.\frac{\partial u(x,t)}{\partial t}\right|_{t=0}=0 \quad , \quad u(0,t)=0 \quad , \quad \left.\frac{\partial u(x,t)}{\partial x}\right|_{x=h}=kT(h,t)$$

The temperature field is described by the heat conduction equation

$$\frac{\partial^2 T}{\partial x^2}=\frac{1}{a}\frac{\partial T}{\partial t} \tag{2}$$

with the following initial and boundary conditions:

$$T(x,0)=0 \quad , \quad \left.\frac{\partial T(0,t)}{\partial x}\right|_{x=0}=0 \quad , \quad T(h,t)=T_h(t)$$

Notation: u - displacement perpendicular to the slab surfaces; $u \equiv u(x,t)$, x - spatial variable in the direction perpendicular to the surfaces from the bottom surface up; t - time, $c^2=\dfrac{2G(1-\nu)}{\rho(1-2\nu)}$, $k=\dfrac{1+\nu}{1-\nu}\alpha_T$, G - shear modulus, ν - Poisson ratio, α_T - thermal expansibility coefficient, $a=\lambda/(\rho c_p)$ - thermal diffusivity, λ - thermal conductivity, ρ - density, c_p - specific heat.

3 The formulae for thermal conductivity

Let us evaluate the eq. (1) with the use of the Laplace integral transform, [3]. We obtain

$$\frac{d^2\bar{u}}{dx^2}-\frac{s^2}{c^2}\bar{u}=k\frac{d\bar{T}}{dx} \tag{3}$$

where $\bar{u}$ denotes the transformed displacement u, $\bar{T}$ stands for the Laplace tranform of temperature, and s is the transformation parameter. The boundary conditions are transformed in the same way.

The solution of eq. (3) has the following form

$$\bar{u}(x,s)=\frac{c}{s}\int_0^x k\frac{d\bar{T}}{dy}\sinh\left[\frac{s}{c}(x-y)\right]dy+$$

$$+\frac{ck}{s}\frac{\sinh(sx/c)}{\cosh(sh/c)}\left\{\bar{T}(h,s)-\int_0^h\frac{d\bar{T}}{dy}\cosh\left[\frac{s}{c}(h-y)\right]dy\right\} \tag{4}$$

Integrating by parts in formula (4) we arrive to the following form of the transformed displacement:

$$\bar{u}(x,s)=\frac{ck}{s}\frac{\sinh(sx/c)}{\cosh(sh/c)}\left\{\bar{T}(0,s)\cosh(sh/c)-\frac{s}{c}\int_0^h\bar{T}(y,s)\sinh[s(h-y)/c]dy\right\}+$$
$$-\frac{ck}{s}\bar{T}(0,s)\sinh(sx/c)+k\int_0^x\bar{T}(y,s)\cosh[s(x-y)/c]dy \tag{5}$$

Consider now the heat conduction equation (2). Its transformed form reads

$$\frac{d^2\bar{T}}{dx^2}-\frac{s\bar{T}}{a}=0 \quad , \qquad \bar{T}=\bar{T}(x,s) \tag{6}$$

Multiplying both sides of (6) by $\sinh[s(x-y)/c]$ and integrating from *0* to *h* we obtain

$$\int_0^h\bar{T}(y,s)\sinh\left[\frac{s}{c}(h-y)\right]dy=\frac{\frac{d\bar{T}(0,s)}{dx}\sinh\left(\frac{sh}{c}\right)+\frac{s}{c}\bar{T}(0,s)\cosh\left(\frac{sh}{c}\right)-\frac{s}{c}\bar{T}(h,s)}{s^2/c^2-s/a} \tag{7}$$

Similarily, multiplying (6) by $\cosh[s(x-y)/c]$ and and integrating from *0* to *x* we arrive to the formula

$$\int_0^x\bar{T}(y,s)\cosh\left[\frac{s}{c}(x-y)\right]dy=\frac{\frac{d\bar{T}(0,s)}{dx}\cosh\left(\frac{sx}{c}\right)-\frac{d\bar{T}(x,s)}{dx}+\frac{s}{c}\bar{T}(0,s)\sinh\left(\frac{sx}{c}\right)}{s^2/c^2-s/a} \tag{8}$$

Substituting (7) and (8) to (5) we get the following formula for transformed displacement:

$$\bar{u}(x,s)\left(\frac{s}{c^2}-\frac{1}{a}\right)=\frac{ck}{s}\frac{\sinh(sx/c)}{\cosh(sh/c)}\bar{T}(h,s)\frac{s}{c^2}-\frac{ck}{s}\frac{1}{c}\frac{d\bar{T}(x,s)}{dx} \tag{9}$$

Let us substitute $\lambda/(\rho c_p)$ instead of the thermal diffusivity coefficient *a*. Then, for $x = h$ and $\bar{T}_h(s)$ standing for the Laplace transform of the function T_h(t) we obtain

$$\left[\rho c_p\bar{u}(h,s)-\frac{k\lambda}{s}\frac{d\bar{T}(x,s)}{dx}\bigg|_{x=h}\right]\frac{1}{\lambda}-\frac{s}{c^2}\bar{u}(h,s)+\frac{k}{c}tgh\left(\frac{sh}{c}\right)\bar{T}_h(s)=0 \tag{10}$$

Finally, we arrive to the formula for the conductivity coefficient, in which the temperature, its *x*-derivative and the displacement appear in the transformed form:

$$\lambda = \frac{\rho c_p \bar{u}(h,s)}{\frac{s}{c^2}\bar{u}(h,s) + \frac{k}{s}\left.\frac{d\bar{T}(x,s)}{dx}\right|_{x=h} - \frac{k}{c} tgh\left(\frac{sh}{c}\right)\bar{T}_h(s)} \tag{11}$$

Inverting the left side of (10) we arrive to the following formula

$$\lambda = \frac{\rho c_p u(h,t)}{\frac{1}{c^2}\frac{du(h,t)}{dt} + k\int_0^t \left.\frac{\partial T(x,\tau)}{\partial x}\right|_{x=h} d\tau - \frac{k}{c}\left[T_h(t) + 2\sum_{n=1}^{\infty}(-1)^n T_h\left(t - 2n\frac{h}{c}\right)\right]} \tag{12}$$

Hence, in order to calculate the conductivity from the formula (12), as an input data we have to take a temperature at an inner point close to the top slab surface as well as temperature and displacement on the top surface. They can be given as sets of discrete data (e.g. from an experiment) or as continuous functions. In the case of discrete data, the derivatives $\partial T/\partial x$ and $\partial u/\partial t$ have to be calculated numerically. Finally, after numerical integration of the derivative $\partial T/\partial x$ the conductivity can be calculated from the formula (12).

The formula (12) is valid in the case when conductivity is a constant. However, when conductivity depends on temperature, the problem has a slightly different solution. The initial conditions for temperature and displacement cannot be zero in that case. For the non-zero initial conditions the transforms of temperature and displacement have the form in which the terms $su(x,0)+u'_t(x,0)$ and $T(x,0)$ appear. In this case, the calculations will be led in the succesive time intervals, in which temperature changes are small. However, the idea of conductivity coefficient identification in each time interval is the same as for constant conductivity.

4 Numerical experiment

In order to simulate measurement of temperature and displacement, first we have to solve an initial-boundary problem for temperature and displacement. Consider the heat conduction equation (2) with the homogeneous initial condition and the following boundary conditions:

$$T(h,t) = T_h(t) = 100, \quad \partial T(0,t)/\partial x = 0 \tag{13}$$

Its solution in the transformed form reads

$$\bar{T}(x,s) = \frac{100\cosh\left(x\sqrt{s/a}\right)}{s\cosh\left(h\sqrt{s/a}\right)} \tag{14}$$

Hence

$$T(x,t)=100\left[1+2\sum_{n=1}^{\infty}(-1)^n\frac{\cos(\lambda_n x/h)}{\lambda_n}e^{-\lambda_n^2\frac{a}{h^2}t}\right] \tag{15}$$

where $\lambda_n=\pi(2n-1)/2$. Obviously, we obtain $T_h(t)$ = *100* for $x = h$.

Next, consider the following uniaxial strain state problem in displacements governed by the equation (1) subject to the homogeneous initial condition and the following boundary conditions:

$$u(0,t)=0 \quad , \quad \frac{\partial u(h,t)}{\partial x}=kT(h,t)=100k \tag{16}$$

The temperature in $(16)_2$ is given by the formula (15).

The solution of the problem (1), (16) in transformed form reads

$$\bar{u}(x,s)=\frac{100k}{s}\left[\frac{s\sinh(sx/c)/c}{s\left(s/c^2-1/a\right)\cosh(sh/c)}-\frac{\sqrt{s/a}\,\sinh\left(\sqrt{s/a}x\right)}{s\left(s/c^2-1/a\right)\cosh(sh/a)}\right] \tag{17}$$

Hence we can find *u(x,t)* and then *u(h,t)* as well as $\partial u)x,t)/\partial t|_{x=h}$. The displacement *u(x,t)* reads

$$u(x,t)=100kx-200k\sum_{n=1}^{\infty}(-1)^{n+1}\frac{\sin(\lambda_n x/h)}{\lambda_n^2a^2+c^2h^2}\times$$

$$\times\left[ha^2\cos(\lambda_n ct/h)+ach^2\sin(\lambda_n ct/h)/\lambda_n+(h^3c^2/\lambda_n^2)e^{-\lambda_n^2\frac{a}{h^2}t}\right] \tag{18}$$

First we check how the formula (11) which expresses λ in terms of $\bar{u}$ and $\bar{T}$ works. Let c=5971.23 [m/s], k=0.00002228 [1/deg], a=0.0000119 [m^2/s], ρ=7800 [kg/m^3], c_p =428.672 [J/kg deg], h=0.05 m. The exact conductivity value is equal 39.789335 W/(m.deg). Substituting exact values of the transformed temperature, (14), its x-derivative and the transformed displacement, (17), we obtain the exact value of conductivity.

In the case of the formula (12) the results of calculations for the exact input data are presented at Fig. 1. Numerical integrating of the exact data leads to a small inaccuracy in the final result. This inaccuracy tends to zero with time tending to infinity (comp. Fig. 1).

In the next step, simulated experimental data have to be generated. The temperature at the top surface is equal to 100 $^{\circ}$C (comp. cond. $(13)_1$). The internal temperature response (simulated measurements) at the point x = *0.95h* = *0.0475* m is taken as the exact data plus a 5% random noise. The displacement and temperature histories obtained this way are presented in Figs. 2 and 3.

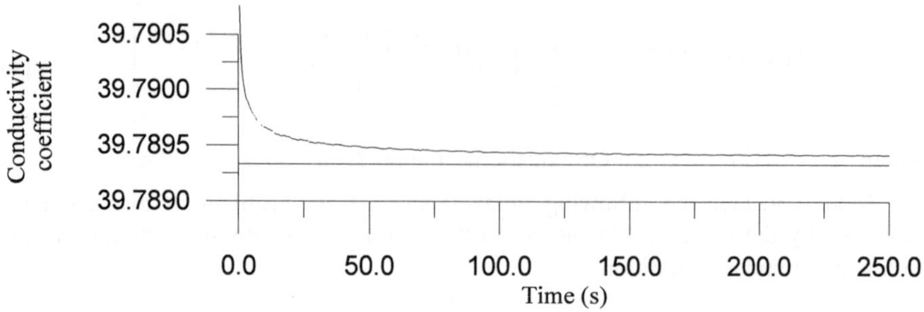

Figure1. The conductivity identification based on the exact data concerning the temperature and displacement.

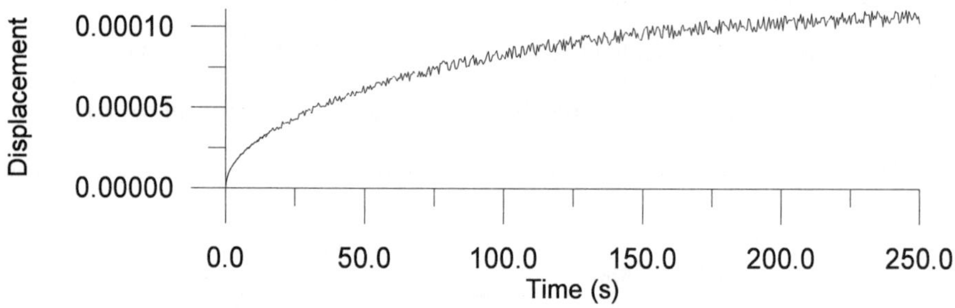

Figure 2. "Measured" displacement of the top surface of the slab.

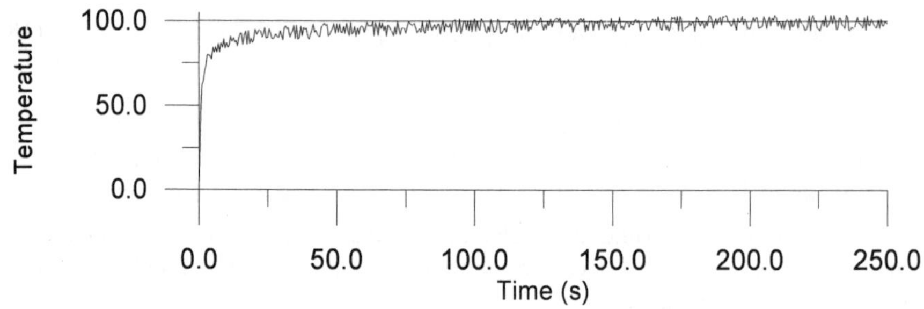

Figure 3. "Measured" temperature at $x = 0.0475$.

Results of calculations for two cases, concerning the techniques of derivative calculation, are presented in Figs 4 and 5. In Fig. 4 the results are obtained for the case when the first derivatives that appear in the formula (12) are approximated with the first backward finite difference. Fig.5 illustrates what happens to the calculated conductivity if the first derivatives are approximated with the central finite difference calculated according to the formulae

$$\frac{du(h,t_k)}{dt} \approx \frac{u(h,t_{k-2}) - 8u(h,t_{k-1}) + 8u(h,t_{k+1}) - u(h,t_{k+2})}{12\Delta t} \tag{19}$$

In both cases average value for conductivity taken for the time interval <100, 250> s differs from the exact value. The average value, its error and the standard deviation is presented in Table 1.

Table 1. Average value of conductivity

	Average value of conductivity	Error %	Standard deviation
backward finite difference	43.23046	8.65	0.68469
central finite difference (19)	42.32706	6.38	0.52616

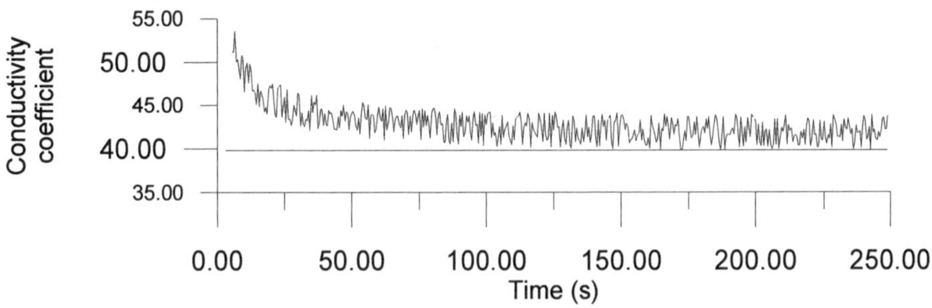

Figure 4. Conductivity obtained in the case of derivatives calculated as first backward finite differences

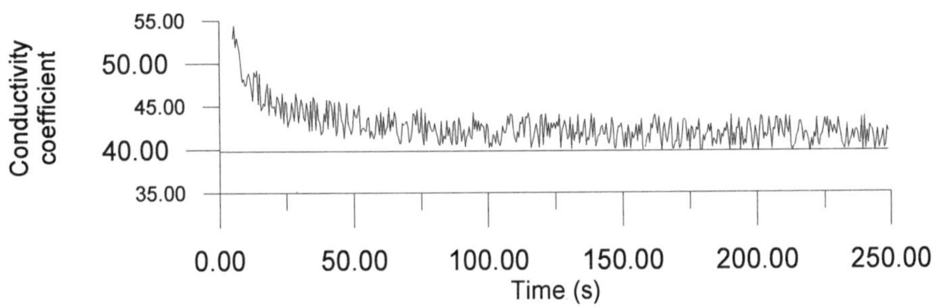

Figure5. Conductivity obtained in the case of derivatives calculated as the central finite differences

The inaccuracy in both cases is of the same order as the inaccuracy of the temperature and displacement. If we add that numerical integration was led in the simplest way (time step × value of integrated function), then we can see that accuracy of the results may be improved. Moreover, in a real experiment measurement errors usually do not exceed 0.1 °C. For such experimental data the identified conductivity would differ from its exact data not more than 0.02-0.03 W/(m.deg).

5 Final remarks

The method of the conductivity coefficient identification presented in the paper is very simple. In the presented numerical experiment the inaccuracy introduced by noisy input data was of the order of the noise itself.

The method is easy to be extended for the case of conductivity depending on temperature. In such case, the formulae (11) and (12) will change slightly, as mentioned at the end of section 3. In that case calculations practically will not differ from those for the constant conductivity.

References

1. Fung Y.C., Fundamentals of the solid mechanics (PWN, Warszawa, 1969) (In Polish)
2. Nowacki W., Thermoelasticity (PWN, Warszawa, 1960) (In Polish)
3. Osiowski J., Outline of the operator calculus (WNT, Warszawa, 1972) (In Polish)

THERMOMECHANICAL COUPLING IN VISCOPLASTICITY: CONSTITUTIVE MODELLING OF SHAPE–MEMORY BEHAVIOUR

P. HAUPT AND D. HELM

Institute of Mechanics, University of Kassel, Mönchebergstraße 7, D-34109 Kassel, Germany

E-mail: haupt@ifm.maschinenbau.uni-kassel.de and helm@ifm.maschinenbau.uni-kassel.de

In this article we present a thermomechanically consistent constitutive model, which is capable of simulating the main effects of shape memory alloys, i.e. the one– and two–way effect, pseudoelasticity and pseudoplasticity as well as the transition between the pseudoelasticity and the pseudoplasticity. The material model is formulated within the phenomenological theory of thermoviscoplasticity. The thermomechanical consistency is proved in the sense of the Clausius–Duhem inequality, a special form of the second law of thermodynamics. The numerical integration of the constitutive equations for various strain and stress–temperature paths demonstrates the ability of the proposed constitutive model to represent the essential phenomena of shape memory behaviour.

1 Introduction

The behaviour of all kinds of materials depends on temperature. A physically consistent representation of this dependence can be achieved if the principles of thermodynamics are observed. In continuum thermomechanics a sufficient basis is provided by the local energy balance (1st Law),

$$(\psi + \theta\eta)^{\cdot} = \frac{1}{\rho}\mathbf{T} \cdot \dot{\mathbf{E}} - \frac{1}{\rho}\operatorname{div}\boldsymbol{q} + r, \tag{1}$$

and the Clausius–Duhem inequality (2nd Law),

$$-\dot{\psi} - \dot{\theta}\eta + \frac{1}{\rho}\mathbf{T} \cdot \dot{\mathbf{E}} - \frac{1}{\rho\theta}\boldsymbol{q} \cdot \operatorname{grad}\theta \geq 0. \tag{2}$$

Here, ψ and η denote respectively the free energy and entropy per unit mass; θ is the absolute temperature, $\boldsymbol{q}$ the heat flux vector, r the volume–distributed heat supply and ρ the mass density. $\mathbf{T}$ is the Cauchy stress tensor and $\mathbf{E}$ the linearized Green strain, i.e., only small deformations are considered.

Material properties are expressed by means of constitutive equations for the free energy,

$$\psi = \hat{\psi}(\mathbf{E}_{\mathrm{e}}, \theta, \mathbf{E}_{\mathrm{i}}, q_1, \ldots, q_n), \tag{3}$$

and the heat flux vector,

$$\boldsymbol{q} = \hat{\boldsymbol{q}}(\ldots; \operatorname{grad}\theta), \tag{4}$$

which must at least depend on the temperature gradient. The free energy depends on the elastic part of the total strain,

$$\mathbf{E}_\mathrm{e} = \mathbf{E} - \mathbf{E}_\mathrm{i}, \tag{5}$$

on the temperature θ and on a set of internal variables. Among these $\mathbf{E}_\mathrm{i}$ is the inelastic part of the total strain; $\mathbf{E}_\mathrm{i}$ and the variables $q_1, \ldots, q_n$ facilitate a phenomenological representation of microstructural changes within the material.

The constitutive description is complete if a set of evolution equations for the internal variables is formulated, specifying the time rates $\dot{\mathbf{E}}_\mathrm{i}$, $\dot{q}_1, \ldots, \dot{q}_n$ as functions of the current values and the total strain $\mathbf{E}(t)$ (or its elastic part $\mathbf{E}_\mathrm{e}$) as well as the temperature $\theta(t)$. Indeed, the entropy inequality (2) is satisfied if the stress and entropy relations hold,

$$\frac{1}{\rho}\mathbf{T} = \frac{\partial\hat{\psi}}{\partial\mathbf{E}_\mathrm{e}}, \qquad \eta = -\frac{\partial\hat{\psi}}{\partial\theta}, \tag{6}$$

and the remaning entropy production is non–negative. This implies the internal dissipation inequality

$$\delta = \left(\frac{1}{\rho}\mathbf{T} - \frac{\partial\hat{\psi}}{\partial\mathbf{E}_\mathrm{i}}\right)\cdot\dot{\mathbf{E}}_\mathrm{i}(t) - \sum_k \frac{\partial\hat{\psi}}{\partial q_k}\dot{q}_k(t) \geq 0, \tag{7}$$

a restrictive condition for all evolution equations.

The energy balance (1) causes a coupling between strain and temperature fields: From eqs. (1), (3), (5) and (6) we infer the generalized equation of heat conduction,

$$\begin{aligned} c_\mathrm{d}\dot{\theta} + \frac{1}{\rho}\operatorname{div}\boldsymbol{q} - r = {} & \underbrace{\frac{1}{\rho}\theta\frac{\partial\mathbf{T}}{\partial\theta}\cdot\dot{\mathbf{E}}_\mathrm{e}}_{w_\mathrm{e}} + \underbrace{\frac{1}{\rho}\mathbf{T}\cdot\dot{\mathbf{E}}_\mathrm{i}}_{w_\mathrm{i}} \\ & \underbrace{-\left(\frac{\partial(\psi+\theta\eta)}{\partial\mathbf{E}_\mathrm{i}}\cdot\dot{\mathbf{E}}_\mathrm{i} + \sum_k \frac{\partial(\psi+\theta\eta)}{\partial q_k}\dot{q}_k\right)}_{w_\mathrm{s}}, \end{aligned} \tag{8}$$

where

$$c_\mathrm{d} = -\theta\frac{\partial^2\hat{\psi}}{\partial\theta^2} \tag{9}$$

is the specific heat at constant deformation. On the left–hand side of the heat conduction equation we have the temperature change and the local heat supply. The right–hand side represents thermomechanical coupling: the first term w_e describes the effect of thermoelastic coupling, the second term w_i is the inelastic part of the stress power, whereas the specific power w_s corresponds to the rate of the internal energy due to the change of the internal variables. Actually, these two terms correspond to a change in the material's microstructure.

A basic assumption of thermoviscoplasticity is the additive decompositon of the free energy into equilibrium and non–equilibrium parts:

$$\psi = \psi_\mathrm{eq} + \psi_\mathrm{ov}. \tag{10}$$

The *equilibrium* part ψ_eq is assumed to be a rate–independent functional of the strain and temperature process and the *non–equilibrium* part ψ_ov a rate–dependent functional, tending asymptotically to zero for slow processes. ψ_eq represents thermodynamic equilibrium states depending on the process history. ψ_ov incorporates the deviation from equilibrium states, which themselves are limits of isothermal static continuations or retarded processes. The decomposition of the free energy implies a corresponding decomposition of the stress tensor into an *equilibrium stress* $\mathbf{T}_\mathrm{eq}$ (with rate–independent hysteresis) and an *overstress* $\mathbf{T}_\mathrm{ov}$ (rate–dependent with relaxation to zero):

$$\mathbf{T} = \mathbf{T}_\mathrm{eq} + \mathbf{T}_\mathrm{ov}. \tag{11}$$

This concept was applied to the representation of the behaviour of metals[3,4] and rubber–like materials [6,7]. In the following it will be employed to model the behaviour of shape–memory alloys, where marked effects of thermomechanical coupling are observed.

2 A Free Energy Function

The phenomena known as shape memory behaviour[2,5,8,9] refer to the equilibrium stress. Depending on temperature, they are related to transitions between *austenitic* and *martensitic* phases or to *twinning* and *detwinning* processes of martensitic crystals. Based on the decomposition of the strain (see figure 1),

$$\mathbf{E} = (1-z)\mathbf{E}_\mathrm{e}^\mathrm{A} + z\mathbf{E}_\mathrm{e}^\mathrm{M} + \mathbf{E}_\mathrm{pp}, \tag{12}$$

we assume the following structure of the constitutive equation for the free energy $\psi = \psi_\mathrm{eq} + \psi_\mathrm{ov}$:

$$\psi = z\left[\psi_\mathrm{e}^\mathrm{M}(\mathbf{E}_\mathrm{e}^\mathrm{M},\theta) + \psi_0^\mathrm{M}\right] + (1-z)\left[\psi_\mathrm{e}^\mathrm{A}(\mathbf{E}_\mathrm{e}^\mathrm{M},\theta) + \psi_0^\mathrm{A}\right] + \psi_\mathrm{pp}(\mathbf{E}_\mathrm{pp},\mathbf{Y}_\mathrm{d}) +$$

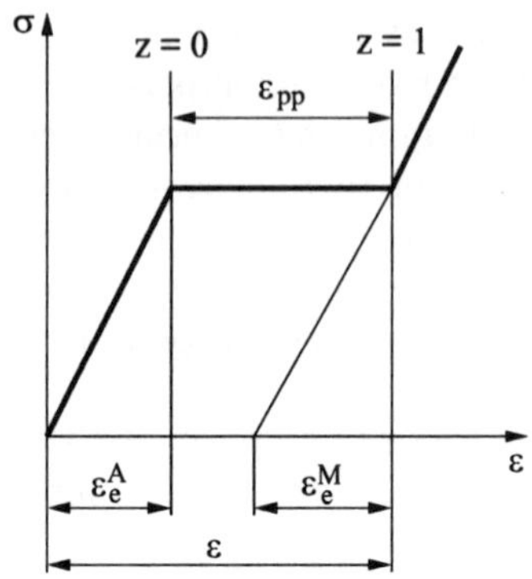

Figure 1. Decomposition of the strain tensor (uniaxial tension)

$$+ \psi_{\mathrm{ov}}(\mathbf{E}, \mathbf{Z}_{\mathrm{d}}) \tag{13}$$

The internal variable z $(0 \le z \le 1)$ denotes the fraction of martensite and $1 - z$ the share of austenite. The superscripts M and A correspond to the two phases and the index pp stands for the pseudoelastic–pseudoplastic part of the strain and free energy, respectively. ψ_0^{A} and ψ_0^{M} correspond to the free energy in the undeformed state; in terms of internal energy and entropy we have

$$\psi_0^{\mathrm{A}} = u_0^{\mathrm{A}} - \theta\eta_0^{\mathrm{A}}, \qquad \psi_0^{\mathrm{M}} = u_0^{\mathrm{M}} - \theta\eta_0^{\mathrm{M}} \tag{14}$$

and

$$\Delta\psi = \Delta u_0 - \theta\Delta\eta_0; \tag{15}$$

the differences in internal energy and entropy,

$$\Delta u_0 = u_0^{\mathrm{M}} - u_0^{\mathrm{A}}, \qquad \Delta\eta_0 = \eta_0^{\mathrm{M}} - \eta_0^{\mathrm{A}}, \tag{16}$$

are regarded as material constants. $\mathbf{Y}_{\mathrm{d}}$ and $\mathbf{Z}_{\mathrm{d}}$ are internal variables of the strain type, in order to allow for kinematic hardening and dissipative effects, respectively. In order to simplify the constitutive modelling as far as possible, the concept of elastoplasticity is applied and isotropic linear elasticity is assumed within the elastic range. Furthermore, the thermoelastic constants of the two phases except $u_0^{\mathrm{A,M}}$ and $\eta_0^{\mathrm{A,M}}$ are assumed to be equal. Then, we have $\mathbf{E}_{\mathrm{e}}^{\mathrm{A}} = \mathbf{E}_{\mathrm{e}}^{\mathrm{M}} = \mathbf{E}_{\mathrm{e}}$ and the decomposition (12) simplifies to

$$\mathbf{E} = \mathbf{E}_{\mathrm{e}} + \mathbf{E}_{\mathrm{pp}}. \tag{17}$$

A very simple free energy relation is assumed to be

$$\rho\psi = \mu \,\mathrm{tr}\left(\mathbf{E}_{\mathrm{e}}^{\mathrm{D}2}\right) + \frac{\kappa}{2}\left(\mathrm{tr}\mathbf{E}_{\mathrm{e}}\right)^2 - 3\alpha\kappa\left(\mathrm{tr}\mathbf{E}_{\mathrm{e}}\right)(\theta-\theta_0) + \rho c_{\mathrm{d}}\left[\theta-\theta_0-\theta\ln\frac{\theta}{\theta_0}\right] + \rho\psi_0^{\mathrm{A}} + \rho z\left(\Delta u_0 - \theta\Delta\eta_0\right) + \mu_{\mathrm{pp}}\mathrm{tr}\left[\left(\mathbf{E}_{\mathrm{pp}}-\mathbf{Y}_{\mathrm{d}}\right)^2\right] + \mu_{\mathrm{ov}}\mathrm{tr}\left[\left(\mathbf{E}^{\mathrm{D}}-\mathbf{Z}_{\mathrm{d}}\right)^2\right]. \tag{18}$$

The coefficients μ, κ, α, c_{d}, μ_{pp}, μ_{ov} as well as Δu_0 and $\Delta\eta_0$ are material constants.

The evolution of the martensitic phase is caused by stress and temperature processes. Therefore, the decomposition

$$z = z^{\mathrm{SIM}} + z^{\mathrm{TIM}} \tag{19}$$

is assumed, where the superscripts SIM and TIM stand for stress– and temperature–induced martensite, respectively.

The free energy (18) leads to the stress relations

$$\mathbf{T}_{\mathrm{eq}} = \rho\frac{\partial\psi}{\partial\mathbf{E}_{\mathrm{e}}} = 2\mu\mathbf{E}_{\mathrm{e}}^{\mathrm{D}} + \kappa\left(\mathrm{tr}\mathbf{E}_{\mathrm{e}}\right)\mathbf{I} - 3\alpha\kappa(\theta-\theta_0)\mathbf{I}, \tag{20}$$

$$\mathbf{T}_{\mathrm{ov}} = \rho\frac{\partial\psi}{\partial\mathbf{E}} = 2\mu_{\mathrm{ov}}\left(\mathbf{E}^{\mathrm{D}}-\mathbf{Z}_{\mathrm{d}}\right) \tag{21}$$

and the entropy relation

$$\eta = -\frac{\partial\psi}{\partial\theta} = c_{\mathrm{d}}\ln\frac{\theta}{\theta_0} + \eta_0^{\mathrm{A}} + z\Delta\eta_0 + \frac{3\alpha\kappa}{\rho}\left(\mathrm{tr}\mathbf{E}_{\mathrm{e}}\right). \tag{22}$$

Together with the definition of the backstress,

$$\mathbf{X} = \rho\frac{\partial\psi}{\partial\mathbf{E}_{\mathrm{pp}}} = 2\mu_{\mathrm{pp}}\left(\mathbf{E}_{\mathrm{pp}}-\mathbf{Y}_{\mathrm{d}}\right), \tag{23}$$

the dissipation inequality (7) takes the special form

$$\delta = \frac{1}{\rho}\left(\mathbf{T}_{\mathrm{eq}}^{\mathrm{D}}-\mathbf{X}\right)\cdot\dot{\mathbf{E}}_{\mathrm{pp}} - \left(\Delta u_0-\theta\Delta\eta_0\right)\left[\dot{z}^{\mathrm{SIM}}+\dot{z}^{\mathrm{TIM}}\right] + \frac{1}{\rho}\mathbf{X}\cdot\left(\dot{\mathbf{E}}_{\mathrm{pp}}-\frac{1}{2\mu_{\mathrm{pp}}}\dot{\mathbf{X}}\right) + \frac{1}{\rho}\mathbf{T}_{\mathrm{ov}}\cdot\left(\dot{\mathbf{E}}^{\mathrm{D}}-\frac{1}{2\mu_{\mathrm{ov}}}\dot{\mathbf{T}}_{\mathrm{ov}}\right) \geq 0. \tag{24}$$

This inequality must be guaranteed for all thermomechanical processes.

3 Evolution Equations

In order to represent the pseudoelastic and pseudoplastic behaviour, a flow rule is constructed corresponding to rate–independent plasticity. To this end, various functions and criteria are introduced:

Yield functions

$$f_1 = \|\mathbf{T}^{\mathrm{D}}_{\mathrm{eq}} - \mathbf{X}\| - \sqrt{\frac{2}{3}} k_1(\theta) \tag{25}$$

$$f_2 = \|\mathbf{T}^{\mathrm{D}}_{\mathrm{eq}} - \mathbf{X}\| - \sqrt{\frac{2}{3}} |k_2(\theta)|, \quad -k_1(\theta) \leq k_2(\theta) \leq k_1(\theta) \tag{26}$$

Loading functions

$$B_{1,2} = \dot{f}_{1,2}\Big|_{\dot{\mathbf{E}}_{\mathrm{pp}} = \dot{\mathbf{X}} = \mathbf{0}} = 2\mu \mathbf{N} \cdot \dot{\mathbf{E}}^{\mathrm{D}} - \mathrm{sgn}\,[k_{1,2}(\theta)] \sqrt{\frac{2}{3}} \frac{\mathrm{d}k_{1,2}(\theta)}{\mathrm{d}\theta} \dot{\theta} \tag{27}$$

$$\mathbf{N} = \frac{\partial f_{1,2}}{\partial \mathbf{T}_{\mathrm{eq}}} = \frac{\mathbf{T}^{\mathrm{D}}_{\mathrm{eq}} - \mathbf{X}}{\|\mathbf{T}^{\mathrm{D}}_{\mathrm{eq}} - \mathbf{X}\|} \tag{28}$$

Limit function

$$g(\mathbf{E}_{\mathrm{pp}}, \theta) = \|\mathbf{E}_{\mathrm{pp}}\| - \sqrt{\frac{3}{2}} \gamma^{\max}_{\mathrm{pp}}(\theta) \tag{29}$$

A model structure with two yield functions and one limit function was first used by Bertram[1]. Using the above definitions the following flow rule is proposed:

$$\dot{\mathbf{E}}_{\mathrm{pp}} = \begin{cases} \lambda_{\mathrm{pp}1} \mathbf{N} \text{ for } f_1 = 0,\ B_1 > 0 \text{ and } g \leq 0 & (30) \\ \lambda_{\mathrm{pp}2} \mathbf{N} \text{ for } f_2 = 0,\ k_2(\theta) > 0,\ B_2 < 0 \text{ and } \mathbf{E}_{\mathrm{pp}} \cdot \mathbf{N} > 0 & (31) \\ \lambda_{\mathrm{pp}2} \mathbf{N} \text{ for } f_2 = 0,\ k_2(\theta) < 0,\ B_2 > 0 \text{ and } \mathbf{E}_{\mathrm{pp}} \cdot \mathbf{N} < 0 & (32) \\ \quad \mathbf{0} \quad \text{for all other cases} & (33) \end{cases}$$

The different multipliers $\lambda_{\mathrm{pp}i}$ $(i = 1, 2)$ are determined in the usual way by evaluating the consistency conditions $(\dot{f}_{1,2} = 0)$. The distinction of different cases produces a representation of the pseudoelastic behaviour (eqs. (30), (31), (33)) and the pseudoplastic behaviour as well as the transition from pseudoelasticity to pseudoplasticity (eqs. (30), (32), (33)). For pseudoelasticity we have $\lambda_{\mathrm{pp}1} > 0$ and $\lambda_{\mathrm{pp}2} < 0$, whereas the other cases are characterized

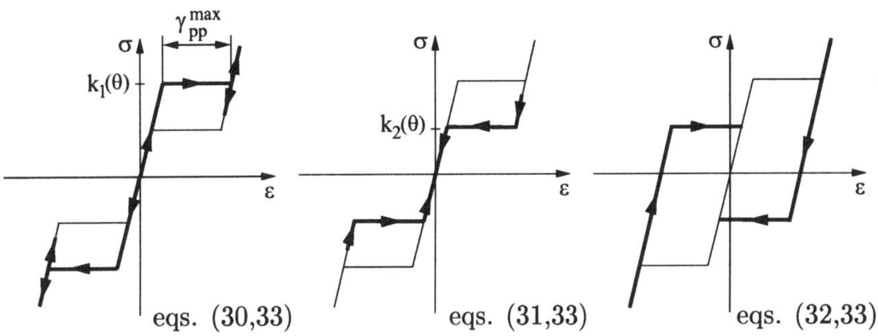

Figure 2. Different cases in the flow rule (30–33)

by $\lambda_{pp1} > 0$ and $\lambda_{pp2} > 0$. The properties of the flow rule (30–33) are schematically illustrated in figure 2.

The evolution of the martensitic phase parallel to the evolution of the pseudoelastic–pseudoplastic strain depends on temperature and the initial phases. This is incorporated by means of the following evolution equation for the stress–induced martensite z^{SIM} (see figure 3 for more details):

$$\dot{z}^{\mathrm{SIM}} = \begin{cases} \frac{(\mathbf{T}_{\mathrm{eq}}^{\mathrm{D}} - \mathbf{X}) \cdot \dot{\mathbf{E}}_{\mathrm{pp}}}{k_1(\theta)\gamma_{\mathrm{pp}}^{\mathrm{max}}(\theta)} & \text{for } f_1 = 0,\ B_1 > 0 \text{ and } z < 1 \quad (34) \\ \frac{(\mathbf{T}_{\mathrm{eq}}^{\mathrm{D}} - \mathbf{X}) \cdot \dot{\mathbf{E}}_{\mathrm{pp}}}{k_2(\theta)\gamma_{\mathrm{pp}}^{\mathrm{max}}(\theta)} & \text{for } f_2 = 0,\ k_2(\theta) > 0,\ B_2 < 0 \text{ and } z > 0 \quad (35) \\ \frac{(\mathbf{T}_{\mathrm{eq}}^{\mathrm{D}} - \mathbf{X}) \cdot \dot{\mathbf{E}}_{\mathrm{pp}}}{k_2(\theta)\gamma_{\mathrm{pp}}^{\mathrm{max}}(\theta)} & \text{for } f_2 = 0,\ k_2(\theta) < 0,\ B_2 > 0,\ -k_1(\theta) < k_2(\theta) \text{ and } z > 0 \quad (36) \\ 0 & \text{for all other cases} \quad (37) \end{cases}$$

In addition to the influence of the stress, the amount of martensite depends on the temperature history. A rough diagram of this functional dependence is set out in figure 4.

A simple way of realising the evolution of the temperature–induced martensite z^{TIM} is achieved as follows:

$$\dot{z}^{\mathrm{TIM}} = \begin{cases} \frac{\dot{\theta}}{M_{\mathrm{f}} - M_{\mathrm{s}}} & \text{for } \dot{z}^{\mathrm{SIM}} = 0,\ \dot{\theta} < 0,\ \theta \leq M_{\mathrm{s}}(\tau_{\mathrm{eff}}) \text{ and } z < 1 \\ -\frac{\dot{\theta}}{A_{\mathrm{f}} - A_{\mathrm{s}}} & \text{for } \dot{z}^{\mathrm{SIM}} = 0,\ \dot{\theta} > 0,\ \theta \geq A_{\mathrm{s}}(\tau_{\mathrm{eff}}) \text{ and } z > 0 \\ 0 & \text{for all other cases} \end{cases} \quad (38)$$

The characteristic temperatures A_{s} (austenite start), A_{f} (austenite finish), M_{s} (martensite start) and M_{f} (martensite finish) depend on the state of stress.

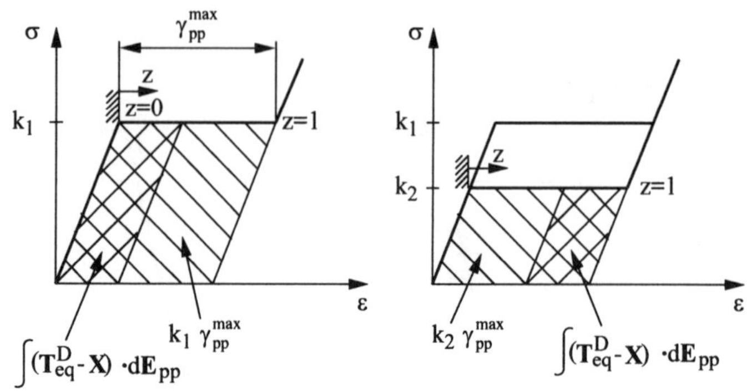

Figure 3. Calculation of the stress induced martensite fraction

The simplest way to allow for experimental results is by means of linear functions of the effective stress $\tau_{\text{eff}} = \left\|\mathbf{T}^{\text{D}}_{\text{eq}} - \mathbf{X}\right\|$ (see figure 4):

$$M_{\text{f}}(\tau_{\text{eff}}) = M_{\text{f}_0} + \frac{\tau_{\text{eff}}}{a} \qquad A_{\text{s}}(\tau_{\text{eff}}) = A_{\text{s}_0} + \frac{\tau_{\text{eff}}}{a} \tag{39}$$

$$M_{\text{s}}(\tau_{\text{eff}}) = M_{\text{s}_0} + \frac{\tau_{\text{eff}}}{a} \qquad A_{\text{f}}(\tau_{\text{eff}}) = A_{\text{f}_0} + \frac{\tau_{\text{eff}}}{a} \tag{40}$$

The internal strain–like variable $\mathbf{Y}_{\text{d}}$ and the corresponding stress represent kinematic hardening. An expedient evolution law is the Armstrong–Frederick equation

$$\dot{\mathbf{X}} = 2\mu_{\text{pp}}\dot{\mathbf{E}}_{\text{pp}} - b\,\dot{s}_{\text{pp}}\,\mathbf{X} \tag{41}$$

with the material constant $b > 0$ and the rate $\dot{s}_{\text{pp}}$ of the pseudoelastic–pseudoplastic arclength (accumulated strain),

$$\dot{s}_{\text{pp}} = \sqrt{\frac{2}{3}}\left\|\dot{\mathbf{E}}_{\text{pp}}\right\|. \tag{42}$$

The simplest evolution equation for the overstress $\mathbf{T}_{\text{ov}}$ is given by

$$\dot{\mathbf{T}}_{\text{ov}} = 2\mu_{\text{ov}}\dot{\mathbf{E}}^{\text{D}} - \frac{2\mu_{\text{ov}}}{\eta_{\text{ov}}}\mathbf{T}_{\text{ov}}, \tag{43}$$

where the viscosity η_{ov} ($\eta_{\text{ov}} > 0$) may depend on temperature and stress. The evolution law (43) for the overstress $\mathbf{T}_{\text{ov}}$ corresponds to a Maxwell element.

These evolution equations complete the definition of the constitutive model.

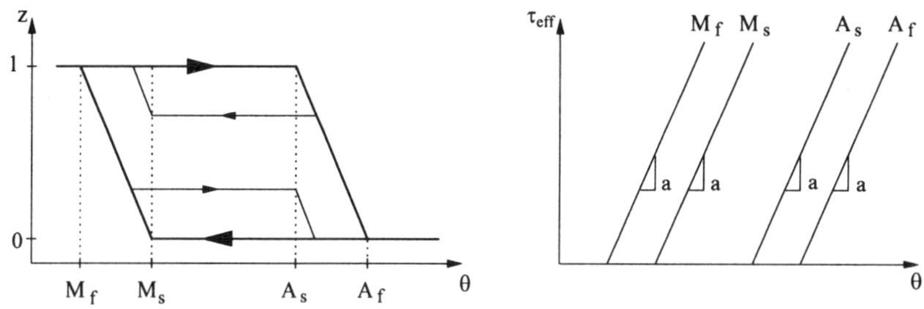

Figure 4. Fraction of martensite as a stress–dependent functional of temperature history

4 Thermodynamic Consistency

It can be shown that all evolution equations are fully compatible with the dissipation inequality (24) in the sense of sufficient conditions.

First of all, it is obvious that the evolution equations (41) and (43) for the backstress and the overstress lead to positive contributions of entropy production. With the help of the flow rule (30)–(33) and the evolution equation (34)–(37) the first two terms of (24) transform into

$$\frac{1}{\rho}\left(\mathbf{T}^{\mathrm{D}}_{\mathrm{eq}} - \mathbf{X}\right) \cdot \dot{\mathbf{E}}_{\mathrm{pp}} - (\Delta u_0 - \theta\Delta\eta_0)\,\dot{z}^{\mathrm{SIM}} =$$

$$\lambda_{\mathrm{pp}i}\sqrt{\frac{2}{3}}|k_i(\theta)|\left[1 - \frac{\Delta u_0 - \theta\Delta\eta_0}{\frac{1}{\rho}k_i(\theta)\gamma_{\mathrm{pp}}^{\max}(\theta)}\right] \qquad (i = 1, 2). \tag{44}$$

In the case of (30) and (34) (transition from austenite to martensite) we have $\lambda_{\mathrm{pp}1} > 0$ and thus

$$1 - \frac{\Delta u_0 - \theta\Delta\eta_0}{\frac{1}{\rho}k_1(\theta)\gamma_{\mathrm{pp}}^{\max}(\theta)} \geq 0, \tag{45}$$

whereas in the case of (31) and (35) (transition from martensite to austenite) we are led to $\lambda_{\mathrm{pp}2} < 0$ and, together with $k_2(\theta) > 0$, to

$$1 - \frac{\Delta u_0 - \theta\Delta\eta_0}{\frac{1}{\rho}k_2(\theta)\gamma_{\mathrm{pp}}^{\max}(\theta)} \leq 0. \tag{46}$$

Furthermore, the inequality

$$1 - \frac{\Delta u_0 - \theta\Delta\eta_0}{\frac{1}{\rho}k_2(\theta)\gamma_{\mathrm{pp}}^{\max}(\theta)} \geq 0 \tag{47}$$

must be fulfilled for $k_2(\theta) < 0$ with $\lambda_{\text{pp}2} > 0$. Summarizing, the quantities k_1, k_2, $\gamma_{\text{pp}}^{\max}$, Δu_0 and $\Delta\eta_0$ are restricted by the inequalities

$$\frac{1}{\rho}k_2(\theta)\gamma_{\text{pp}}^{\max}(\theta) \leq \Delta u_0 - \theta\Delta\eta_0 \leq \frac{1}{\rho}k_1(\theta)\gamma_{\text{pp}}^{\max}(\theta). \tag{48}$$

The third term of the dissipation inequality (24) is investigated on the basis of the evolution equation (38), which implies

$$\begin{aligned} &-(\Delta u_0 - \theta\Delta\eta_0)\dot{z}^{\text{TIM}} \\ &= -(\Delta u_0 - \theta\Delta\eta_0)\begin{cases} \frac{\dot{\theta}}{M_{\text{f}}-M_{\text{s}}} & \text{for } \dot{\theta}<0,\, \theta\leq M_{\text{s}}(\tau_{\text{eff}}),\,\ldots \\ -\frac{\dot{\theta}}{A_{\text{f}}-A_{\text{s}}} & \text{for } \dot{\theta}>0,\, \theta\geq A_{\text{s}}(\tau_{\text{eff}}),\,\ldots \end{cases} \end{aligned} \tag{49}$$

leading to the restrictions

$$\Delta u_0 - \theta\Delta\eta_0 \leq 0 \qquad \text{for} \qquad \theta \leq M_{\text{s}}(\tau_{\text{eff}}) \text{ and} \tag{50}$$

$$\Delta u_0 - \theta\Delta\eta_0 \geq 0 \qquad \text{for} \qquad \theta \geq A_{\text{s}}(\tau_{\text{eff}}). \tag{51}$$

These two inequalities hold if and only if both constants are negative:

$$\Delta u_0 < 0, \qquad \Delta\eta_0 < 0. \tag{52}$$

Supposing that the dissipation is equal for phase transformations in both directions, the equation

$$\frac{1}{\rho}\left[k_2(\theta) + \frac{1}{2}\Delta k(\theta)\right]\gamma_{\text{pp}}^{\max}(\theta) = \Delta u_0 - \theta\Delta\eta_0 \tag{53}$$

with $\Delta k(\theta) = k_1(\theta) - k_2(\theta)$ may be assumed instead of the inequality $(48)_1$.

The relations (48), (52) and (more specifically) (53) are sufficient for the thermodynamic consistency of the proposed constitutive model.

5 Equation of Heat Conduction

If all constitutive assumptions are inserted into the energy balance (1) or (8), a generalized equation of heat conduction is derived. As a result, this heat conduction equation reads as follows:

$$\begin{aligned} c_{\text{d}}\dot{\theta} + \frac{1}{\rho}\operatorname{div}\boldsymbol{q} - r = &\underbrace{-\frac{3\alpha\kappa}{\rho}\theta\left(\text{tr}\dot{\mathbf{E}}_{\text{e}}\right)}_{w_{\text{e}}} + \underbrace{\frac{1}{\rho}\mathbf{T}_{\text{eq}}\cdot\dot{\mathbf{E}}_{\text{pp}}}_{w_{\text{pp}}} + \underbrace{\frac{\eta_{\text{ov}}}{2\rho\mu_{\text{ov}}}\text{tr}\left(\mathbf{T}_{\text{ov}}^2\right)}_{w_{\text{ov}}} \\ &- \underbrace{\left\{\frac{1}{\rho}\mathbf{X}\cdot\dot{\mathbf{E}}_{\text{pp}} - \frac{b\dot{s}_{\text{pp}}}{2\rho\mu_{\text{pp}}}\text{tr}\left(\mathbf{X}^2\right) + \Delta u_0\dot{z}\right\}}_{\dot{w}_{\text{s}}}. \end{aligned} \tag{54}$$

Thermomechanical coupling means that heat is produced during the deformation processes. The heat production $u = \rho(w_\mathrm{e} + w_\mathrm{pp} + w_\mathrm{ov} - w_\mathrm{s})$ on the right–hand side of (54), giving rise to a change in temperature, consists of several parts. The term w_pp is the power of the equilibrium stress at the pseudoelastic–pseudoplastic strain rate and the term w_ov the power of the overstress. The stress power w_e corresponds to the thermoelastic effect; the power w_s is absorbed or released for changes in the microstructure.

Eq. (53), as a more special formulation of $(48)_1$, infers a reformulation of the heat conduction equation: if

$$\begin{aligned}\Delta u_0 \dot{z} &= \frac{1}{\rho}\left[k_2(\theta) + \frac{1}{2}\Delta k(\theta)\right]\gamma_\mathrm{pp}^\mathrm{max}(\theta)\dot{z}^\mathrm{SIM} + \theta\Delta\eta_0\dot{z} \\ &= \frac{1}{\rho}\left[k_2(\theta) + \frac{1}{2}\Delta k(\theta)\right]\sqrt{\frac{2}{3}}\mathrm{sgn}\left[k_i(\theta)\right]\lambda_{\mathrm{pp}i} + \theta\Delta\eta_0\dot{z} \end{aligned} \tag{55}$$

is inserted into (54) ($i = 1$ corresponds to the transition from austenite to martensite; $i = 2$ to the transition from martensite to austenite), we obtain

$$\begin{aligned} c_\mathrm{d}\dot{\theta} + \frac{1}{\rho}\,\mathrm{div}\,\boldsymbol{q} - r = &-\frac{3\alpha\kappa}{\rho}\theta\left(\mathrm{tr}\dot{\mathbf{E}}_\mathrm{e}\right) + \frac{\eta_\mathrm{ov}}{2\rho\mu_\mathrm{ov}}\mathrm{tr}\left(\mathbf{T}_\mathrm{ov}^2\right) + \frac{b\dot{s}_\mathrm{pp}}{2\rho\mu_\mathrm{pp}}\mathrm{tr}\left(\mathbf{X}^2\right) \\ &+ \frac{1}{\rho}\frac{1}{2}\Delta k(\theta)\dot{s}_\mathrm{pp}(t) - \theta\Delta\eta_0\dot{z}, \end{aligned} \tag{56}$$

We see that the the dissipation term $\frac{1}{2}\Delta k(\theta)\dot{s}_\mathrm{pp}(t)$ does not depend on the direction of the phase transformation. In the last term of (56) the quantity $-\theta\Delta\eta_0$ is the *latent heat* due to the phase transition.

6 Numerical Results

The constitutive equations were integrated numerically, using a Runge–Kutta method. Some results are depicted in figures 6–14 below. The calculations are based on assumed values of the material parameters, i.e. the 2nd yield limit

$$k_2(\theta) = -100\,\mathrm{MPa} + \langle\theta - 307.5\mathrm{K}\rangle\left(\rho\frac{\Delta u_0 - \theta\Delta\eta_0}{\gamma_\mathrm{pp}^\mathrm{max}(\theta)} - 100\,\mathrm{MPa}\right) \tag{57}$$

and the other parameters listed in tables 1, 2 and 3. The dependence of the yield limits k_1 and k_2 on temperature is illustrated in figure 5.

Figures 6–10 show the response of the material model to a single strain cycle at different temperatures ($\dot{\varepsilon} = \pm 0.001\,\frac{1}{\mathrm{s}}$; $\varepsilon_\mathrm{max} = \pm 0.08$). Due to the choice of material parameters, the pseudoplastic behaviour without phase transitions ($z = 1$) occurs at temperature $\theta = 240\,\mathrm{K}$ (see figure 6). The

Parameter	κ	μ	ρ	α	Δu_0	$\Delta\eta_0$
Value	70000	25000	6400	$1.06 \cdot 10^{-5}$	$-28 \cdot 10^3$	-87.5
Unit	MPa	MPa	$\frac{\text{kg}}{\text{m}^3}$	$\frac{1}{\text{K}}$	$\frac{\text{J}}{\text{kg}}$	$\frac{\text{J}}{\text{kg K}}$

Table 1. Material parameters of thermoelasticity

Parameter	μ_{pp}	b	M_{f0}	M_{s0}	A_{s0}	A_{f0}	Δk	$k_1(\theta)$
Value	2100	80	275	295	320	340	200	$\Delta k + k_2(\theta)$
Unit	MPa	–	K	K	K	K	MPa	MPa

Parameter	$k_2(\theta)$	$\gamma_{\text{pp}}^{\max}(\theta)$	a
Value	$k_2(\theta)$	0.07	$-\sqrt{\frac{2}{3}}\rho\frac{\Delta\eta_0}{\gamma_{\text{pp}}^{\max}}$
Unit	MPa	–	$\frac{\text{N}}{\text{m}^2\,\text{K}}$

Table 2. Material parameters of pseudoelasticity and -plasticity

Parameter	μ_{ov}	η_{ov}
Value	55000	20000
Unit	MPa	$\frac{\text{MPa}}{\text{s}}$

Table 3. Material parameters of the overstress

deformation mechanism in this case is based on the orientations and reorientations of the martensitic crystals. The strain cycle at the higher temperature $\theta = 315\,\text{K}$ leads to the transition between pseudoplasticity and pseudoelasticity (see figure 7). The austenite transforms into martensite, and vice versa in view of the austenite phase in the initial configuration (see figure 8). Figures 9 and 10 show the simulation of pseudoelasticity at $\theta = 350\,\text{K}$. Here, the stress induced phase transformations occur in loading and unloading.

Figures 11–14 illustrate the numerical results for the one– and two–way shape memory effect due to thermomechanical processes. The initial temperature is $\theta = 240\,\text{K}$ and the first mechanical loading is stress controlled with $\dot{\sigma} = \pm 1\,\frac{\text{MPa}}{\text{s}}$ and $\sigma_{\max} = 300\,\text{MPa}$. After that, the temperature changes at the

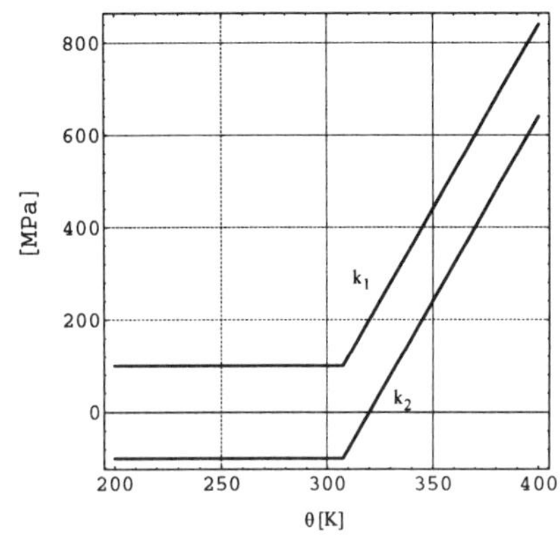

Figure 5. Yield limits as function of temperature

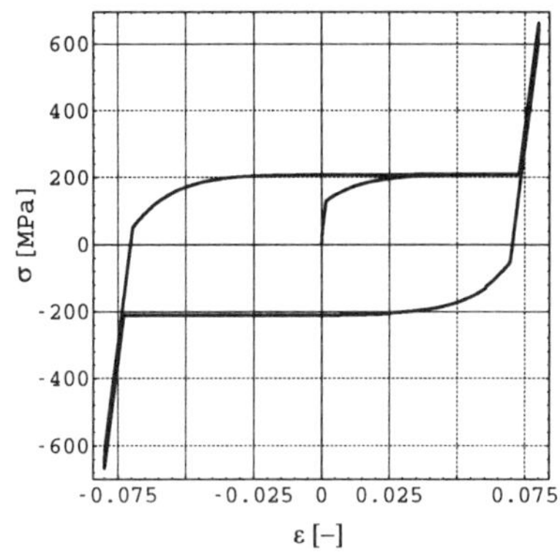

Figure 6. Pseudoplasticity (strain control) $\theta = 240\,\mathrm{K}$

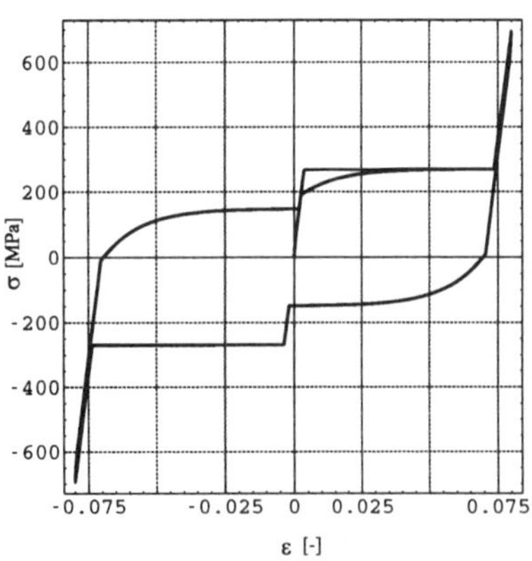

Figure 7. Transition range (strain control) $\theta = 315\,\mathrm{K}$

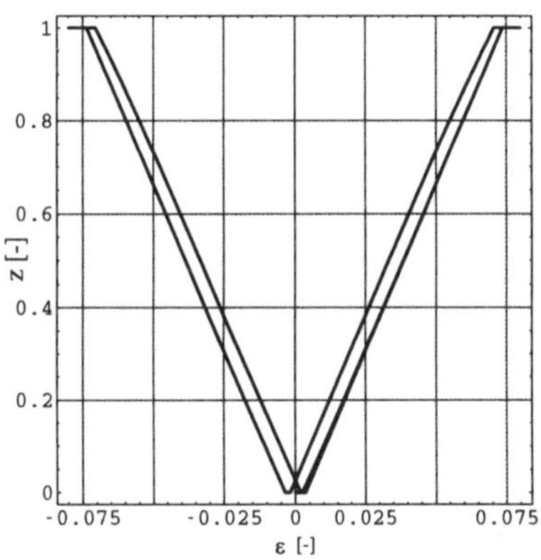

Figure 8. Fraction of martensite in the transition range

rate $\dot{\theta} = 0.3\,\frac{\mathrm{K}}{\mathrm{s}}$ to $\theta_{\max} = 360\,\mathrm{K}$. The initially twinned martensite structure detwins under the mechanical loading applied. In the subsequent heating and cooling processes phase transformation between martensite and austenite and the retransformation takes place. At zero stress, the shape begins to change when heat is applied. This one–way shape memory effect is depicted in figure

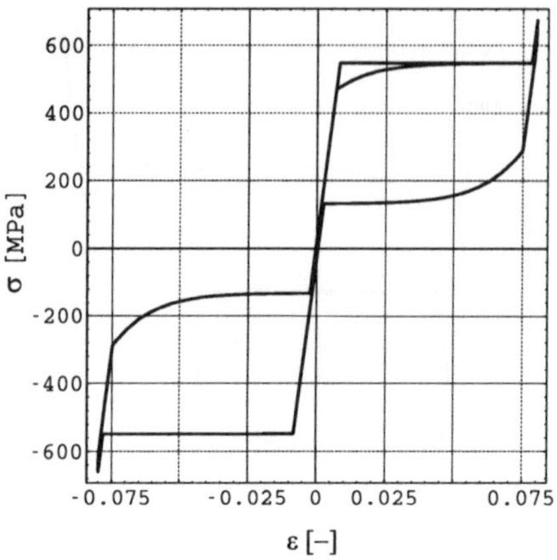

Figure 9. Pseudoelasticity (strain control) $\theta = 350\,\mathrm{K}$

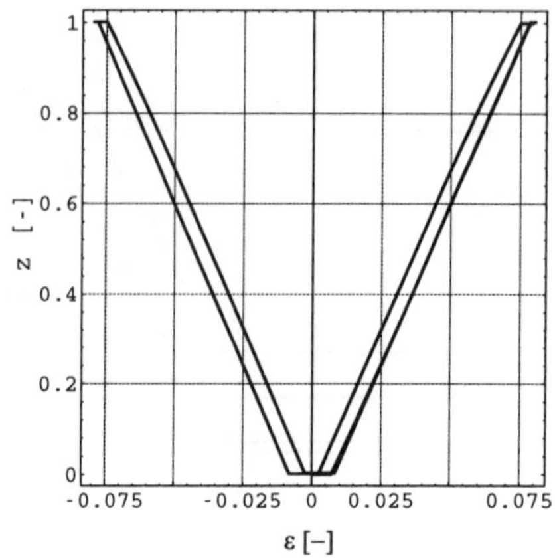

Figure 10. Fraction of martensite in the case of pseudoelasticity

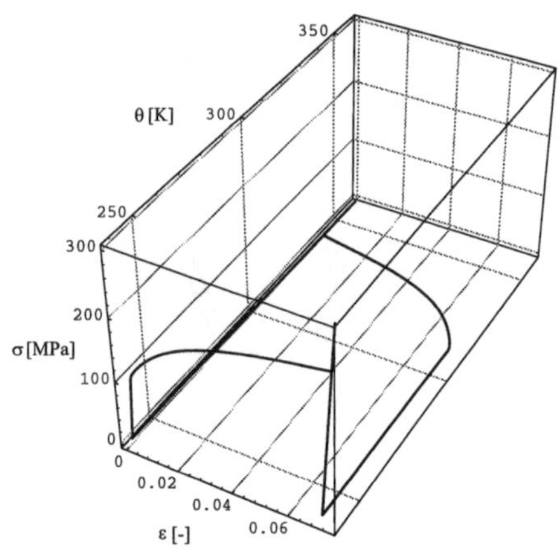

Figure 11. One–way shape memory effect (stress and temperature control)

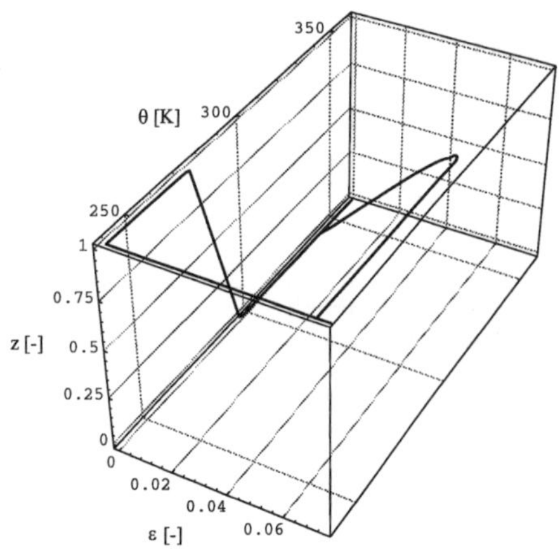

Figure 12. Fraction of martensite in the case of the shape memory effect

11 and the corresponding change in the martensite fraction in figure 12. On the other hand, if the effective stress is high enough, the shape–changing takes places during heating and cooling down. This behaviour, the so–called the two–way shape memory effect is illustrated in figures 13 and 14.

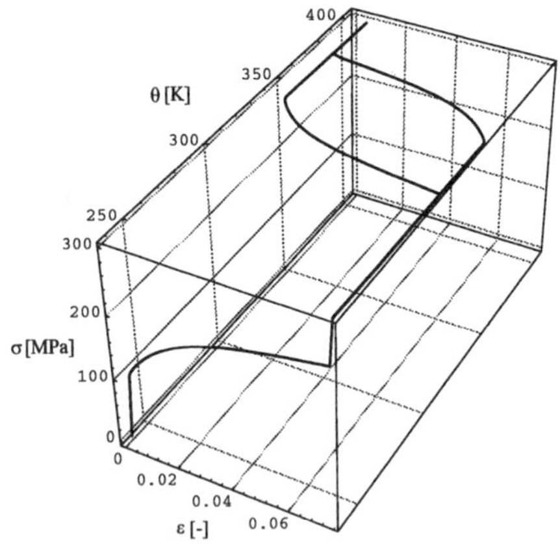

Figure 13. Two–way shape memory effect (stress and temperature control)

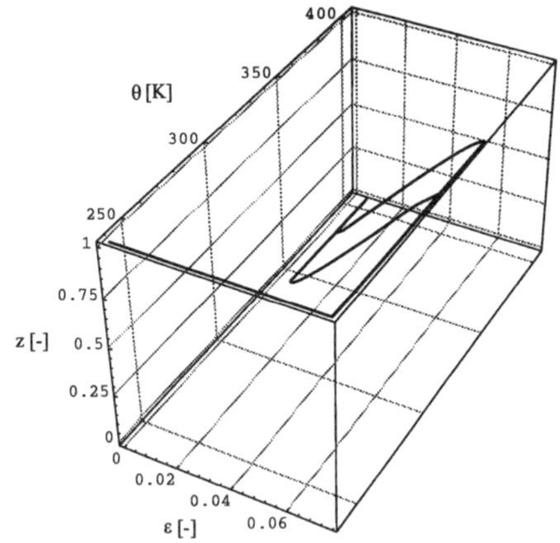

Figure 14. Fraction of martensite in the case of the two–way shape memory effect

7 Conclusion

The numerical simulations show that the essential phenomena of shape memory behaviour are qualitatively represented by the constitutive model. A quantitative representation should be based on material parameters, which are identified from test data. Experimental identification coupled with verification of the constitutive theory are the aims of future research.

Acknowledgment

We gratefully acknowledge the Deutsche Forschungsgemeinschaft (DFG) for the support of this work.

References

1. A. Bertram, Nuclear Engineering and Design **74**, 173 (1982).
2. H. Funakubo, Gordon and Breach Science Publishers(1987).
3. P. Haupt, Proceedings of the ASME Material Division **69–1**, 503 (1995).
4. P. Haupt and A. Lion, Cont. Mech. and Thermodyn. **7**, 73 (1995).
5. Y. Huo and I. Müller, Cont. Mech. Thermodyn. **5**, 163 (1993).
6. A. Lion, Acta Mechanica **123**,1 (1997).
7. A. Lion, Journal of the Mechanics and Physics of Solids **45**, 1805 (1997).
8. N. Nakanishi, Archives of Mechanics **35**, 37 (1983).
9. K. Tanaka *et al.*, Mechanics of Materials **19**, 271 (1995)

EXPLOITING THE DISSIPATION INEQUALITY IN CASE OF CONSTRAINED STATE SPACE VARIABLES

H. HERRMANN, W. MUSCHIK, C. PAPENFUSS, G. RÜCKNER*
INSTITUT FÜR THEORETISCHE PHYSIK
TECHNISCHE UNIVERSITÄT BERLIN
HARDENBERGSTR. 36
D-10623 BERLIN GERMANY

Abstract
We derive the proposition: The result of exploitation the dissipation inequality by Liu's procedure is independent of taking into account constraining balance equations or algebraic constraints before or after having applied the procedure.

1 Introduction

We are dealing with continuum theory local in position and time. In the (classical) five-field theory the wanted fields are the fields of mass density $\rho(\mathbf{x},\mathbf{t})$, momentum density $\rho(\mathbf{x},\mathbf{t})\mathbf{v}(\mathbf{x},\mathbf{t})$, and energy density $\rho e(\mathbf{x},\mathbf{t})$. In extended thermodynamics the fluxes of these quantities are also considered as wanted fields, so for instance in fourteen-field theory we have the additional wanted fields of heat flux $\mathbf{q}(\mathbf{x},\mathbf{t})$, and the symmetric part of the stress tensor $\mathbf{t}(\mathbf{x},\mathbf{t})$. The considerations in this paper are independent of the number of wanted fields. The wanted fields are not specified and abreviated by $\mathbf{a}$.

The balance equations are not a closed system of differential equations; constitutive equations are needed. Constitutive quantities depend on the state space variables in a material dependent manner. The domain of the constitutive mappings, the state space contains the wanted field, in many cases (space and time) derivatives of them, and possibly additional (for instance internal) variables. Constitutive equations are restricted by the second law of thermodynamics.

Starting out with Coleman-Mizel's formulation of the second law *All solutions of the balance equations satisfy the dissipation inequality* a technique to evaluate these restrictions after having chosen the state space has been introduced by Liu [2]. The so called Liu equations and a residual dissipation inequality are derived. Dependent on the chosen state space, the balance equations may represent constraints for the set of state variables, or they may determine the higher variables outside the state space. There is also the possibility that there are algebraic constraints between variables, so for instance

*E-MAIL: MUSCHIK@PHYSIK.TU-BERLIN.DE

the normalization of the 4-velocity in relativistic theories. If there are constraints for state variables, there are different possibilities of taking them into consideration:

- the balance equations can be inserted before applying Liu's procedure for exploiting the dissipation inequality,
- the balance equations can be inserted after having performed Liu's procedure,
- the balance equations can be differentiated, thus loosing their character as constraints and fitting in the usual scheme.

The state variables satisfy balance equations of first order in time and space. If now $\partial_t \varrho$, $\partial_j \varrho$, $\partial_j v_k$ are included in the state space, the balance equation of mass becomes a constraint for the set of the state variables. If $\partial_t \varrho$ is not included in the state space, the balance of mass does not represent any constraint for the state variables, but it determines $\partial_t \varrho$ as a higher derivative outside the state space. It is demonstrated that such constraints can be taken into account in different ways yielding always the same result which is independent of applying Liu's procedure before or after having taken into account the constraining balances. Also in the case of differentiating the constraining balance equation, we can rediscover the original result. In any case, the entropy density is independent of the time derivatives included in the state space, and the entropy production density does not depend on the higher derivatives outside the state space.

2 Balances and State Space

The n quantities $\boldsymbol{a}$ satisfy the n balance equations

$$\partial_t \boldsymbol{a} + \nabla \cdot \mathcal{F} = \boldsymbol{b}. \tag{1}$$

Here $\mathcal{F}$ are the $3n$ fluxes belonging to $\boldsymbol{a}$, and $\boldsymbol{b}$ the volume quantities influencing $\partial_t \boldsymbol{a}$. The system of balances (1) is supplemented by the *dissipation inequality*

$$\partial_t S + \nabla \cdot \boldsymbol{h} \geq 0. \tag{2}$$

To perform the derivatives ∂_t and ∇ we need a *statespace* which consists of m variables z_j and spans the domain of the constitutive mappings. By using the chain rule we obtain from (1) and (2) the *balances on the state space*

$$\frac{\partial \boldsymbol{a}}{\partial \boldsymbol{z}} \cdot \partial_t \boldsymbol{z} + \frac{\partial \mathcal{F}^\top}{\partial \boldsymbol{z}} : \nabla \boldsymbol{z} = \boldsymbol{b}, \tag{3}$$

$$\frac{\partial S}{\partial \boldsymbol{z}} \cdot \partial_t \boldsymbol{z} + \frac{\partial h}{\partial \boldsymbol{z}} : \nabla \boldsymbol{z} \geq 0. \tag{4}$$

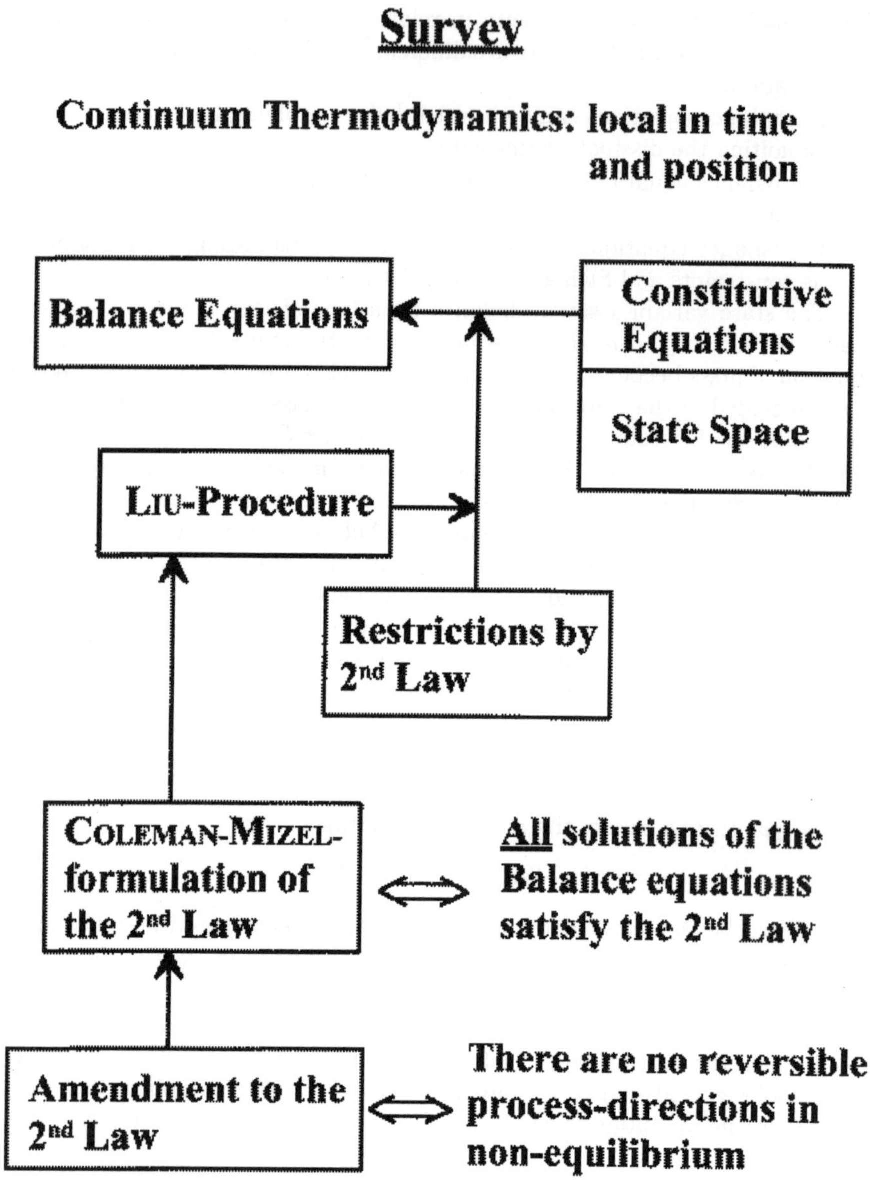

Figure 1. In addition to the balance equations constitutive equations are needed with constitutive functions defined on the state space. Constitutive equations are restricted by the second law of thermodynamics.

These balances write in matrix formulation

$$A_{\Delta\gamma}x_\gamma \; + \; B_\Delta \; = 0, \tag{5}$$

$$\alpha_\gamma x_\gamma \; + \; \beta \; \geq 0 \tag{6}$$

(Δ = number of the balance, γ = number of the higher derivative). Here the *higher derivatives* are

$$x_\gamma^\top \; = \; (\partial_t \boldsymbol{z}, \nabla \boldsymbol{z}), \tag{7}$$

and the constitutive quantities $A_{\Delta\gamma}, \alpha_\gamma, B_\Delta, \beta$ do not depend on the higher derivatives

$$A_{\Delta\gamma} \; = \; \left(\frac{\partial \boldsymbol{a}}{\partial \boldsymbol{z}}, \frac{\partial \mathcal{F}^\top}{\partial \boldsymbol{z}}\right), \; \alpha_\gamma^\top \; = \; \left(\frac{\partial S}{\partial \boldsymbol{z}}, \frac{\partial h}{\partial \boldsymbol{z}}\right), \; B_\Delta \; = \; (-\boldsymbol{b}), \; \beta \; = \; (0). \tag{8}$$

3 Example I: Mass Balance as a Constraint

The mass balance

$$\partial_t \varrho \; + \; \nabla \cdot \varrho \boldsymbol{v} \; = \; \partial_t \varrho \; + \; \boldsymbol{v} \cdot \nabla \varrho \; + \; \varrho \mathbf{1} : \nabla \boldsymbol{v} \; = \; 0 \tag{9}$$

contains the first derivatives $\partial_t \varrho, \nabla \varrho$, and $\nabla \boldsymbol{v}$. Therefore it depends on the choice of the state space, if the mass balance represents a contraint for the state variables or not. If we consider the state space

$$\boldsymbol{z}^I \; = \; \{\varrho, \boldsymbol{v}, \nabla \boldsymbol{v}, \nabla \varrho, ...\} \tag{10}$$

(9) is the equation for determining the higher derivative $\partial_t \varrho$. If also $\partial_t \varrho$ is included in the state space

$$\boldsymbol{z}^{II} \; = \; \{\varrho, \boldsymbol{v}, \nabla \boldsymbol{v}, \partial_t \varrho, \nabla \varrho, ...\} \tag{11}$$

the mass balance (9) represents a constraint for the state variables, because now $\partial_t \varrho$ is not anymore a higher derivative with respect to (10), but a state variable in (11). The higher derivatives are by this choice of the state space

$$x_\gamma^\top \; = \; (\partial_t \boldsymbol{v}, \partial_t \nabla \boldsymbol{v}, \nabla\nabla \boldsymbol{v}, \partial_t \partial_t \varrho, \partial_t \nabla \varrho, ...). \tag{12}$$

We investigate constraints in a more general connection in chapter 6.

4 Example II: Normalization of the 4-velocity in relativistic continuum mechanics

In relativistic continuum mechanics the state space includes the 4-velocity $\mathbf{u}$:

$$z = \{\mathbf{u}, \ldots\} \quad . \tag{13}$$

Only three of the four components of the 4-velocity are independent, and we have the normalization as an algebraic constraint:

$$u^\alpha u_\alpha = -c^2 \tag{14}$$

5 Second Law and Liu's Proposition

We now have to exploit the second law (2). There are different procedures to do that [1]. Here we refer to Liu's procedure [2] which is the most correct one. There is also a flood of different formulations of the second law [3]. Here we refer to that of Coleman-Mizel [4] which runs as follows: *Each solution x_γ of the balance equations (1) satisfies the dissipation inequality (2).* This special formulation of the second law can be proved by the non-reversible direction axiom [5] stating that there are no reversible process directions in non-equilibrium. The CM-formulation allows to prove Liu's

□Proposition [2,5]: There exist state space functions $\Lambda_\Delta(z)$, so that Liu's equations

$$\Lambda_\Delta(z) A_{\Delta\gamma}(z) \ = \ \alpha_\gamma(z) \tag{15}$$

and the reduced dissipation inequality

$$-\Lambda_\Delta(z) B_\Delta(z) \ + \ \beta \ \geq \ 0 \tag{16}$$

are valid. □

As (15) and (16) show, the constitutive equations $A_{\Delta\gamma}, B_\Delta(z), \alpha_\gamma$, and β are not independent of each other: after eliminating the fuctions $\Lambda_D(z)$ we obtain the restrictions on the constitutive functions imposed by the dissipation inequality. This dependence describes the class of materials which are compatible with the second law for a given choice of the state space spanned by the state variables z.

6 Constraints

We now introduce a simple abstract state space (for more extended state spaces of discrete systems see [7])

$$z = \{\boldsymbol{u},\boldsymbol{r},\boldsymbol{w}\} \tag{17}$$

and consider different cases for exploiting the dissipation inequality by Liu's procedure:

1. Without taking into consideration any constraint
2. Taking into account constraints *before* applying Liu's procedure
$$z = \{\boldsymbol{u},\boldsymbol{r},\boldsymbol{w} = \boldsymbol{g}(\boldsymbol{u},\boldsymbol{r})\} \tag{18}$$
3. Taking into account constraints *after* having performed Liu's procedure
$$\longrightarrow \quad \begin{array}{l} A_{\Delta\gamma}x_\gamma + B_\Delta = 0, \\ \alpha_\gamma x_\gamma + \beta \geq 0 \end{array} \quad \longrightarrow \text{Liu equations} \leftarrow \boldsymbol{w} = \boldsymbol{g}(\boldsymbol{u},\boldsymbol{r}) \tag{19}$$
4. Differentiating the constraints to obtain higher derivatives
$$\partial_t \boldsymbol{w} = \dot{\boldsymbol{u}}\partial_{\boldsymbol{u}}\boldsymbol{g} + \dot{\boldsymbol{r}}\partial_{\boldsymbol{r}}\boldsymbol{g}, \tag{20}$$
$$\nabla \boldsymbol{w} = \nabla \boldsymbol{u}\partial_{\boldsymbol{u}}\boldsymbol{g} + \nabla \boldsymbol{r}\partial_{\boldsymbol{r}}\boldsymbol{g}, \tag{21}$$
and including them into the application of Liu's procedure.

The balances (1) and (2) become

$$\partial_t \boldsymbol{a}(\boldsymbol{u},\boldsymbol{r},\boldsymbol{w}) + \nabla\cdot\mathcal{F}(\boldsymbol{u},\boldsymbol{r},\boldsymbol{w}) = \boldsymbol{b}, \tag{22}$$
$$\partial_t S(\boldsymbol{u},\boldsymbol{r},\boldsymbol{w}) + \nabla\cdot\boldsymbol{h}(\boldsymbol{u},\boldsymbol{r},\boldsymbol{w}) \geq 0 \tag{23}$$

from which we obtain the balances on the state space (3) and (4)

$$\begin{aligned} &\partial_t \boldsymbol{u}\partial_{\boldsymbol{u}}\boldsymbol{a} + \partial_t \boldsymbol{r}\partial_{\boldsymbol{r}}\boldsymbol{a} + \partial_t \boldsymbol{w}\partial_{\boldsymbol{w}}\boldsymbol{a} + \\ &\quad + \partial_{\boldsymbol{u}}\mathcal{F}\cdot\nabla\boldsymbol{u} + \partial_{\boldsymbol{r}}\mathcal{F}\cdot\nabla\boldsymbol{r} + \partial_{\boldsymbol{w}}\mathcal{F}\cdot\nabla\boldsymbol{w} = \boldsymbol{b}, \end{aligned} \tag{24}$$
$$\begin{aligned} &\partial_t \boldsymbol{u}\partial_{\boldsymbol{u}}S + \partial_t \boldsymbol{r}\partial_{\boldsymbol{r}}S + \partial_t \boldsymbol{w}\partial_{\boldsymbol{w}}S + \\ &\quad + \partial_{\boldsymbol{u}}\boldsymbol{h}\cdot\nabla\boldsymbol{u} + \partial_{\boldsymbol{r}}\boldsymbol{h}\cdot\nabla\boldsymbol{r} + \partial_{\boldsymbol{w}}\boldsymbol{h}\cdot\nabla\boldsymbol{w} \geq 0. \end{aligned} \tag{25}$$

In matrix formulation (24) and (25) write

$$x_\gamma = (\partial_t \boldsymbol{u}, \partial_t \boldsymbol{r}, \partial_t \boldsymbol{w}, \nabla\boldsymbol{u}, \nabla\boldsymbol{r}, \nabla\boldsymbol{w}), \tag{26}$$
$$A_{\Delta\gamma} = (\partial_{\boldsymbol{u}}\boldsymbol{a}, \partial_{\boldsymbol{r}}\boldsymbol{a}, \partial_{\boldsymbol{w}}\boldsymbol{a}, \partial_{\boldsymbol{u}}\mathcal{F}, \partial_{\boldsymbol{r}}\mathcal{F}, \partial_{\boldsymbol{w}}\mathcal{F}), \tag{27}$$
$$\alpha_\gamma^\top = (\partial_{\boldsymbol{u}}S, \partial_{\boldsymbol{r}}S, \partial_{\boldsymbol{w}}S, \partial_{\boldsymbol{u}}\boldsymbol{h}, \partial_{\boldsymbol{r}}\boldsymbol{h}, \partial_{\boldsymbol{w}}\boldsymbol{h}), \tag{28}$$
$$B_\Delta = (-\boldsymbol{b}). \tag{29}$$

Liu's equations (15) result in

$$
\begin{aligned}
\boldsymbol{\Lambda}_1 \cdot \partial_{\boldsymbol{u}}\boldsymbol{a} &= \partial_{\boldsymbol{u}} S, &(30)\\
\boldsymbol{\Lambda}_1 \cdot \partial_{\boldsymbol{r}}\boldsymbol{a} &= \partial_{\boldsymbol{r}} S, &(31)\\
\boldsymbol{\Lambda}_1 \cdot \partial_{\boldsymbol{w}}\boldsymbol{a} &= \partial_{\boldsymbol{w}} S, &(32)\\
\boldsymbol{\Lambda}_1 \cdot \partial_{\boldsymbol{u}}\mathcal{F} &= \partial_{\boldsymbol{u}}\boldsymbol{h}, &(33)\\
\boldsymbol{\Lambda}_1 \cdot \partial_{\boldsymbol{r}}\mathcal{F} &= \partial_{\boldsymbol{r}}\boldsymbol{h}, &(34)\\
\boldsymbol{\Lambda}_1 \cdot \partial_{\boldsymbol{w}}\mathcal{F} &= \partial_{\boldsymbol{w}}\boldsymbol{h}. &(35)
\end{aligned}
$$

(20) and (21) Here according to #1 no constraints have been taken into account.

Taking the constraints (18) and (19) according to #2 and #3 into consideration we obtain the following equations

$$
\begin{aligned}
\partial_{\boldsymbol{u}} S + \partial_{\boldsymbol{w}} S \partial_{\boldsymbol{u}}\boldsymbol{g} &= \boldsymbol{\Lambda}_1 \cdot (\partial_{\boldsymbol{u}}\boldsymbol{a} + \partial_{\boldsymbol{w}}\boldsymbol{a}\partial_{\boldsymbol{u}}\boldsymbol{g}), &(36)\\
\partial_{\boldsymbol{r}} S + \partial_{\boldsymbol{w}} S \partial_{\boldsymbol{r}}\boldsymbol{g} &= \boldsymbol{\Lambda}_1 \cdot (\partial_{\boldsymbol{r}}\boldsymbol{a} + \partial_{\boldsymbol{w}}\boldsymbol{a}\partial_{\boldsymbol{r}}\boldsymbol{g}), &(37)\\
\partial_{\boldsymbol{u}}\boldsymbol{h} + \partial_{\boldsymbol{w}}\boldsymbol{h}\partial_{\boldsymbol{u}}\boldsymbol{g} &= \boldsymbol{\Lambda}_1 \cdot (\partial_{\boldsymbol{u}}\mathcal{F} + \partial_{\boldsymbol{w}}\mathcal{F}\partial_{\boldsymbol{u}}\boldsymbol{g}), &(38)\\
\partial_{\boldsymbol{r}}\boldsymbol{h} + \partial_{\boldsymbol{w}}\boldsymbol{h}\,\partial_{\boldsymbol{r}}\boldsymbol{g} &= \boldsymbol{\Lambda}_1 \cdot (\partial_{\boldsymbol{r}}\mathcal{F} + \partial_{\boldsymbol{w}}\mathcal{F}\partial_{\boldsymbol{r}}\boldsymbol{g}). &(39)
\end{aligned}
$$

Consequently we obtain the same set of equations (37) to (40) independently if we take the constraints into account before or after the application of Liu's procedure.

Differentiation of the constraint according to #4 results in the additional differential equations ((20)) and ((21)) to be taken into account together with the balance equation ((23)). Consequently three functions Λ_1 to Λ_3 appear. Then applying Liu's procedure results in

$$
\begin{aligned}
\partial_{\boldsymbol{u}} S &= \boldsymbol{\Lambda}_1 \cdot \partial_{\boldsymbol{u}}\boldsymbol{a} + \boldsymbol{\Lambda}_2 \cdot \partial_{\boldsymbol{u}}\boldsymbol{g}, &(40)\\
\partial_{\boldsymbol{r}} S &= \boldsymbol{\Lambda}_1 \cdot \partial_{\boldsymbol{r}}\boldsymbol{a} + \boldsymbol{\Lambda}_2 \cdot \partial_{\boldsymbol{r}}\boldsymbol{g}, &(41)\\
\partial_{\boldsymbol{w}} S &= \boldsymbol{\Lambda}_1 \cdot \partial_{\boldsymbol{w}}\boldsymbol{a} - \boldsymbol{\Lambda}_2, &(42)\\
\partial_{\boldsymbol{u}}\boldsymbol{h} &= \boldsymbol{\Lambda}_1 \cdot \partial_{\boldsymbol{u}}\mathcal{F} + \boldsymbol{\Lambda}_3 \cdot \partial_{\boldsymbol{u}}\boldsymbol{g}, &(43)\\
\partial_{\boldsymbol{r}}\boldsymbol{h} &= \boldsymbol{\Lambda}_1 \cdot \partial_{\boldsymbol{r}}\mathcal{F} + \boldsymbol{\Lambda}_3 \cdot \partial_{\boldsymbol{r}}\boldsymbol{g}, &(44)\\
\partial_{\boldsymbol{w}}\boldsymbol{h} &= \boldsymbol{\Lambda}_1 \cdot \partial_{\boldsymbol{w}}\mathcal{F} - \boldsymbol{\Lambda}_3. &(45)
\end{aligned}
$$

We obtain back the equations (36) to (39) by performing the following procedures

$$
\begin{aligned}
&(40) + \partial_{\boldsymbol{u}}\boldsymbol{g}\cdot(42),\\
&(41) + \partial_{\boldsymbol{r}}\boldsymbol{g}\cdot(42),\\
&(43) + \partial_{\boldsymbol{u}}\boldsymbol{g}\cdot(45),\\
&(44) + \partial_{\boldsymbol{r}}\boldsymbol{g}\cdot(45).
\end{aligned}
$$

Consequently we proved that differentiation of the constraints results in the same equations as without differentiating them.

7 Conclusions

We have shown that the three different possibilities of taking into account constraints between state space variables in Liu's procedure, as well as not taking them into account in the procedure at all, results in the same set of restrictions on the constitutive equations. Therefore the practically most convenient choice of disregarding the constraints in the exploitation of the dissipation inequality is possible.

Acknowledgements

We thank the Draloric Company, D-95100 Selb, and the Isolde-Dietrich-Stiftung for financial support.

References

1. W. Muschik, C. Papenfuss, H. Ehrentraut: Concepts of Continuum Thermodynamics, Kielce University of Technology 1996, ISBN 83-905132-7-7, chap. 4
2. I.S. Liu: Method of Lagrange multipliers for exploitation of the entropy principle, Arch. Rat. Mech. Anal. 46 (1972) 131
3. W. Muschik: Formulations of the second law – Recent Developments, J. Phys. Chem. Solids 49 (1988) 709
4. B.D. Coleman, V.J. Mizel: Existence of caloric equations of state in thermodynamics, J. Chem. Phys. 40 (1964) 1116
5. W. Muschik, H. Ehrentraut: An amendment to the second law, J. Non-Equilib. Thermodyn. 21 (1996) 175
6. W. Muschik: Fundamentals of Non-Equilibrium Thermodynamics, in: Non-Equilibrium Thermodynamics with Application to Solids (W. Muschik Ed), Springer Wien, 1993, chap.9
7. W. Muschik, R. Domínguez-Cascante: On extended thermodynamics of discrete systems, Physica A 233 (1996) 523

THE STABILITY OF THE METHOD OF FICTITIOUS SOURCES IN AN INVERSE HEAT CONDUCTION PROBLEM FOR A LAYER

J. JANKOWSKI
Poznań University of Technology, Institute of Applied Mechanics
ul. Piotrowo 3, 60-965 Poznań, Poland
e-mail: jankowsk@sol.put.poznan.pl

M. CIAŁKOWSKI
Poznań University of Technology, Chair of Thermal Engineering
ul. Piotrowo 3, 60-965 Poznań, Poland
e-mail: cialkows@sol.put.poznan.pl

The paper presents a method of an approximate solution of a parabolic initial boundary value problem. The method consists in introducing some fictitious heat sources outside the region under consideration. The intensities of the sources are derived to verify boundary conditions. The method is applied to the inverse problem. The stability of the solution is also considered.

1 Method of fictitious sources

Temperature distribution field inside a bounded region D satisfies diffusion equation with initial and boundary conditions (first and second kind) is

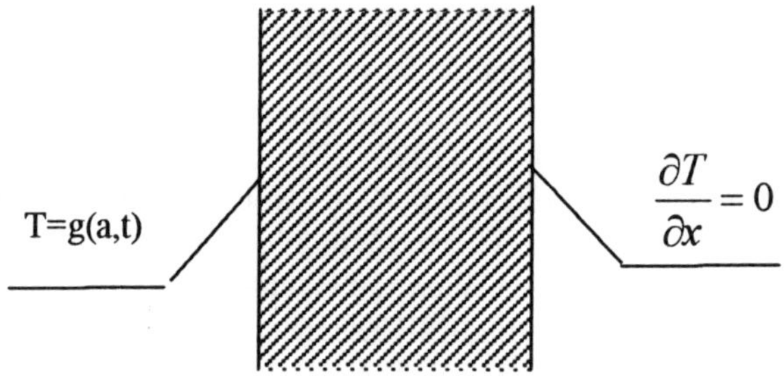

Figure 1. Boundary conditions for a layer.

searching for. The considerations are taken, for simplicity, only for one-dimensional region $D = (a, b)$.

Let us consider the following boundary value problem

$$T_{,xx}(x,t) = T_{,t}(x,t), \quad T(a,t) = h(t), \quad T_{,x}(b,t) \equiv 0, \quad T(x,0) = \varphi(x) \qquad (1.1)$$

where $x \in D, t > 0$.

The method of fictitious sources consist in finding a solution of certain nonhomogeneous Cauchy problem in an unbounded region,

$$\Theta_{,xx} = \Theta_{,t} + f(x,t), \quad \Theta(x,0) = \varphi(x), \quad x \in (-\infty,\infty), \quad t > 0, \qquad (1.2)$$

where the heat sources f are located outside the region D.

$$f(x,t) = w(t)\delta(x-c) + v(t)\delta(x-d), \quad c < a, \quad d > b \qquad (1.3)$$

The heat sources are chosen in such a way, that the solution of the Cauchy problem satisfies the boundary conditions as well. Then the solution of the problem (1.1) has the form:

$$T(x,t) = \int_0^t \int_{-\infty}^{\infty} E(x-y,t-s) f(y,s) dy ds + \int_{-\infty}^{\infty} E(x-y,t)\varphi(y) dy, \quad x \in (a,b), \quad t > 0 \quad (1.4)$$

where E is the fundamental solution of parabolic equation.

The main difficulty lies in the exact calculation of the source intensity $f(y,s)$.

The calculated temperature has got any singularities only at the points where the "fictitious" sources are located. That means that the temperature and its gradient are continuous functions of spatial variable and time.

2 Applications of the method

The method of fictitious sources is applicable to the direct and inverse problems for the one- two- and three-dimensional regions with arbitrary geometry. Also the mixed boundary conditions may be taken into account.

3 Heat sources in an infinite region

Heat conduction in one-dimensional region leads to the well-known Fourier law:

$$q = -\lambda \frac{\partial T}{\partial x}, \qquad (3.1)$$

where T -temperature field, q -rate heat transfer, λ -heat conductivity.

Heat energy balance has the form:

$$c\rho\frac{\partial T}{\partial t}+\frac{\partial q}{\partial x}=W\ . \tag{3.2}$$

Above W are heat sources, c -specific heat, ρ -mass density.

One-dimensional Cauchy problem for temperature field generates from (3.1) and (3.2) :

$$\begin{aligned} &c\rho\frac{\partial T}{\partial t}=\frac{\partial}{\partial x}(\lambda\frac{\partial T}{\partial x})+W, \quad t>0, \\ &T=0, \quad t=0. \end{aligned} \tag{3.3}$$

It transforms under the assumption $c, \rho, \lambda,$ to be constant as follows:

$$\begin{aligned} &\frac{\partial^2 \vartheta}{\partial x^2}=\frac{\partial \vartheta}{\partial \tau}-\frac{\omega}{\lambda}, \quad \tau>0, \\ &\vartheta=0, \quad \tau=0. \end{aligned} \tag{3.4}$$

We introduce the notation: $\vartheta(x,\tau)\equiv T(x,\frac{\tau}{\kappa}), \omega(x,\tau)\equiv W(x,\frac{\tau}{\kappa}), \kappa=\frac{\lambda}{c\rho}$.

The solution of the Cauchy problem (3.4) takes the form ([3],s.363):

$$\vartheta(x,\tau)=\frac{1}{\lambda}\int_0^{\tau}\int_{-\infty}^{\infty}\frac{\omega(y,\sigma)}{\sqrt{4\pi(\tau-\sigma)}}exp(-\frac{(x-y)^2}{4(\tau-\sigma)})dyd\sigma, \quad x\in(-\infty,\infty), \tau>0. \tag{3.5}$$

Return to the origin notation we have the solution of problem (3.3):

$$T(x,t)=\frac{1}{\lambda}\int_0^{\infty}\int_{-\infty}^{\infty}\frac{W(y,s)}{\sqrt{4\pi\kappa(t-s)}}exp(-\frac{(x-y)^2}{4\kappa(t-s)})dyds, \tag{3.6}$$

where $x\in(-\infty,\infty)$

We will use this temperature field distribution in an unbounded region to solve the initial-boundary value problem for the layer.

4 Inverse problem

Let us consider a layer of thickness (b-a) thermally insulated on the right side. Let the given temperature $g(t)$ will be at a fixed point x* inside the layer.

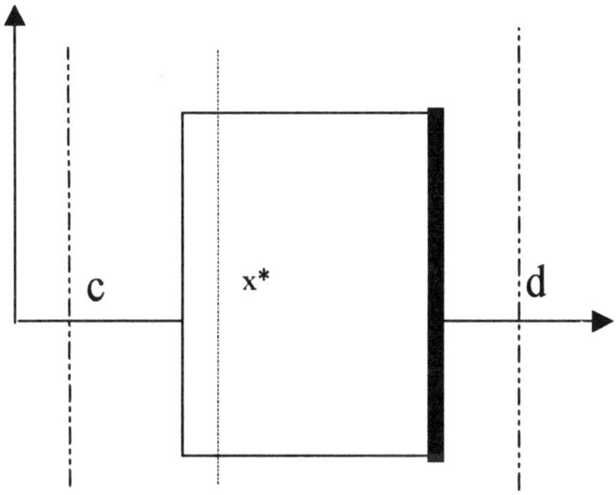

Figure 2. The location of "fictitious sources" and the given temperature.

Now the problem is:

$$T_{,xx}(x,t)=T_{,t}(x,t),\quad T(x^*,t)=g(t),\quad T_{,x}(b,t)\equiv 0,\quad T(x,0)\equiv 0 \tag{4.1}$$

where $x \in D, t > 0$.

The exact solution has got the form (3.6) in D.

An approximate solution of the problem may be obtained by discretization of the source in time and space. The "fictitious" heat sources may be expressed by the formula:

$$W(x,t)=\sum_{l=1}^{m}\sum_{j=1}^{2}W_l^j\delta(x-c_j)\eta(t-\tau_l) \tag{4.2}$$

Now the solution (3.6) is found to be:

$$T(x,t)=\frac{1}{2\lambda}\sqrt{\frac{\kappa}{\pi}}\sum_{l=1}^{m}\sum_{j=1}^{2}W_l^j\eta(\kappa t-\tau_{l-1})F_{x\,j}^{t\,l} \tag{4.3}$$

where

$$F_{x\,j}^{t\,l}=\int_{\tau_l/\kappa}^{t}\frac{1}{\sqrt{t-s}}exp(-\frac{(x-c_j)^2}{4\kappa(t-s)})ds \tag{4.4}$$

This symbol is in fact a function of the distance between the heat sources and the place, where the temperature is derived.

5 Stability of solution

Strictly speaking the values $2\lambda\sqrt{\pi/\kappa}\,g(t_k)$ are not exact so the error δg^k will be created. The corresponding change of derived temperature δT will be as follows:

$$\delta T(x,t) = \sum_{l=1}^{m}\sum_{j=1}^{2}\delta W_l^j F_{x\,j}^{t\,l} \tag{5.1}$$

In particular for m=3 we can the solution error expressed by δg^k in explicit way:

$$\delta T(x,t) = \delta g^1\left(\frac{P_{22x1}^{11t1}}{M_{12}^{11}} + \frac{M_{21}^{12}P_{21x2}^{22t2} + N_{21}^{12}P_{11x2}^{22t2}}{M_{12}^{11}M_{12}^{22}} + \right.$$

$$\left. + \frac{(M_{21}^{13}M_{12}^{22} + N_{21}^{12}N_{12}^{23} + M_{21}^{23})P_{21x2}^{33t3} + (N_{21}^{13}M_{12}^{22} - N_{21}^{12}M_{12}^{23} - M_{21}^{12}N_{21}^{23})P_{11x2}^{33t3}}{M_{12}^{11}M_{12}^{22}M_{12}^{33}}\right) + \tag{5.2}$$

$$+\delta g^2\left(\frac{P_{22x1}^{22t2}}{M_{12}^{22}} + \frac{M_{21}^{23}P_{21x2}^{33t3} + N_{21}^{23}P_{11x2}^{33t3}}{M_{12}^{22}M_{12}^{33}}\right) + \delta g^3\frac{P_{22x1}^{33t3}}{M_{12}^{33}}$$

where

$$M_{ij}^{kl} = F_{ii}^{kk}F_{jj}^{lk} - F_{ij}^{kk}F_{ji}^{lk} \qquad N_{ij}^{kl} = F_{ij}^{kk}F_{ii}^{lk} - F_{ii}^{kk}F_{ij}^{lk}$$
$$P_{ijxl}^{kktk} = F_{ij}^{kk}F_{xl}^{tk} - F_{il}^{kk}F_{xj}^{tk} \qquad F_{ij}^{kl} = F_{x_i^*\,j}^{t_k l} \tag{5.3}$$

The coefficient P_{ijxl}^{kktk} takes the values grater than 1, for x<x* and less than 1, for x>x*. It means that the boundary temperature deriving from the inverse problem has the worst values in the whole region. However, the amplification of error is linear, so the stability of the solution is kept within whole layer. But the solution error generates in the linear form by P_{ijxl}^{kktk}. That means the stability of solution is kept during the numerical processes. Then in the vicinity of the measuring point the method is such exact and stable as in the case of direct solution.

Acknowledgement

The paper has been supported by the State Committee for Scientific Research 116/T10/97/13 grant.

References

1. J. Stefaniak, *J. Tech. Phys.*, **26**, 349-358 (1985).
2. J. Stefaniak, J. Jankowski, *J. Tech. Phys.*, **28**, 87-96 (1987).
3. H. Marcinkowska, *Wstęp do teorii równań różniczkowych cząstkowych* (PWN, Warszawa 1972).

NONCLASSICAL HEAT TRANSPORT AND NONEQUILIBRIUM THERMODYNAMICS

DAVID JOU
Departament de Física Estadística, Universitat Autònoma de Barcelona
08193 Bellaterra, Catalonia, Spain
and
Institut d'Estudis Catalans, Carme 47
08001 Barcelona, Catalonia, Spain

Several kinds of nonclassical heat transport are presented, as thermal waves, ballistic phonon propagation, phonon hydrodynamics, and flux limiters. After having discussed the basic phenomenology and its mathematical description through suitably generalized transport equations, we discuss their thermodynamic foundations, which require to go beyond the local-equilibrium hypothesis, and which open a number of basic questions in the frontiers of nonequilibrium thermodynamics.

1 Introduction

The classical description of heat transport is based on the well-known Fourier's law (1810). However, in spite of its success to describe a wide diversity of situations of practical interest, the classical description is not sufficient to cover all the presently known phenomenology of heat transport. Amongst the very interesting phenomena not accounted for in the classical theory there are thermal waves, ballistic transport, phonon hydrodynamics, and flux limiters, whose description requires the use of generalized transport equations including memory, nonlocal effects, and nonlinear effects. These effects are of especial relevance in fast transport in miniaturized devices at high power, and thus have strong connections with present technological efforts.

An example of a transport equation allowing for the description of some of the mentioned phenomena is the so-called Guyer-Krumhansl (GK) equation (1964), which has the form

$$\tau \frac{\partial \boldsymbol{q}}{\partial t} + \lambda \nabla T + \boldsymbol{q} = \ell^2 \left(\nabla^2 \boldsymbol{q} + 2 \nabla \nabla . \boldsymbol{q} \right) \tag{1}$$

where $\boldsymbol{q}$ is the heat flux, λ the thermal conductivity, τ the relaxation time of the heat flux and l a correlation length; in dielectric solids, τ and l are given by $\tau = \tau_R$ and $l^2 = (1/5)c_S{}^2\tau_N\tau_R$, with c_s the Debye phonon velocity, and τ_N and τ_R the respective relaxation times for the "normal" momentum-preserving and resistive collisions. Expression (1) shows memory effects, through the relaxation term, and non-local effects, through the terms containing the gradients of the heat flux. It reduces to

Fourier's law when both effects are ignored, and to the so-called Maxwell-Cattaneo-Vernotte (MCV) equation (1865, 1948, 1958) when memory effects are kept but nonlocal one are ignored. Equation (1) describes successfully second sound and phonon hydrodynamics. However, it yields an infinite speed for ballistic propagation and it does not include non-linear effects, necessary for the description of flux-limiters. Here, we will propose a generalization of it which is more satisfactory in some of these aspects.

The consequences of transport equation (1) have been explored in much detail. However, no comparable attention has been paid to its thermodynamic implications, which are certainly of much interest, as they require a theory no longer based on the local-equilibrium hypothesis. Extended irreversible thermodynamics is an especially well suited framework for this purpose. This lecture aims to discuss both the dynamical and thermodynamical aspects of this more general description of heat transport, and to point out several of the fundamental questions that are raised at the frontiers of our present understanding of nonequilibrium thermodynamics.

2 Phenomenology of nonclassical heat transport

a) Classical heat transport. The classical Fourier's law describes the diffusive transport of heat. Its formulation was very influential not only in the theory of transport processes but also in pure mathematics, giving raise to the methods of Fourier transforms. Fourier's law is very successful in most of practical situations, and an enormous literature has grown from it. Furthermore, it is the simplest constitutive equation compatible with the classical local-equilibrium version of nonequilibrium thermodynamics. However, it has some drawbacks which make it unable to describe the behaviour of heat transport at high frequencies or at short wavelengths, which, anyway, it is not expected to describe because it is beyond its domain of validity. This makes worth the analysis of more general transport equations able to cope with these situations.

b) Thermal waves and second sound. Fourier's law predicts an infinite speed for the propagation of thermal pulses. This is not what is experimentally observed. Indeed, when a heat pulse is produced at the boundary of a small sample, it advances with finite speed towards the opposite boundary of the sample, where it is detected. The first description of this phenomenon was based on the MCV equation, which adds a relaxation term to the classical Fourier's law for the heat flux, namely

$$\tau \frac{\partial \boldsymbol{q}}{\partial t} + \boldsymbol{q} = -\lambda \nabla T \tag{2}$$

When this equation is introduced into the energy-balance equation, it yields for the speed of pulses $v = (\lambda/\rho c \tau)^{1/2}$, which diverges when τ tends to zero. In solids, this speed takes the value $v = (1/3)^{1/2}c_S$, with c_S the phonon speed. Many measurements of the second sound were carried out in the decades of 1960 and 1970 in several materials (mainly dielectric crystals) at low temperature (mostly between 20 K and 4 K).

Some successful practical applications may be found in situations with high local heating rate, as in short-pulse (nanoseconds to femtoseconds) laser heating of metals, used in the fabrication of microstructures, synthesis of advanced materials, measurement of thin-film properties and the analysis of structure transformations. Other examples of high heating rates are high-speed grinding, high-speed friction or fast explosions. A quantum version of (2) has been proposed in solid state physics, in the hydrodynamic description of nuclear collisions and in the analysis of the primordial Universe.

There are, however, several facts that make the MCV equation insufficient. On the one side, it is observed that a rectangular heat pulse does not maintain a rectangular form, but it becomes slightly open and rounded off during the propagation, in contrast with the MCV predictions. On the other side, one observes in the signal arriving to the detector a more complex behaviour than that predicted by the MCV equation: it is composed of a precursor moving at the phonon speed, which corresponds to the so-called ballistic phonon propagation, followed by a second peak moving at the second-sound speed, and finally, a wide signal corresponding to heat difusion. The relative importance of the different parts of the signal depends on the length of the sample on which the experiments are performed -usually at very low temperatures, where the relaxation times are long enough to be observable. These two drawbacks of the MCV equation can be partially superseded by adding to it non-local terms, as in the GK equation. Such terms may arise from different microscopic models, and they reflect the microstructure of the system.

c) Phonon hydrodynamics. The presence of nonlocal terms allows the GK equation to descibe a regime known as phonon hydrodynamics, where (1) reduces to

$$\lambda \nabla T = \ell^2 \nabla^2 \boldsymbol{q} \tag{3}$$

This form is similar to the equation describing Poiseuille flow of Newtonian fluids provided that $\boldsymbol{q}$, ∇T, and $l^{\,2}/\lambda$ are regarded as the velocity $\boldsymbol{v}$, the pressure gradient ∇p, and the viscosity η, respectively. This analogy with hydrodynamics is the reason of the name of phonon hydrodynamics. Poiseuille flow of phonons, first observed in solid He^4, in 1964, is found in a cylindrical heat conductor when the mean free path of the resistive collisions is much longer than the radius but the mean free path of momentum-conserving collisions is much shorter than it. In this regime, the total heat flowing along a cylinder in response to a temperature gradient is not proportional to the transverse area of the cylinder, but to the fourth power of its radius, in analogy with the total flow rate of fluid in the mentioned Poiseuille flow.

d) Subsystems with different temperatures. Another situation of interest is found when considering complex systems consisting of several subsystems, each of them at its own temperature. The equilibration rate between the different subsystems introduces a characteristic time in these problems. These subsystems may be, for instance, the electrons and the lattice in a metal submitted to a short-pulse laser heating, where the electron temperature is much higher than the lattice one during some time after recpetion of the pulse, to which the electrons are much more sensitive than the lattice. This situation may be described by assuming the following evolution equations for the electron and lattice temperatures T_e and T_l respectively

$$c_e \frac{\partial T_e}{\partial t} = \nabla.(\lambda \nabla T_e) - G(T_e - T_l) \tag{4a}$$

$$cl \frac{\partial Tl}{\partial t} = G(T_e - T_l) \tag{4b}$$

with G the electron-phonon coupling term describing the energy transfer from electrons to the lattice, and c_e and c_l the specific heats of electrons and lattice per unit volume. Equation (4b) yields $T_e = T_l + (c_l/G)\partial T_l/\partial t$. Introducing this result into (4a) yields

$$\nabla^2 T_l + \frac{c_l}{G}\frac{\partial \nabla^2 T_l}{\partial t} = \frac{c_l + c_e}{\lambda}\frac{\partial T_l}{\partial t} * \frac{c_e c_l}{\lambda G}\frac{\partial^2 T_l}{\partial t^2}. \tag{5}$$

This evolution equation for the lattice temperature (an analogous equation is also obtained for the electron temperature) may also be obtained by introducing (1) into the balance energy equation and making suitable identifications. Other systems where a two-temperature description is convenient are, for instance, heterogeneous systems, where the liquid and solid phases may also have different temperatures, or in polyatomic gases, which may have different temperatures for translational and internal degrees of freedom. It is interesting to note that an equation of the form (5) may also be obtained from the following constitutive equation, also known as dual-phase-lag equation,

$$\tau \dot{\boldsymbol{q}} + \boldsymbol{q} = -\lambda\left(\nabla T + \tau' \dot{\nabla T}\right) \tag{6}$$

which includes relaxation effects in both the heat flux and the temperature gradient, and whose physical consequences are described in much detail in the recent monograph by Tzou.

Notwithstanding its success in describing several nonlocal effects, the GK equation has also some drawbacks: it yields an infinite speed of propagation for the ballistic phonon flow, and it does not include nonlinear terms, which makes it unable to cope with the reduction of the value of the heat flux with respect to the predictions of Fourier's law at high values of the temperature gradient.

3 Entropy and entropy flux

Up to now we have dealt with the purely dynamical aspects of transport problems. Now, we explore their implications for thermodynamics. Indeed, it is well known that the classical Fourier transport law is compatible with the classical local-equilibrium thermodynamics. Thus, it is logical to ask whether a generalization of the transport equations implies also a generalization of the underlying thermodynamic formalism, though this aspect is often neglected in the literature.

To understand in the simplest way that the use of generalized transport equations like (1) requires a generalization of basic thermodynamic concepts, we assume the simplest situation in which it reduces to the MCV equation, which yields for the evolution of temperature profile the following telegrapher's equation

$$\tau \frac{\partial^2 T}{\partial t^2} + \frac{\partial T}{\partial t} - \chi \nabla^2 T = 0 \tag{7}$$

where χ stands for the thermal diffusivity, namely $\chi = \lambda/\rho c$. It is easy to show that if one assumes that a sinusoidal temperature profile is imposed as the initial condition in an isolated solid, and if the evolution of the temperature profile is described by (7), the total entropy of the system defined as the volume integral of the local-equilibrium entropy is no longer monotonous, but it decreases at some intervals. Therefore, it is clear that the MCV equation is inadmissible in the framework of the local-equilibrium hypothesis.

However, this transport equations becomes compatible with a monotonous increase of the entropy if instead of the local-equilibrium entropy one uses an extended entropy defined as

$$s = s_{eq} - \frac{\tau_1 v}{2\lambda T^2} \boldsymbol{q}.\boldsymbol{q} \tag{8}$$

where s_{eq} is the local-equilibrium entropy. The nonclassical terms in $\boldsymbol{q}.\boldsymbol{q}$ vanish in the limit when the relaxation time of the heat flux tends to zero, and one recovers both the classical Fourier law and the classical local-equilibrium entropy.

We are also in conditions to take into account the non-local terms in (1). It may be seen that these terms modify the form of the entropy flux, which becomes

$$J^s = \theta^{-1}q + \frac{\ell^2}{\lambda\theta^2}(\nabla q).q + \frac{\ell^2}{\lambda\theta^2}(\nabla.q)q \tag{9}$$

In the limit where the correlation length tends to zero, the nonclassical terms in (9), i.e. the terms containing the correlation length and the gradient of the heat flux, vanish, and one recovers the classical local-equilibrium expression.

Thus, it follows that *if the transport equation (1) describing the dynamics of heat transport is to be used, one must generalize both the entropy and the entropy flux.* This provides an extension of the local-equilibrium theory when the relaxation time and correlation length of the heat flux (or other fluxes) do not vanish. This has far reaching implications for nonequilibrium thermodynamics. On the one side, it is clear that one cannot trust the local-equilibrium hypothesis as a sound basis to explore the restrictions of the second law on the transport equations, but there is an intimate connection and parallel development between thermodynamics and dynamics, in which the entropy is not given *a priori*, but it must be discovered in parallel to the transport equations.

On the other side, the temperature defined from the derivative of the entropy with respect to internal energy $\theta^{-1} = (\partial s/\partial u)$, is no longer equal to the local-equilibrium temperature T, because θ depends not only on u but also on q, whereas T only depends on u. Then, one is led to ask what temperature is measured by the thermometer in presence of a heat flux. It follows that, in the presence of a heat flux, equipartition is no longer valid, and that the average value of the molecular translation speed perpendicular to the heat flux is not the same as that along the heat flux, but somewhat lower. Thus, the temperature indicated by a thermometer will depend in general to the degrees of freedom of the system to which the thermometer is most sensitive. Indeed, in equilibrium, all degrres of freedom have the same temperature, and the final reading of the thermometer does not depend on its sensitivity to the different degrees of freedom, but this is no longer valid out of equilibrium. It would be naive, thus, to pretend, as it is almost always done in computer simulations of molecular dynamics, that the temperature indicated by a thermometer is directly related to the average of the translational kinetic energy of the particles. We believe, indeed, that the phenomenology of temperature in nonequilibrium situations, even in steady-state situations, is far from being completely understood at present, though it is, of course, a fundamental question in any theory of nonequilibrium thermodynamics.

A further argument in favor of the generalized entropy is that it is possible to obtain a variational formulation of the GK equation by using the principle of minimum entropy production, provided one uses the generalized entropy instead of the local-equilibrium entropy. Finally, the entropy (8) may be given a microscopic basis from kinetic theory, information theory, fluctuation theory, and by other methods in nonequilibrium statistical mechanics, which have been studied in much detail.

4 Higher-order fluxes

Equation (1), which we have used as the basis of the description of several aspects of nonclassical heat transport, yields an infinite speed for ballistic phonon transport, in contrast to experiments. To obtain a more satisfactory equation and a deeper understanding of the complexities of dynamics and thermodynamics in nonequilibrium situations, we must use a more general framework which takes as independent variables not only the lowest-order fluxes, but also higher-order fluxes. In fact, from a phenomenological point of view it would be sufficient to incorporate as a further independent variable the flux of the heat flux, a second-order tensor. However, we provide here the most general view. Indeed, it turns out the the relaxation time of the higher-order fluxes is of the same order of that of the heat flux. Therefore, when one must consider the heat flux as an independent quantity, namely, at frequencies comparable to the inverse of the relaxation time of the heat flñux, all the higher-order fluxes should in fact be considered as independent variables too.

.Thus, instead of taking as single non-equilibrium variable the heat flux $\boldsymbol{q}$, we introduce higher-order variables $\boldsymbol{J}^{(1)}, \boldsymbol{J}^{(2)},\ldots, \boldsymbol{J}^{(n)}$, where $\boldsymbol{J}^{(n)}$, a tensor of order n, is the flux of $\boldsymbol{J}^{(n-1)}$. In microscopic terms, $\boldsymbol{J}^{(1)} = \boldsymbol{q}$ is a part of the third-order moment of the velocity distribution function, and $\boldsymbol{J}^{(n)}$ is, correspondingly, a part of its (n+2)th-order moment. Up to the nth-order moment, the extended Gibbs equation takes the form

$$\mathrm{d}s = T^{-1}\mathrm{d}u - \alpha_1 v \boldsymbol{J}^{(1)}.\mathrm{d}\boldsymbol{J}^{(1)} - \ldots - \alpha_n v \boldsymbol{J}^{(n)} \otimes \mathrm{d}\boldsymbol{J}^{(n} \tag{10a}$$

and the extended entropy flux reads as

$$\boldsymbol{J}^s = \beta_0 \boldsymbol{J}^{(1)} + \beta_1 \boldsymbol{J}^{(2)}.\boldsymbol{J}^{(1)} + \ldots + \beta_{n-1} \boldsymbol{J}^{(n)} \otimes \boldsymbol{J}^{(n-1} \tag{10b}$$

where all the coefficients α_i and β_i are generally temperature dependent. The symbol $\otimes$ denotes the total contraction of the corresponding tensors. The above expression suggests the following set of evolution equations for the $\boldsymbol{J}^{(n)}$:

$$\begin{aligned} \tau_1 \dot{\boldsymbol{J}}^{(1)} &= -\left(\boldsymbol{J}^{(1)} + \lambda \nabla T\right) + \frac{\beta_1 \tau_1}{\alpha_1} \nabla \boldsymbol{J}^{(2)} \\ \tau_n \dot{\boldsymbol{J}}^{(n)} &= -\boldsymbol{J}^{(n)} + \frac{\beta_n \tau_n}{\alpha_n} \nabla \boldsymbol{J}^{(n+1)} + \frac{\beta_{n-1} \tau_n}{\alpha_n} \nabla \boldsymbol{J}^{(n-1)} \qquad (n = 2, 3,\ldots, N) \end{aligned} \tag{11}$$

with $\mu_n \geq 0$, as required from the positiveness of the entropy production.

Since $J^{(n)}$ is the flux of $J^{(n-1)}$, the coefficient of $\nabla . J^{(n)}$ in the evolution equation of $J^{(n-1}$ must be equal to –1. As a consequence, $\rho\beta_n = -\alpha_n$, a relation which reduces considerably the number of independent parameters. The GK equation arises in the particular case where all the $J^{(n)}$ with $n > 2$ vanish and when the relaxation time of the flux of the heat flux $J^{(2)}$ is zero. In the Fourier space, the hierarchy of equations (11) yields a generalized transport law with (ω, k)-dependent coefficients which may be written in a continued-fraction expansion. It may be shown that if all the higher-order fluxes are assumed to have the same relaxation time, the speed of propagation of the ballistic pulses becomes

$$v_{p\infty}^2 = \frac{\chi}{\tau_\infty - \left(\ell_\infty^2/\chi\right)}. \tag{12}$$

By identifying this expression for the phase velocity with the standard expression $v_p = (\chi/\tau_{eff})^{1/2}$, one may introduce a "renormalized" or "effective" relaxation time. When the values for the correlation length are introduced in (12), one obtains for the value of the ballistic phonon flow $v_s \approx 1{,}006 c_S$. This value is very close to the expected value c_S. Furthermore, this value for the effective relaxation time is closer to the experimental results obtained from ultrasound propagation in rarefied gases than the usual Grad´s relaxation time appearing in the first-order theory

It can thus be stated that the introduction of the whole set of higher-order fluxes and the elimination of the fast variables through the definition of an effective relaxation time yields results which are in better agreement with the experimental results than a simple theory restricted to the first thirteen moments. The above procedure presents the advantage that it does not ignore *a priori* all the other fluxes. By considering only the first-order fluxes as independent variables, one obtains nevertheless a satisfactory qualitative understanding of high-frequency phenomena, in contrast with the classical theory of irreversible processes.

The hierarchy (11) has been recently generalized, by including terms in which a nonvanishing temperature gradient couples the fluxes of different orders. This yields in an elegant form a so-called flux-limiter, i.e. a nonlinear generalization of Fourier's law in which the effective thermal conductivity depends not only on the temperature but also on the temperature gradient, in such a way that for high values of the gradient the heat flux is considerably reduced with respect to the predictions of Fourier's law. This reduction is important in the analysis of radiation hydrodynamics or the interaction between a laser and a plasma in inertial nuclear fusion.

5 Conclusions

We could summarize the main contents of this lecture by outlining the following points:

- *Nonclassical heat transport is an open challenge for researchers interested in transport processes.* Such nonclassical transport has practical interest, because memory effects become important for the analysis of high-frequency perturbations and nonlocal and nonlinear effects are especially relevant in small systems, where the dimensions of the device become comparable to the mean-free-path of the energy carriers, and where small temperature differences give raise to very high temperature gradients.
- *Nonclassical heat transport is incompatible with local-equilibrium thermodynamics, and thus requires to go beyond the frontiers of the classical theory.* Thus, one is faced with fundamental questions as the definition and the meaning of entropy and of absolute temperature in nonequilibrium situations, or with the choice of variables defining the nonequilibrium state of the systems. In fact, these questions arise not only in the particular frame of extended irreversible thermodynamics, but are common to most of the theories, either macroscopic or microscopic, attempting to go beyond the local-equilibrium approximation. Extended irreversible thermodynamics is useful in identifying these questions, but, since it is a limited theory, it cannot pretend to give definitive answers to the questionsit opens.

Acknowledgments

I acknowledge many years of very fruitful discussions on these and many other topics with Profs J. Casas-Vázquez, from Autonomous University of Barcelona, and G. Lebon, of Liège University. I acknowledge Prof. B. Maruszewski for his invitation to lecture in this symposium. This work has been supported by a EU grant in the framework of the Program of Human Capital and Mobility (European Thermodynamic Network ERB-CHR-XCT 920 007) and by the Dirección General de Investigación Científica y Técnica of the Spanish Ministry of Education under grant PB94-0718.

References

Since the existing bibliography on these topics is very wide, we have restricted this list to a few general references where the reader will find the essential ideas and an ample bibliographical survey.

Review articles

Dreyer W and Struchtrup H *Continuum Mech. Thermodyn.* **5** 1(1993)
Joseph D D and Preziosi L *Rev. Mod. Phys* .**61**, 41(1990);. **62**, 375 (1991)
Jou D, Casas-Vázquez J and Lebon G *Rep. Prog. Phys.* **51**, 1104 (1988); *J. Non-Equilib. Thermodyn.* **17** (1992) 383;. **21** (1996) 103

Books

Jou D, Casas-Vázquez J and Lebon G *Extended irreversible thermodynamics* (Springer Verlag, Berlin, 1996, 2nd edition)

Muller I and Ruggeri T *Extended thermodynamics* (Springer, New York, 1993)

Sieniutycz S *Conservation laws in variational thermohydrodynamics* (Kluwer, Dordrecht, 1994)

Sieniutycz S and Salamon P ed *Extended Thermodynamic Systems* (Taylor and Francis, New York, 1992)

Tzou D Y *Macroscale to microscale heat transport: the lagging* (Taylor and Francis, New York, 1997)

Wilmanski K, Thermomechanics of continua (Springer, Berlin, 1998)

CONCEPTS OF THEORETICAL FOUNDATIONS FOR BIOMECHANICS BASED ON MULTISCALE MODELLING

J. KACZMAREK
Institute of Fluid-Flow Machinery, Polish Academy of Sciences, 80-952 Gdańsk, ul. J.Fiszera 14, Poland

In this paper a concept of multiscale modelling in biomechanics is discussed. Motivations for theoretical ideas follow from analysis of the cytogel behaviour. As a result two elements of the description are discussed. The first one, more physical, consists in introducing of idea of molecular catalysts mechanics. The second one, more mathematical, is related to collection of dynamical systems with dimensional reduction. The dimensional reduction procedure realizes transition between scales. Continuum mechanics widely applied in biomechanics is viewed to be a particular case of the collection of dynamical system approach. Concept introduced in this paper creates possibility of immersion of biomechanical modelling in well determined physical foundations owing to multiscale description.

1 Introduction

The most important medium responsible for biological motion is the cytogel. Behaviour of this medium is characterized by complicated forms of deformation induced by chemical reactions. The main components of the cytogel are myosin and actin fibers. They are crosslinked by gelation factors in the gel state. Cytogel can undergo solation as a result of varying calcium concentration. Calcium concentration is also the main factor which decide on contraction of actin-myosin units [1], [2]. In general, all processes related to motion of cytogel are very complicated. However, the calcium concentration can be viewed as a component especially responsible for controllability.

Summing up these considerations, we notice that the cytogel have unusual properties with respect to possible motion and its controllability. This attracts special attention. Modelling of such a medium is considered in literature. The continuum mechanics is applied to this end [3].

However, satisfactory mathematical models of behaviour of such a complicated medium which take into account deep physical foundations have to be very complicated. Nevertheless, simplified and general theoretical concepts should be discussed in order to go towards a satisfactory description.

In this paper a concept of theoretical foundations for large class of problems related to biomechanics is suggested. The starting point for presented ideas is intention of mathematical modelling of the cytogel. To this end two main concepts are introduced. The first one, more physical, is an idea of molecular catalysts (MC) mechanics. The second one, more mathematical, is collection of dynamical systems with dimensional reduction.

2 Collection of dynamical systems with the dimensional reduction

Let us consider a dynamical system in a general form

$$\dot{\boldsymbol{\phi}} = L(\boldsymbol{\phi},\, \mathbf{f}) \,, \tag{1}$$

where $\boldsymbol{\phi} \in \mathcal{M}_\phi$, $\mathbf{f} \in \mathcal{F}$ are variables of this system and external interactions respectively. Let us introduce also sets of processes defined on the time interval T, $\boldsymbol{\phi}(t) \in \mathcal{M}_{\phi T}$, $\mathbf{f}(t) \in \mathcal{F}_{\phi T}$.

The dynamical system (1) is predestined to description of the matter on the most elementary level.

We assume that we are able to distinguish a collection of dynamical subsystems in the framework of the elementary dynamical system (1). Elements of the collection of dynamical systems are subsystems with separated subgroup of variables $\boldsymbol{\phi}_i = \{\phi_{i\alpha}\}$, $i \in I_N$, $\alpha \in I_A$, where i stands for number of subsystem and α stands for number of variable within it.

In order to obtain a simplified description based on (1) we introduce a dimensional reduction procedure. To this end let us consider also variables of the dimensionally reduced system as $\mathbf{d} = \{\mathbf{d}_i\} \in \mathcal{M}_d$ which are defined with the help of $\boldsymbol{\phi}_i$.

The dimensional reduction procedure is defined as $DR = \{\pi_T,\ SDS(\mathbf{C}),\ app\}$, where $\pi_T : \mathcal{M}_{\phi T} \to \mathcal{M}_{dT}$ transforms processes of the elementary dynamical system EDS, in this case (1), into the dimensionally reduced one. $SDS(\mathbf{C})$ is the skeletal dynamical system which represents set of dynamical systems for $\mathbf{d}$ variable and parameterized by constants $\mathbf{C}$. *app* is a set of approximation procedures which should be applied in order to obtain the reduced dynamical system $RDS = SDS(\bar{\mathbf{C}})$ by option of best constants $\bar{\mathbf{C}}$ from all admissible $\mathbf{C}$.

Balance of equations are the starting point for determination of SDS. Let us consider a general form of balance of mass and energy for collection of dynamical systems.

Let us discuss a global form of these balance equations. Let $I_N \subset I_S$ be sets of indexes of the collection of dynamical systems. I_S represents indexes of an extended dynamical system which comprise (1). Then, $I_E = I_S - I_N$ stands for indexes related to an external dynamical system. This system, in turn, allows to model external interactions. M_r is mass assigned to r-th subsystem, C_r is production of mass and J_{ij} is interchange of mass from j-th to i-th subsystem. We have $J_{ij} + J_{ji} = 0$. The balance of mass is postulated in the form

$$\sum_{r\in I_N} (\dot{M}_r - C_r + \sum_{j\in I_S} J_{rj}) = 0 \tag{2}$$

with additional conditions related to external fluxes of mass

$$J_{rj} = \bar{J}_{rj} \;,\; r \in I_N,\; j \in I_E \;, \tag{3}$$

where $\bar{J}_{rj}$ is given.

Let E_r stands for energy of the r-th subsystem, R_r be production of energy and W_{ij} be interchange of energy from j-th to i-th subsystem. We have the action and reaction law as $W_{ij} + W_{ji} = 0$. The global form of the balance of energy is assumed by

$$\sum_{r\in I_N} (\dot{E}_r - R_r + \sum_{j\in I_S} W_{rj}) = 0 \tag{4}$$

with additional relations in the form

$$W_{rj} = \bar{W}_{rj}\ ,\ r \in I_N,\ j \in I_E \tag{5}$$

for given value of $\bar{W}_{rj}$.

More detailed considerations and in particular local forms of these equations generating the SDS are discussed in the next Section.

We assume also that there is a possibility of assigning geometrical objects of miscellaneous dimensions to each distinguished subsystems. With the help of them their space properties can be analysed in detail and in an averaged manner.

3 Concept of molecular catalysts mechanics

The myosin and actin units create deformable systems which are stable with respect to composition during their interactions with chemical reactions. Properties of molecules like these lead to rather universal in biology notion of molecular catalysts. In particular, special mechanical properties suggest necessity of discussion of molecular catalysts mechanics as mechanics of stable with respect to composition systems of molecules in interaction with chemical reactions.

Collection of dynamical systems with dimensional reduction seems to be appropriate mathematical approach to model of this kind of mechanics.

Bearing in mind that chemically reacting particles represent an environment for MC, we distinguish two parts of the dynamical system on the elementary level

$$\dot{\phi} \equiv \{\dot{\phi}_{MC},\ \dot{\phi}_R\} = \{C(\phi_{MC},\ \phi_R,\ \mathbf{f}),\ R(\phi_{MC},\ \phi_R,\ \mathbf{g})\}\ . \tag{6}$$

The first one, MC-system describes catalysts molecule which can be divided into subsystems. The second one R-system is related to chemically interacting particles. Properties of these two systems differs considerably. Thus, two kinds of DR should be applied and next interactions between these systems should be postulated.

Let MC be divided into subsystems indexed by $h \in I_C$. Then, the following new variables are introduced $\mathbf{d}_C = \{\mathbf{d}_{Ch} = \{\mathbf{x}_h,\ \theta_h,\ \boldsymbol{\xi}_h\};\ h \in I_C\}$, where $\mathbf{x}_h$ stands for slowly varying part of position assigned to the dynamical subsystem, θ_h is responsible for averaged fluctuations of positions of particles in the subsystem, $\boldsymbol{\xi}_h$ stands for internal state variables, $I_C \subset I_{CS}$.

Let us introduce also functions $a_{h\alpha} = a_{h\alpha}(\mathbf{x}_l)$, $l \in I_h^{a\alpha}$ where $I_h^{a\alpha}$ is a defined set of indexes related to subsystems interacting with the h-th one. In this case such an interaction is described by means of variables $\mathbf{x}_l$. This function can have the form $a_h = \mathbf{x}_h - \mathbf{x}_l$ for instance, $I_h^a = \{h,\ l\}$. Similar functions $b_{h\beta} = b_{h\beta}(\theta_h)$ are introduced for variable θ_h.

The energy function is postulated in the form $E = \sum_h (V_h + H_h\theta_h + K_h)$, where $V_h = V_h(\mathbf{C}_V,\ a_{h\alpha},\ \theta_h,\ \boldsymbol{\xi}_h)$ stands for potential energy with respect to variables $a_{h\alpha}$, H_h are forces conjugate with variables θ_h, $K = K(C_K,\ \dot{\mathbf{x}},\ \dot{a}_{h\alpha})$ is the kinetic energy term, $R_h = f_{hi}\dot{x}_{hi} + r_h$, $W_{hm} = Q_{hm} + \bar{W}_{hm}$. Summation convention is not applied to indexes related to I_C, I_{SC}. Consequently, the balance of energy is assumed in the following form

$$\sum_{h \in I_C} (\dot{V}_h + \dot{H}_h\theta_h + H_h\dot{\theta}_h + \dot{K}_h - f_{hj}\dot{x}_{hj} - r_h + \sum_{m \in I_{CS}} Q_{hm} + \sum_{m \in I_{CS}} \bar{W}_{hm}) = 0\ . \tag{7}$$

During this approach we assume that balance of mass for MC-system takes the form $\dot{m}_h = 0$. It means that no supply and fluxes of mass between subsystems are considered. Then, masses are constant coefficients in functions V and K.

Let us assume that $I_C = \{h\}$. Then, taking into account variables introduced and independence of time processes, we obtain from (7) the following equations

$$\sum_{\alpha} \mathcal{V}_{hj\alpha} + K_{hj} + \sum_{\alpha} \Phi_{hj\alpha} - f_{hj} = 0 \,, \tag{8}$$

$$\frac{\partial V_h}{\partial \theta_h} + H_h = 0 \,, \tag{9}$$

$$\dot{H}_h \theta_h - r_h - \sum_{m \in I_{CS}} Q_{hm} + \frac{\partial V_h}{\partial \xi_{hp}} \dot{\xi}_{hp} = 0 \tag{10}$$

for each $h \in I_C$. Additionally for $m \in I_{CS} - I_C$, we have

$$\sum_{\alpha} (\mathcal{V}_{mj\alpha} + \Phi_{mj\alpha}) - P_{mj} = 0 \,, \tag{11}$$

$$Q_{hm} = \bar{Q}_{hm} \,, \tag{12}$$

where $\mathcal{V}_{hj\alpha} = \sum_{h'} \frac{\partial V_{h'}}{\partial a_{h'\alpha}} \frac{\partial a_{h'\alpha}}{\partial x_{hj}}$, $K_{hj} = \frac{\partial^2 K_h}{\partial \dot{x}_{hj} \partial \dot{x}_{hn}}(0) \ddot{x}_{hn}$, $\Phi_{hj\alpha} = \sum_{h'} \frac{\partial^2 K_{h'}}{\partial \dot{a}_{h'\alpha} \partial \dot{a}_{h'\alpha}}(0) \ddot{a}_{h'\alpha} \frac{\partial a_{h'\alpha}}{\partial x_{hj}}$ and $P_{mj} = \sum_{h',n} \frac{\partial \bar{W}_{h'n}}{\partial \dot{x}_{mj}}(0)$ where $\bar{W}_{hm}$ and $\bar{Q}_{hm}$ are given. The summation convention is not applied to α in these expressions.

Particles taking part in chemical reactions are described by the second part of (6). The skeletal dynamical system is derived from the balance of mass equation (2) in this case. The variables related to this part of SDS are assumed in the form $\mathbf{d}_R = \{\mathbf{d}_{Rl} = \{M_l,\ \boldsymbol{\eta}_l\},\ l \in I_R\}$, where M_l is mass assigned to l-th subsystem and $\boldsymbol{\eta}_l$ represents internal variables which are able to describe structure of distribution of particles in l-th subsystem. $I_R \subset I_{RS}$. $\mathbf{d} = \{\mathbf{d}_C,\ \mathbf{d}_R\} \in \mathcal{M}_d$.

The equation (2) applied to $I_R = \{n\}$ gives

$$M_n - C_n + \sum_{q \in I_{RS} - \{n\}} J_{nq} = 0 \,. \tag{13}$$

Furthermore, for $q \in I_{RS} - \{n\}$, we assume the additional condition

$$J_{nq} = \bar{J}_{nq} \,, \tag{14}$$

where $\bar{J}_{nq}$, is a given value. Let us note that space distribution of mass can be analysed with the help of geometrical objects assigned to each subsystem.

In order to solve equations (8)-(14), some constitutive relations have to be introduced. They are assumed in the following form

$$V_h = V_h(\mathbf{C}_V,\ a_{h\alpha},\ \theta_h,\ b_{h\beta},\ \boldsymbol{\xi}_h) \,,\ h \in I_C \,, \tag{15}$$

$$Q_{hl} = Q_{hl}(\mathbf{C}_Q,\ a_{h\alpha},\ \theta_h,\ b_{h\beta},\ \boldsymbol{\xi}_h) \,,\ h,\ l \in I_C \,, \tag{16}$$

$$J_{np} = J_{np}(\mathbf{C}_J,\ M_n,\ \boldsymbol{\eta}_n)\ ,\ n,\ p \in I_R\ , \tag{17}$$

$$\dot{\xi}_h = A_\xi(\mathbf{C}_\xi,\ a_{h\alpha},\ \theta_h,\ b_{h\beta},\ \boldsymbol{\xi}_h)\ ,\ h \in I_C\ , \tag{18}$$

$$\dot{\eta}_n = A_\eta(\mathbf{C}_\eta,\ M_n,\ \boldsymbol{\eta}_n)\ ,\ n \in I_R\ . \tag{19}$$

They are valid for isolated MS and R systems. Interactions between them are introduced by means of expression

$$\bar{Q}_{hm} = \bar{Q}_{hm}(\bar{\mathbf{C}},\ M_h,\ \boldsymbol{\eta}_h)\ ,\ h \in I_C,\ m \in I_R\ . \tag{20}$$

Let us note that h and m are connected with different kind of subsystems. Precise interpretation of this can be attained with the help of geometrical objects assigned to each subsystem considered. Then, their space placements can be discussed and as a result geometrical interpretation of efflux $\bar{Q}_{hm}$ is possible.

Equations (8)-(20) create the skeletal dynamical system for our averaged model of mechanical behaviour of molecular catalysts molecule in chemical environment.

We discuss one equation of type of (13). This is motivated by fact that one component (calcium) is responsible for controllability. In general, the number of such equations is equal to number of reacting components.

We can construct two kind of processes defined on $\mathcal{M}_{dT}$. The first one, $\pi_T(\phi(t))$ is obtained by solution of (6) with initial conditions defined for $\phi(t_0)$. The second one, $\mathbf{d}(\mathbf{C},\ t)$ is obtained by solution of equations of SDS (8)-(20) with assumed constants. With the help of a metric γ defined on $\mathcal{M}_{dT}$ we introduce the function $h(\phi(t_0),\ \mathbf{f}) = inf_{\mathbf{C}}\gamma(\mathbf{d}(\mathbf{C},\ t),\ \pi_T(\phi(t)))$. We are able to identify the best constants $\bar{\mathbf{C}}$ by means of this function. Then, the reduced dynamical system $RDS = SDS(\bar{\mathbf{C}})$ is determined.

Introduced description is predestined to describe single large molecule. Then, subsystems are related to group of atoms. Consequently, we are able to describe some distinguished properties of such a molecule in a simplified form.

4 Continuum modelling as a particular case of the dimensional reduction

Model given in the previous section is designed to describe behaviour of single or system of molecules. Such a model is related to scale for which the atoms are considerable objects. On the other hand the cytogel in larger scale is appropriate medium for continuum modelling. Then, introduced hitherto description is deemed to be a foundation for the continuum one. Transition between scales is realized by means of the dimensional reduction. Therefore, the continuum theory has to have finite-dimensional fields [4].

General idea of continuum with finite-dimensional fields consists in fulfilling the balance equations for a finite family of subbodies of the body $\mathcal{B}$ only [4]. In classical continuum such equations are satisfied for all subbodies.

Above assumptions creates situation similar to previously discussed for collection of dynamical systems. Distinguishing of subsystems rests on defining a family of subbodies

$K_h \in \mathcal{K}_C$, $\bigcup_{h \in I_B} K_h = \mathcal{B}$, $K_h \cap K_l = \emptyset$, $h \neq l, h, l \in I_B$. Then, the dynamical subsystem is represented by group of atoms which are placed in K_h on the most elementary level.

Let us assign χ_h to a center $\mathbf{X}_{Ch}$ of each K_h. Similarly as previously we define a function $a_h = a_h(\chi_l)$, $l \in I_h^a$ defined on values assigned to neighbouring K_l. Then, the continuous deformation function $\alpha(\{\chi_h, a_h\}) = \chi^{\mathcal{K}}(\mathbf{X})$, $\mathbf{X} \in \mathcal{B}$ is assigned to the discrete fields. Then, we are able to introduce the gradient of deformation as $a_{hij} = \frac{\partial \chi_i^{\mathcal{K}}}{\partial X_j}(\mathbf{X}_{Ch})$. The field of temperature can be defined by analogy as $T(\mathbf{X}) = \beta(\{T_h, b_h\})$, where $b_h = b_h(T_l)$, $l \in I_h^b$.

In general, we discuss finite-dimensional fields of miscellaneous kinds owing to finite family of subbodies $\mathcal{K}$ introduced. Continuum fields are obtained as an image of interpolation mappings acting on these discrete fields. Then, they create finite-dimensional spaces of continuum fields.

The balance equations are postulated with the help of (2)-(5). The balance of energy takes the form

$$\dot{E}(\mathcal{B}) + P(\mathcal{B}) + W(\partial\mathcal{B}) - R(\mathcal{B}) = 0 \tag{21}$$

with the additional condition

$$W(\partial\mathcal{B}_s) = \bar{W}(\partial\mathcal{B}_s)\ ,\ for\ each\ \ \partial\mathcal{B}_s \subset \partial\mathcal{B}\ , \tag{22}$$

where $E(\mathcal{B}) = \sum_{h \in I_B} E(K_h)$ is internal energy of the body $\mathcal{B}$, $P(\mathcal{B}) = \sum_{h \in I_B} \dot{K}(K_h)$ is power of inertia forces, $R(\mathcal{B}) = \sum_{h \in I_B} R(K_h)$ is source of energy, efflux of energy has the form $W(\partial\mathcal{B}) = \sum_{h \in I_B} \sum_{j \in I_{BS}} (Q_{hj}(\partial K_{hj})) + p_{hj}(\partial K_{hj})\dot{\chi}_{(hj)}$. $\chi_{(hj)}$ is the deformation function assigned to $\partial K_{hj} = \bar{K}_h \cap \bar{K}_j$. $\partial\mathcal{B}_s$ is admissible boundary part of the body considered.

The balance of mass is assumed by

$$\dot{M}(\mathcal{B}) - C(\mathcal{B}) + J(\partial\mathcal{B}) = 0 \tag{23}$$

with the additional condition

$$J(\partial\mathcal{B}_s) = \bar{J}(\partial\mathcal{B}_s)\ ,\ for\ each\ \ \partial\mathcal{B}_s \subset \partial\mathcal{B}\ , \tag{24}$$

where $M(\mathcal{B}) = \sum_{h \in I_B} M(K_h)$ is mass of the body $\mathcal{B}$, $C(\mathcal{B}) = \sum_{h \in I_B} C(K_h)$ stands for source of mass, $J(\partial\mathcal{B}) = \sum_{h \in I_B} \sum_{j \in I_{BS}} (J_{hj}(\partial K_{hj}))$ represents efflux of mass. Additionally, the inequality following from the balance of entropy is postulated

$$\dot{S}(\mathcal{B}) - N(\mathcal{B}) + H(\partial\mathcal{B}) \geq 0\ , \tag{25}$$

where $S(\mathcal{B}) = \sum_{h \in I_B} S(K_h)$ is entropy, $N(\mathcal{B}) = \sum_{h \in I_B} N(K_h)$ is source of entropy and $H(\partial\mathcal{B}) = \sum_{h \in I_B} \sum_{j \in I_{BS}} (H_{hj}(\partial K_{hj}))$ represents efflux of entropy.

Variables of the new dynamical system are given by $\bar{\mathbf{d}} = \{\bar{\mathbf{d}}_C, \bar{\mathbf{d}}_R\}$, $\bar{\mathbf{d}}_C = \{\chi_h, \dot{\chi}_h, T_h, \boldsymbol{\mu}_h\}$, $h \in I_B \equiv I_{BC}$, $\bar{\mathbf{d}}_R = \{\varrho_n, \boldsymbol{\nu}_n\}$, $n \in I_{BR}$, where $\boldsymbol{\mu}$, $\boldsymbol{\nu}$ are internal state variables. ϱ_n represents density $\varrho_n = M_n/V(K_n)$.

We introduce also the free energy $\Psi(K_h) = \Psi_h$ by means of expression $E_h = \Psi_h + S_h T_h$. Let us assume that $\Psi_h = \Psi_h(a_h, T_h, \boldsymbol{\mu}_h)$. Let us introduce also a discrete differentiation operation D_{hi}, $i = 1, 2, 3$ which depends on kind of division of $\mathcal{B}$ on subbodies. Then, using also a discrete counterpart of the Green theorem, we can derive the following equations for system of molecular catalysts (C-system)

$$- D_{hi}(T_{hni}) - f_{hn} + M_h\ddot{\chi}_{hn} - D_{hm}(i_{hmp}\ddot{a}_{hnp}) = 0 \,, \tag{26}$$

$$\dot{S}_h T_h + D_{hi}(q_{hi}) - R_{eh} + S_h^d \dot{T}_h - T_{hni}^d \dot{a}_{hni} + \frac{\partial \Psi_h}{\partial \mu_{hk}} \dot{\mu}_{hk} = 0 \tag{27}$$

and the boundary conditions

$$T_{s_h ni} a_{s_h ji} n_{s_h j} - p_{s_h n} = 0 \,, \quad q_{s_h i} n_{s_h i} = \bar{q}_{s_h} \,. \tag{28}$$

Furthermore, using equation (23) we obtain

$$\frac{\partial \varrho_g}{\partial t} + D_{gi}(j_{gi}) - c_{gi} = 0 \,, \quad j_{s_g i} n_{s_g i} = \bar{j}_{s_g} \tag{29}$$

as equations describing evolution of the R-system.

The constitutive equations has to take into account interactions between the both systems. Let us introduce notations $\mathcal{H}_C = \{a_h,\ T_h,\ b_h,\ \boldsymbol{\mu}_h\}$, $\mathcal{H}_R = \{\varrho_g,\ r_g,\ \boldsymbol{\nu}_g\}$, $\mathcal{R}_C = \{\Psi_h,\ \mathbf{T}_h,\ S_h,\ q_{hj}\}$, $\mathcal{R}_R = \{j_{gk},\ c_g\}$, where $h \in I_{BC}$, $g \in I_{BR}$. $r_g(\varrho_l)$ is a function defined by analogy to a_h, $l \in I_h^r$. Gradients of ϱ_g can be expressed with the help of r_g.

Constitutive quantities are expressed on different sets $\mathcal{K}_C$ and $\mathcal{K}_R$ of subbodies. On the other hand, interactions between C and R systems have to be expressed by constitutive equations since the scale of averaging is big sufficiently at this level of description. Thus, let $\mathcal{H}_{R|C}$ stands for $\mathcal{H}_R$ transformed to the system $\mathcal{K}_C$ of subbodies and $\mathcal{H}_{C|R}$ means $\mathcal{H}_C$ transformed to $\mathcal{K}_R$. We can consider for instance $\mathcal{H}_{R|C} = \{\varrho_h,\ r_h,\ \boldsymbol{\nu}_h\}$, $\varrho_h = \varrho(\mathbf{X}_{Ch})$ $\nu_h = \nu(\mathbf{X}_{Ch})$ $(r_h)_l = \frac{\partial \varrho}{\partial X_l}(\mathbf{X}_{Ch})$, where $\varrho(\mathbf{X}) = \alpha_\varrho(\{\varrho_g,\ r_g\})$, $\boldsymbol{\nu}(\mathbf{X}) = \alpha_\nu(\{\nu_g\})$. Accordingly, the constitutive equations are postulated in the general form as

$$\mathcal{R}_C = \mathcal{R}_C(\mathbf{C}_{RC},\ \mathcal{H}_C,\ \mathcal{H}_{R|C}) \,, \quad \mathcal{R}_R = \mathcal{R}_R(\mathbf{C}_{RR},\ \mathcal{H}_{C|R},\ \mathcal{H}_R) \tag{30}$$

and additionally evolution equations for internal variables are given by

$$\dot{\mu} = A_\mu(\mathbf{C}_\mu,\ \mathcal{H}_C,\ \mathcal{H}_{R|C}) \,, \quad \dot{\nu} = A_\nu(\mathbf{C}_\nu,\ \mathcal{H}_{C|R},\ \mathcal{H}_R) \,. \tag{31}$$

Finally, we have introduced the skeletal dynamical system on larger scale of averaging, appropriate for continuum description, by equations (26)-(31).

5 Final remarks

In this paper a concept of mathematical modelling in biomechanics motivated by cytogel behaviour is introduced. The main intention of this approach is to give an unified theoretical scheme as far as possible. Complexity of physical phenomena observed in miscellaneous scales creates necessity of taking into account physical foundations in the framework of the theoretical approach. This induces multiscale description which consists of two segments.

The first one is the concept of molecular catalysts mechanics and is based on physical assumptions. This kind of theory is predestined to describe behaviour of single large molecule or a system of such molecules in interaction with chemical reactions.

The second segment is more formal mathematically. It is the collection of dynamical systems with dimensional reduction. This kind of multiscale modelling is appropriate for molecular catalysts mechanics. Transition between scales is realized by means of the dimensional reduction procedure. Such concept allows to formulate problems on the most elementary level and next obtain more simple models which describe especially interesting features. Owing to this, biomechanics usually connected with larger scale of averaging can be placed in system with well defined physical foundations.

Let us notice also that collection of dynamical systems comprise continuum mechanics as a particular case. On the other hand, molecular dynamics can be placed on the most elementary level of this description. This method is widely applied in chemical physics and material science and can be viewed as creating physical foundations. As a result both these important theories can be integrated within one theoretical scheme.

References

[1] R.S.GOODY, K.C.HOLMES, Biochimica et Biophysica Acta, 726, 13-39, (1983).

[2] S.HIGHSMITH, Biochimica et Biophysica Acta, 639, 31-39, (1981).

[3] G.F.OSTER, G.M.ODELL, Physica **12D**, 333-350, (1984).

[4] J.KACZMAREK, *A formulation of continuum mechanics as a dimensional reduction of a finite-dimensional dynamical system*, Arch. Mech., **49**, 49-66, (1997).

MODELING OF ULTRASONIC WAVES IN MAGNETIC LIQUIDS: TWO PHASE APPROACH

M. KACZMAREK
Department of Environmental Mechanics, Pedagogical University in Bydgoszcz
Chodkiewicza 30,85-064 Bydgoszcz, Poland
E-mail: mkk@rose.man.poznan.pl

T. HORNOWSKI, M. LABOWSKI and A. SKUMIEL
Institute of Acoustics, Adam Mickiewicz University
Matejki 48/49, 60-769 Poznan, Poland

Two phase model of dynamical behavior of magnetic liquid influenced by a constant magnetic field is considered. The model takes into account mutual interactions of clusters (skeleton stiffness), interactions of clusters and liquid carrier and transversal isotropy of the system. Rotational degrees of freedom are neglected. Some predictions of the model are discussed and compared to experimental data.

1 Introduction

In the past few decades an increase of technological importance as well as significant interest in theoretical and experimental investigations of magnetic liquids are observed. Magnetic liquids when not subjected to magnetic field are homogeneous colloidal suspensions of ferromagnetic particles in a carrier liquids, such as water, oil or kerosene. Typically, the diameters of solid particles range from 5 to 10 nanometers. Although to prevent coagulation of the particles they are coated with surface-active dispersive medium experimental investigations of magnetic liquids in DC magnetic field show that a certain amount of colloidal particles forms aggregates (clusters) which tend to join into a chain-like structure as long as hundreds of nanometers [1]. This microstructure causes an increase of stiffness of the medium, induces anisotropy of its mechanical properties, and contributes to the dissipation of mechanical energy of internal processes due to interaction of clusters with the surrounding liquid.

A useful way to study mechanical properties of magnetic liquids is based on application of ultrasonic methods. In particular these methods appeared to be very powerful to observe anisotropy of properties of the liquids in variable magnetic field and wide range of temperature. The ultrasonic methods can be also used to study the role of clusters in macroscopic behavior of the liquids.

Numerous models of wave propagation in magnetic liquids have been considered, see e. g. Tarapov [8], Parsons.[6], Taketomi [3], and Gogosov et al.[7]. However, most of them fail to predict properly anisotropic properties of attenuation and phase velocity of ultrasonic waves. The most promising model which offers a description of some characteristic features of attenuation of mechanical waves is given by Taketomi [3]. This model takes into account the consequences of motion of clusters in magnetic fluid and particularly the mutual interaction between clusters and clusters with carrier fluid. Since the Taketomi's approach is based on analysis of dissipated energy of the system it is limited to the study of attenuation and cannot predict dispersion of waves as well as other consequences of the two-phase nature of magnetic liquid (e.g. total number of wave modes in the liquid that can propagate taking into account given anisotropy and stiffness of skeleton).

More thorough two-phase approach applied for the description of ultrasonic waves in magnetic liquids was proposed by Gogosov et al. [7]. The model however treats the magnetic liquid as a suspension in which mutual interactions between clusters are neglected.

The purpose of the present paper is to develop a two phase model of wave propagation in magnetic liquid using the concept of superimposed continua. The basic notions of the model are borrowed from the theory of multiphase media, and particularly from the theory of anisotropic porous materials saturated with fluid [4 9]. The two phases of the medium are: the skeleton made of interacting chains of clusters and the carrier liquid. The mutual interactions of clusters are represented through a non-vanishing stress tensor for the skeleton. The interactions of clusters with liquid are expressed as a sum of viscous and inertial drag forces. The model assumes a transversal isotropy of the medium induced by DC magnetic field. Quantitative predictions of the model are discussed and its numerical predictions are compared with experimental data.

2 Model description

It is assumed that the magnetic liquid consists of phases: solid skeleton (s) and fluid (f) and the motion of the phases is governed by the linearized balances of linear momentum; i.e.

$$\rho^f \frac{\partial \mathbf{v}^f}{\partial t} - \nabla \cdot \mathbf{T}^f - \mathbf{R}^f = 0 \ , \qquad \rho^s \frac{\partial \mathbf{v}^s}{\partial t} - \nabla \cdot \mathbf{T}^s - \mathbf{R}^s = 0 \ , \tag{1}$$

where ρ^f and ρ^s stand for the mass density of fluid and solid phase, $\mathbf{v}^f$ and $\mathbf{v}^s$ are velocities of the fluid and solid skeleton, $\mathbf{T}^f$ and $\mathbf{T}^s$ denote stress tensors, and $\mathbf{R}^f = -\mathbf{R}^s$ are interaction forces between the phases.

The stress tensors and interaction forces are assumed to be constitutive functions that express material properties of the medium. The stress tensor in fluid phase will consist of two parts: hydrostatic component and viscous stress:

$$T^f{}_{ij} = -np\delta_{ij} + \pi_{ij} \quad , \tag{2}$$

where p is the fluid pressure, n is the volume fraction of carrier liquid in the magnetic liquid and π is the viscous stress tensor. The hydrostatic component of the stress tensor in fluid phase $\mathbf{T}^f$ for the transversally isotropic medium reads [9]

$$np = -M(e_{11} + e_{22}) + Qe_{33} + R\epsilon \quad , \tag{3}$$

where axis x_3 is assumed to be perpendicular to the isotropy plane which includes axis x_1 and x_2. e_{ij} denotes the strain tensor for the skeleton and ϵ is the relative volume change or dilatation of fluid that can be expressed by displacements of skeleton, $u^s{}_i$, and fluid, $u^f{}_i$, as following $e_{ij} = \frac{1}{2}(\frac{\partial u^s{}_i}{\partial x_j} + \frac{\partial u^s{}_j}{\partial x_i})$, and $\epsilon = \frac{\partial u^f{}_i}{\partial x_i}$. The coefficients M and Q are coupling coefficients relating deformations of solid to the hydrostatic stress in fluid while R is a measure of the pressure required to change the volume of fluid phase while the total volume of the medium remains constant. The viscous components of the stress tensor in fluid phase are assumed to be dependent on rate of deformation of fluid $d_{ij} = \frac{1}{2}(\frac{\partial v^f{}_i}{\partial x_j} + \frac{\partial v^f{}_j}{\partial x_i})$ and for given symmetry are following:

$$\pi_{11} = 2\mu d_{11} + \alpha(d_{11} + d_{22} + d_{33}) \quad , \tag{4}$$

$$\pi_{22} = 2\mu d_{22} + \alpha(d_{11} + d_{22} + d_{33}) \quad , \tag{5}$$

$$\pi_{33} = 2\gamma d_{33} + \beta(d_{11} + d_{22} + d_{33}) \quad , \tag{6}$$

$$\pi_{13} = \pi_{31} = 2\delta d_{13} \ , \quad \pi_{23} = \pi_{32} = 2\delta d_{23} \ , \quad \pi_{12} = \pi_{21} = 2\mu d_{12} \ , \tag{7}$$

where μ, α, γ, β, and δ are the viscosity constants.

The stress tensor of skeleton depends on deformation of skeleton and dilatation of fluid and its components for transversally isotropic case take on the form [9]

$$T_{11} = 2Ne_{11} + A(e_{11} + e_{22}) + Fe_{33}) + M\epsilon \quad , \tag{8}$$

$$T_{22} = 2Ne_{22} + A(e_{11} + e_{22}) + Fe_{33} + M\epsilon \quad , \tag{9}$$

$$T_{33} = 2Ce_{33} + F(e_{11} + e_{22}) + Q\epsilon \quad , \tag{10}$$

$$T_{13} = T_{31} = 2Le_{13} \ , \quad T_{23} = T_{32} = 2Le_{23} \ , \quad T_{12} = T_{21} = 2Ne_{12} \ , \tag{11}$$

where N, A, F, C, and L are the elastic constants characterizing mechanical responce of skeleton.

The interaction force between fluid and skeleton is assumed to be a sum of viscous (frictional) force and inertial force

$$R^s{}_1 = -b_t(v^f{}_1 - v^s{}_1) - c_t\frac{\partial}{\partial t}(v^f{}_1 - v^s{}_1) \quad R^s{}_2 = -b_t(v^f{}_2 - v^s{}_2) - c_t\frac{\partial}{\partial t}(v^f{}_2 - v^s{}_2) \ , \tag{12}$$

$$R^s{}_3 = -b_n(v^f{}_3 - v^s{}_3) - c_n\frac{\partial}{\partial t}(v^f{}_3 - v^s{}_3) \ , \tag{13}$$

where b_t, b_n, and c_t, c_n are coefficients characterizing viscous and inertial couplings in transversally isotropic medium.

Combining the above introduced constitutive functions and balances of linear momentum for solid and fluid phase one obtains equations for macroscopic displacements of phases which in particular govern the propagation of harmonic waves in infinite medium. In order to discuss properties of the solution of the system of equations in a form of plane harmonic waves the following functions are considered: $u^f{}_i = b^i e^{i\omega(t-q_\alpha x_\alpha)}$, and $u^s{}_i = a^i e^{i\omega(t-q_\alpha x_\alpha)}$ where $\mathbf{a} = \{a_1, a_2, a_3\}$ and $\mathbf{b} = \{b_1, b_2, b_3\}$ are amplitudes of displacement of solid and fluid, ω is the angular frequency, and $\mathbf{q} = \{q_1, 0, q_3\}$ is the slowness vector which is related to complex wave vector, $\mathbf{k}$, according to the relation $\mathbf{k} = \omega\mathbf{q}$. The slowness vector is chosen in such way that the plane of constant phase is parallel to the axis x_2. Plugging the assumed solutions into the motion equations of phases one obtains six homogeneous equations with unknown parameters a_i and b_i:

$$D_{11}a_1 + D_{12}a_3 + D_{13}b_1 + D_{14}b_3 = 0 \quad D_{31}a_1 + D_{32}a_3 + D_{33}b_1 + D_{34}b_3 = 0 \ , \tag{14}$$

$$D_{41}a_1 + D_{12}a_3 + D_{13}b_1 + D_{14}b_3 = 0 \quad D_{61}a_1 + D_{62}a_3 + D_{63}b_1 + D_{64}b_3 = 0 \ , \tag{15}$$

$$D_{51}a_2 + D_{52}b_2 = 0 \ , \qquad D_{21}a_2 + D_{22}b_2 = 0 \ , \tag{16}$$

where D_{ij} are combinations of material constants, angular frequency and components of slowness vector. Since the equations which contain a_2 and b_2 are independent of the other equations one can notice that the waves which are related to the motion of particles in direction x_2 are not coupled with waves which are disturbances in the plane determined by x_1 and x_3. The set of four homogeneous equations (14) and (15) has non-trivial solution if the appropriate determinant of coefficients is equal to zero, $det D^I = 0$. This condition yields a dispersion equation corresponding to the waves which are disturbances in the plane of x_1 and x_3. Similarly, the uniqness of the solution of equations (16) demands that the corresponding determinant is equal to zero, $det D^{II} = 0$. This condition yields a dispersion equation for waves which are disturbances in direction x_2.

Introducing into the dispersion equations a complex value slowness parameter, α, such that $q_1 = \alpha sin\Theta$, and $q_3 = \alpha cos\Theta$, where Θ is the angle between the magnetic field and the direction of ultrasonic wave, polynomials in α are obtained. The roots of the polynomials with positive values of real components of α determine the number of wave modes predicted by the model as well as the velocities and coefficients of attenuation of the waves. In general, the model predicts four types of waves which are related to disturbances in plane x_1, x_3. Two of the modes are quasi-longitudinal waves for which the motion of particles of phases takes place in or out of phase. The remaining waves are the shear like waves resulting from the stiffness of solid skeleton and viscosity of carrier liquid.

3 Discussion of results

In order to discuss properties of the proposed two-phase model of magnetic liquid some predictions of the model will be analyzed and compared with experimental data. Since the model disregards rotational motion of clusters the analysis is limited to waves which propagate along the magnetic field. The experimental data are obtained using the Matec 2470 Automatic Attenuation Recorder which measures the logarithmic difference (in dB) between two selected echoes. When A and B are two consecutive echoes the log(A/B) output is proportional to the absorption coefficient of the sample. The variations of log(A/B), e.g. as a function of the magnetic field, give direct changes in the absorption coefficient. The accuracy of determination of the changes in absorption coefficient is better than 1%. The absolute values of the absorption coefficient are accurate within 5 %. The ultrasonic cell used in measurements is made of brass which remains fixed in the gap between the electromagnetic pole pieces. The magnetic field is measured to within 0.5 % with a F. W. Bell Gaussmeter model 9200. Other details of the ultrasonic setup used are reported by Skumiel at al. [5]. The measurements are carried out in magnetic liquid denoted as EMG-605 (produced by Ferrofluids Inc.) consisting of magnetic particles of Fe_3O_4 suspended in water. The volume concentration of the particles and initial viscosity of the liquid are 3.5 % and 0.5 Ns/m. The measurements are performed at temperature 50^oC , frequency 1.18 MHz, and for different strength of magnetic field. The results for wave velocity and absorption coefficient (attenuation) are shown in Fig. 1. Fig. 2 shows the predicted by the two-phase model values of attenuation, and velocity of the fast quasi-longitudinal wave propagating in direction x_3 versus volume fraction of solid skeleton (clusters). To obtain the numerical results it is assumed that the increase of magnetic field causes proportional growth of number of clusters while the radius of clusters remains constant and equal to 1000 nm.

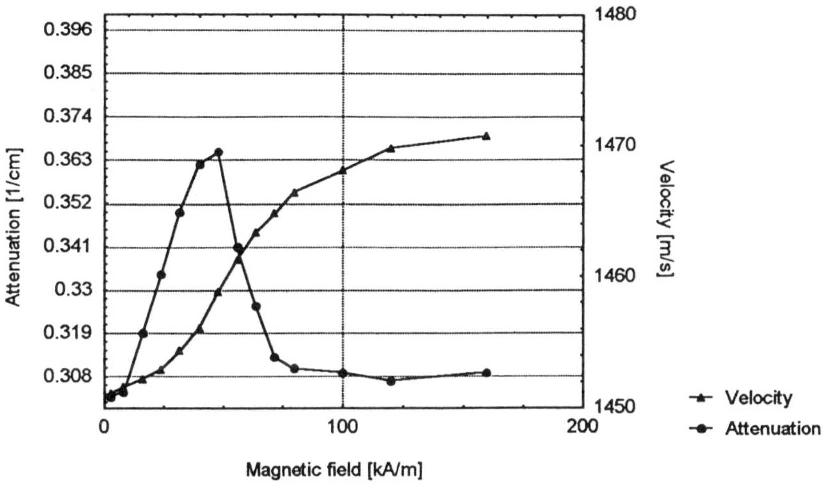

Fig. 1: Dependence of the absorption coefficient (attenuation) and wave velocity on magnetic field for $\Theta = 0$, $T = 50^oC$, and f = 1,18 MHz.

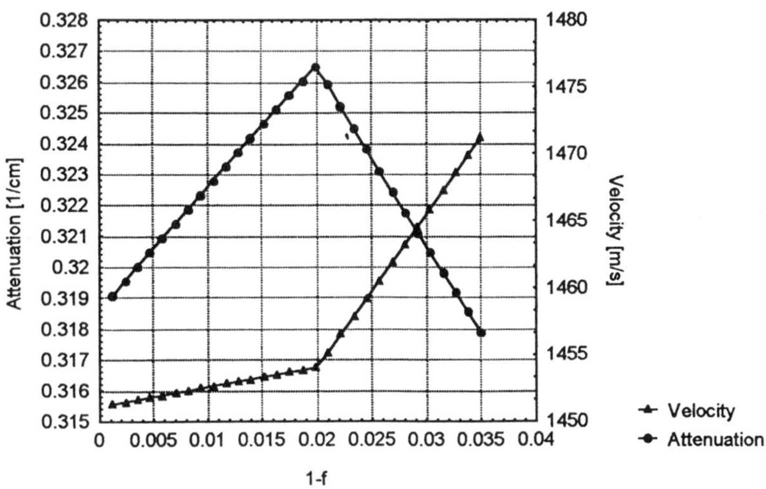

Fig. 2: Predictions of the two-phase model of magnetic liquid.

For volume fractions of clusters lower than 0.02 the interaction of clusters is assumed to be negligible (there is no load bearing skeleton). When the volume fraction approaches the limit 0.02 clusters start to interact proportionally to the volume fraction of the cluster above the limit and this causes the appearance of stiffness of the skeleton in direction x_3. No further increase of stiffness of the skeleton takes place when magnetic saturation is reached (1-f approaches the maximum value).

The numerical values of other parameters of the model were chosen in such a way that the orders of attenuation and velocity predicted by the model agree with the appropriate data obtained from the experiments. Although further refinement of the model is necessary to include e.g. changes in viscosity of carrier liquid and frequency dependence of interaction between phases, one should notice from the comparison of the results given in Fig. 1 and 2 that there is a good qualitative agreement of theoretical results and experimental data. An analysis of other predictions of the model in relation to the measurements to evaluate e.g. the effects of anisotropy of the material and the role of frequency of ultrasonic waves is necessary and will be approached in the nearest future.

Acknowledgments

The studies were supported by KBN Research Grant No. 2PO3B 07912.

References

1. I. Nakatani, Proc. Int. Symp. On Hydrodynamics of Magnetic Liquids and its Application, Sendai, p.9 (1997),
2. D. A. Kruger, IEEE Trans. on Magnetics, MAG-16 2, 251-253 (1980),
3. S. Taketomi, J.Phys. Soc. Japan 55, 838-844 (1983),
4. R. M. Bowen, Theory of mixtures, in Continuum Physics, vol. III, A. C. Erring (ed.), New York: Academic Press, pp. 1-127, (1976),
5. A. Skumiel, M. abowski, and T. Hornowski, J.Phys.D:Appl.Phys. 30, 25 (1997),
6. J. D. Parsons, J. Phys. D. Appl. Phys., 8, 1219 (1975),
7. W. W. Gogosov, W. W. Martynov, S. I. Curikov, and G. A. Saposnikova, Magn. Gidrodin., 2, 19, (1987),
8. I. E. Tarapov, Magnetohydrodynamics, 1,1 (1972),
9. M. D. Sharma and M. L. Gogna, J. Acoust. Soc. Am., 90, 25, (1991).

DEFORMATION AND FRACTURE OF POLYMERS: RELATIVISTIC APPROACH

G.M.KERCH
Centre of Science and Engineering EKOTRA, Nicgales str.4-235,LV-1035, Riga, Latvia,
E-mail:garry.k@usa.net

Relativistic phenomena can play the important role in processes of fracture of solids, especially when velocity of kinetic units on microscopic or mesoscopic level approaches some limiting velocity. The role of the value of limiting velocity in manifestation of relativistic effects was considered and the relationship between structure, value and sign of the thermal expansion coefficient on the one hand and weight changes of polymers during the process of fracture on the other hand was discussed.

1 Introduction

It has been shown[1-3] that equations of motion of crystal dislocations can be brought into a form analogous to those of a particle in special relativity. Dislocations suffer Lorentz contraction in the direction of motion and the total energy is also given by the relativistic equation. But in all equations velocity of light is replaced by velocity of sound. Velocity of elastic waves is the limiting velocity for mechanical displacements in solids. Mott [4] found that the velocity of a crack in solids should asymptotically approach a terminal velocity. It has been shown recently [5] that comparatively large weight changes of viscose fibers as a result of fracture can be observed due to relativistic reasons.

The purpose of this paper is to demonstrate the role of finite limiting velocity in special relativity and in derivation of Lorentz transformations in order to show how the relativistic approach can be applied to mechanical phenomena in solids.

2 Results and discussions

Let us consider a system of parallel layers moving with some constant velocities relatively to each other. The layer 2 moves with velocity V relative to the layer 1 and the layer 3 moves with velocity v relative to the layer 2. If in the framework of the system the relative velocity of mutual displacements is limited and the value of the limiting velocity is equal to some arbitrary constant c, then one can arrive at a conclusion that velocity u of the layer 3 relative to the layer 1 cannot be equal to the sum V + v. In alternative case, when magnitude of velocity is not limited, one can increase step by step the value of velocity of each next layer by certain value Δv until infinity. So the simple arithmetical law of addition of velocities does not stand within the framework of a system provided that the value of velocity of mutual mechanical displacements is limited.

The limiting velocity should not by all means be equal to velocity of light. All above mentioned is valid for any other value of limiting velocity. So $u \neq V + v$ and we must assume

$$u = A(V + v) = Av(1+ V/v), \tag{1}$$

$$v = A(u - V) = Au(1 - V/u). \tag{2}$$

As a matter of fact Eqs.(1) and (2) demonstrate viewpoints of two different observers. Combining Eqs.(1) and (2) yields

$$A^2 = \frac{1}{(1+V/v)(1-V/u)}. \tag{3}$$

If A=1, then classical law of addition of velocities $u = V + v$ is valid. Let us denote $A^2 = 1/1\text{-}q$ and after simple transformations of Eq.(3) we obtain

$$1 - V/u = (1 - q)/(1 + V/v),$$

$$u = (v + V)/(1 + qv/V). \tag{4}$$

The boundary conditions, due to the existence of limiting velocity c, are

$$c = A(V + c),\ A = 1/(1+V/c). \tag{5}$$

According to Eq.(4) we obtain

$$u_{v=c} = (c + V)/(1 + qc/V). \tag{6}$$

Eq.(6) is consistent with boundary conditions (5) provided $q=V^2/c^2$ .Finally, the law of addition of velocities takes the form

$$u = (v+V)/(1 + Vv/c^2). \tag{7}$$

The difference between the relative velocity u of layers 3 and 1 and the arithmetical sum of velocities V+v increases with decreasing the limiting velocity of mechanical displacements c within the system. In the case $c = \infty$ the law $u = V+v$ is valid. So any limitation in the magnitudes of velocities of mutual displacements leads to the law of addition of velocities in the form of Eq.(7).One can assume that infinite velocity is impossible in systems of finite size and the Eq.(7) will be valid in all such systems. The consequence of the limitation of the magnitudes of velocities is that "rigid bodies" and "incompressible fluids" have become impossible objects. For, by definition, they would transmit signals instantaneously. One can conclude that rigidity of a body will increase with increasing the value of speed limit. It is consistent with well known formula

$$E = \rho c^2 \tag{8}$$

where E is the elasticity modulus,ρ is the density and c is the velocity of sound. Eq.(7) can be rewritten as

$$\frac{dx'}{dt'} = \frac{V + \frac{dx}{dt}}{1 + \frac{dx}{dt}\frac{V}{c^2}} = \frac{dx + V\,dt}{dt + dx\frac{V}{c^2}} \tag{9}$$

Hence

$$x' = \gamma(x + Vt),$$

$$t' = \gamma(t + x\frac{V}{c^2}). \tag{10}$$

By symmetry, we must have

$$x = \gamma(x' - Vt'),$$

$$t = \gamma(t' - x'V/c^2). \tag{11}$$

Taking into account Eqs.(11) , the Eqs.(10) can be written as

$$x' = \gamma[\gamma(x' - Vt') + \gamma V(t' - x'V/c^2)] = \gamma^2 x'(1 - V^2/c^2),$$

$$t' = \gamma[\gamma(t' - x'V/c^2) + \gamma V/c^2 (x' - Vt')] = \gamma^2 t'(1 - V^2/c^2). \qquad (12)$$

Hence

$$\gamma = 1/(1 - V^2/c^2)^{1/2}, \qquad (13)$$

and Eqs.(10) - (11) are the Lorentz transformations. But instead of velocity of light we must use the limiting velocity c characteristic for the system under consideration.

Derivations of Lorentz transformations that dispense with postulate of the invariance of speed of the light are known for a long time[6-9] and underline a general character of special relativity as a universal theory describing the structure of space-time. The apparently ad hoc and privileged selection of a specific physical process of the light-signal propagation as a fundamental phenomenon underpinning the basis of special relativity obviously restricts the area of application of the theory. It has been noted [9] that special relativity rules all classes of natural phenomena, not only electromagnetic interactions, which have no privilege other than a historical one. So, it is a more general approach if the bases of Einsteinian relativity are taken as the special relativity principle and the existence of limiting speed, instead of the relativity principle and some specific experimental observation such as the law of light propagation.

Consequently the acceptance of the relativity principle must necessarily lead to the manifestation of relativistic effects in solids and velocity of elastic waves should be considered as the limiting velocity for mechanical displacements.

Kobayashi et al[10] experimentally demonstrated that cracks in amorphous brittle materials always travel at velocities smaller than the Rayleigh wave speed. Relativistic effects have been observed in a number of works[5,11-20] dealing with dynamical fracture. Yoffe[11] showed that stresses in the neighborhood of the crack adopt a universal form near the tip, and this universal singularity contracts in the direction of motion as the cracks approach the speed of sound and that at around 60% of the Rayleigh wave speed a crack should become unstable, since the maximum tensile stress would no longer be directly ahead of the crack, but would instead be off at an angle. Crack tip instabilities have been discussed recently by Marder and Gross[12] and Marder and Fineberg [13].The energy needed to form a new crack surface increases drastically[14] , the velocity of cracks begins to oscillate[15] and the fracture surface shows periodic structure as the velocity of a crack approaches some characteristic limiting velocity. It has been observed that effective surface energy[16], stress intensity factor[17,18], dynamic fracture toughness[19,20] also increase sharply as the limiting speed is approached.

If an amount of energy ΔU be given to a body, the inertial mass of the body increases by an amount $\Delta U / c^2$. It was found that inertia is not a fundamental property of matter but a property of energy. In the case if ΔU is internal energy changes of a solid the c should be equal to velocity of sound. So comparatively large changes of the weight of polymers are expected to be observed under the comparatively small changes of internal energy.

Analytical balance was used to measure weights of polymer samples before and after stretching and fracture with accuracy 0.05 mg.

Table 1: Weight changes in polymer samples

No	Weight before fracture, 10^{-6} kg	Weight after fracture, 10^{-6} kg	Weight changes, 10^{-6} kg
Low density polyethylene (compression molded sheets)			
1	5590.6	5589.6	-1.0
2	5491.3	5490.6	-0.7
3	5488.2	5487.2	-1.0
4	5439.5	5438.3	-1.2
5	5444.1	5443.3	-0.8
Poly(vinyl chloride)			
6	3170.4	3166.3	-4.1
7	3164.7	3161.2	-3.5
8	3180.2	3175.9	-4.3
9	3083.9	3079.7	-4.2
Oriented low density polyethylene			
10	195.6	196.1	0.5
11	206.0	206.8	0.8
12	173.6	174.4	0.8
13	182.2	183.3	1.1
14	194.2	195.1	0.9

The results are shown in Table 1. The values of mass decrease or increase depend on the changes of internal energy of a body after stretching and fracture. The increase of weight has been observed for polymer with negative linear thermal expansion coefficient. The comparatively great changes of masses are due to low value of limiting velocity. The value of weight changes is too high to be attributed to the absorption or desorption of gases or to other experimental errors. One can calculate from the data of Table 1 internal energy changes $\Delta U = \Delta mc^2$, which vary between -0.5 and 2.0 J. These values are consistent with experimental data reported by Godovsky.[21]

It has been shown[22] that negative linear thermal expansion coefficient and evolution of heat under stretching of an oriented polyethylene may be associated with crystalline structure characterized by c-axis orientation of crystallites and positive linear thermal expansion coefficient and absorption of heat under stretching of an oriented polyethylene may be associated with crystalline structure characterized by a-axis orientation of crystallites.

So one can expect positive weight changes as a result of fracture for oriented polymers with negative linear thermal expansion coefficient and weight losses for polymers with positive linear thermal expansion coefficient.

3 Conclusions

The realization of the significance of the limiting velocity and relativistic ideas about interrelation between energy and inertia of the processes and their sharp increase in the neighborhood of the limiting velocity gives the opportunity to develop a new approach to consideration of mechanical deformation and fracture of solids. Manifestation of relativistic effects is more pronounced in polymers which are characterized by the great value of thermal expansion coefficient and comparatively low value of the velocity of elastic waves

References

1. J.I.Frenkel in *Vvedenie v teoriju metalov*, (Nauka, Leningrad, 1972).
2. F.C Frank, *Proc. Phys. Soc.* **62**, 131(1949).
3. J. D.Eshelby, *Proc. Phys. Soc.* **62**, 307 (1949).
4. N. Mott,*Engineering* **165**,16 (1947).
5. G.Kerch, *Int.J.Fracture* **83**,L33 (1997).
6. N.Umov in *Selected Works*,(Gosizdat,Moscow,1950).
7. Y.Terletskii in *Paradoxes in the Theory of Relativity*,(Plenum,NY, 1968).
8. W.Rindler in *Essential Relativity*,(Springer-Verlag,New York,1979).
9. J-M.Levy-Leblond,*Am.J.Phys.* **44**,271 (1976).
10. A.Kobayashi *et al,J.Appl.Pol.Sci.***18**,1625 (1974).
11. E.Yoffe,*Phil.Mag.* **42**,739 (1951).
12. M.Marder and S.Gross,*J.Mech.Phys.Solids* **43**,1(1995).
13. M.Marder and J.Fineberg,*Physics Today*,N9,24 (1996).
14. A.Green and P.Pratt, *Eng.Fracture Mech.* **6**,71(1974).
15. J.Fineberg *et al,Phys.Rev.* B **45**,5146 (1992).
16. L.Rose,*Int.J.Fracture* **12**,799 (1976).
17. J.Dally *et al,Int.J.Fracture* **27**,159 (1985).
18. W.Knauss and K.Ravi-Chandar, *Int.J.Fracture* **27**,127 (1985).
19. T.Kanazawaet *et al,Exp.Mech.* **21**,76.
20. A. Rosakis.and A. Zehnder,*Int.J.Fracture* **27**,169 (1985).
21. J.K.Godovsky in *Teplofizika polimerov*,(Chimija, Moscow,1982).
22. G.M.Kerch and L.A.Irgen, *Mehanika polimerov*, N3, 547 (1975).

MATRIX ANALYSIS APPROACH TO DETERMINATION OF ENTROPIES FOR HYPERBOLIC SYSTEMS OF CONSERVATION EQUATIONS

W. LARECKI

Institute of Fundamental Technological Research, Polish Academy of Sciences, 00-049 Warszawa, Swietokrzyska 21, Poland

E-mail: wlarecki@ippt.gov.pl

For many problems of continuum phisics, the governing equations take the form of the system of conservation equations while the thermodynamic restrictions can be expressed as a requirement of consistency with the additonal conservation equation ("balance of entropy"). Liu's method of Lagrange multipliers or, the more general, the "main dependency relation" of Friedrichs have been employed for analysing consistency of the system of conservation equations with the "balance of entropy". An alternative approach is based on the fact that the system of conservation equations in a normal form is symmetrizable by premultiplication by Hessian iff it is consistent with the additional conservation equation, so the conditon that the system of conservation equations admitts Hessian as symmetrizer can be equivalently analysed. In this paper, the respecive procedure is developed in detail for hyperbolic systems of conservation equations in several space dimensions. Since such systems involve real simple matrices with real eigenvalues, the representation of the family of left symmetric symmetrizers of simple matrices is derived and the condition that $m \geq 2$ simple matrices have common left symmetric symmetrizers is obtained. Then, Frobenius theorem is employed to determine the conditon that Hessians are contained in the family of common left symmetric symmetrizers. The application of developed procedure to the equations of isentropic flow of barotropic fluid is presented.

1 Introduction

Hyperbolic systems of conservastion equations are often encountered in continuum physics, especially in the context of wave phenomena. Then, consistency with the additional conservation equation ("balance of entropy") provides criteria for the uniqueness of weak solutions (shock waves) and makes it possible to obtain equivalent symmetric hyperbolic systems, for which the Cauchy problem is well posed. Liu's method of Lagrange multipliers [1] or, the more general, the "main dependency relation" of Friedrichs [2] have been used to study the consistency of the systems of conservation equations with the "balance of entropy" and, consequently, for derivation of equivalent symmetric hyperbolic systems. In [3], an alternative approach has been proposed, which is based on the fact known from [4] that the system of conservation equations in a normal (Cauchy) form is symmertizable by premultiplication by Hessian matrix iff it implies the "balance of entropy". Hence, the conditions imposed on the system of conservation equations by consistency with "balance of entropy" can be obtained from the requirement that it admits Hessian matrix as a symmetrizer. The respective general procedure has been proposed in [3] and the objective of this paper is to develop this procedure in detail for hyperbolic systems, not necessarily in the normal form. As an illustrative physical example, the application of developed procedure to the equations of isentropic flow of barotropic fluid is presented.

2 Consistency of Hyperbolic Systems with "Balance of Entropy"

The hyperbolic (in the time direction t) system of N conservation equations for N unknowns $\mathbf{u}^T(t,\mathbf{X}) = [u^1(t,\mathbf{X}),\ldots,u^N(t,\mathbf{X})]$ is considered

$$\begin{aligned} &\mathbf{A}^0(\mathbf{u})\partial_t\mathbf{u} + \mathbf{A}^j(\mathbf{u})\partial_j\mathbf{u} = \mathbf{d}(\mathbf{u},t,\mathbf{X}) \ , \quad j = 1,2,\ldots,m \ , \\ &\mathbf{A}^0(\mathbf{u}) = [\nabla_{\mathbf{u}}\mathbf{f}^0(\mathbf{u})], \quad \mathbf{A}^j(\mathbf{u}) = [\nabla_{\mathbf{u}}\mathbf{f}^j(\mathbf{u})] \ , \end{aligned} \tag{1}$$

where $\mathbf{X} := [X^j] \in R^m$ are spatial coordinates, $t \in (0,I], I < \infty$ is the time, $\mathbf{u} \in \mathcal{D} \subset R^N, \mathcal{D}$ - open bounded convex domain, $\mathbf{f}^0, \mathbf{f}^j : \mathcal{D} \to R^N$ are twice continously differentiable, $\partial_t := \dfrac{\partial}{\partial t}$, $\partial_j := \dfrac{\partial}{\partial X^j}$, and the usual summation convention over repeated lower and upper indices is assumed. Here, $\nabla_{\mathbf{v}}$ denotes gradient with respect to components of vector $\mathbf{v}$ and $(.)^T$ denotes the transpose. $\mathbf{A}^0$ and $\mathbf{A}^j$ are the $N \times N$ matrices. Hyperbolicity of (1) requires $\det\left[\nabla_{\mathbf{u}}\mathbf{f}^0(\mathbf{u})\right] \neq 0$ for all $\mathbf{u} \in \mathcal{D}$, and therefore (1) can be transformed into the equaivalent hyperbolic conservative system in the normal (Cauchy) form

$$\partial_t\mathbf{w} + \tilde{\mathbf{A}}^j(\mathbf{w})\partial_j\mathbf{w} = \tilde{\mathbf{d}}(\mathbf{w},t,\mathbf{X}) \ , \tag{2}$$

with the aid of the invertible transformation of dependent variables $\mathbf{w} = \tilde{\mathbf{w}}(\mathbf{u}) = \mathbf{f}^0(\mathbf{u})$ and $\mathbf{u} = \tilde{\mathbf{u}}(\mathbf{w}) = \mathbf{f}^{0^{-1}}(\mathbf{w})$. Hence, $\tilde{\mathbf{f}}^j(\mathbf{w}) = \mathbf{f}^j(\mathbf{f}^{0^{-1}}(\mathbf{w}))$ and, consequently,

$$\begin{aligned} &\tilde{\mathbf{A}}^j(\mathbf{w}) := [\nabla_{\mathbf{w}}\tilde{\mathbf{f}}^j(\mathbf{w})] = [\nabla_{\mathbf{u}}\mathbf{f}^j(\tilde{\mathbf{u}}(\mathbf{w}))][\nabla_{\mathbf{w}}\tilde{\mathbf{u}}(\mathbf{w})] = \mathbf{A}^j(\tilde{\mathbf{u}}(\mathbf{w}))[\mathbf{A}^0(\tilde{\mathbf{u}}(\mathbf{w}))]^{-1} \\ &\tilde{\mathbf{d}}(\mathbf{w},t,\mathbf{X}) = \mathbf{d}(\tilde{\mathbf{u}}(\mathbf{w}),t,\mathbf{X}). \end{aligned} \tag{3}$$

Hyperbolicity of (1) (and of (2), (3)) means that the eigenvalue problem $\det(\tilde{\lambda}\mathbf{I} - n_j\tilde{\mathbf{A}}^j) = 0$ has N real roots $\overset{(K)}{\tilde{\lambda}}(\mathbf{w};\mathbf{n})$ and there are N linearly independent eigenvectors $\overset{(K)}{\tilde{\mathbf{r}}}(\mathbf{w};\mathbf{n})$ such that $[\overset{(K)}{\tilde{\lambda}}\,\mathbf{I} - n_j\tilde{\mathbf{A}}^j]\overset{(K)}{\tilde{\mathbf{r}}} = \mathbf{0}$, for all $\mathbf{n} = [n_1,\ldots,n_m]$. Hence, the $N \times N$ matrix

$\mathbf{L}(\mathbf{w};\mathbf{n}) := [\overset{(1)}{\tilde{\mathbf{r}}}(\mathbf{w};\mathbf{n}),\ldots,\overset{(N)}{\tilde{\mathbf{r}}}(\mathbf{w};\mathbf{n})]$ has the inverse $\mathbf{S}(\mathbf{w};\mathbf{n})$, such that its N rows $\overset{(K)}{\tilde{\mathbf{l}}^T}(\mathbf{w};\mathbf{n})$, called the left eigenvectors, are linearly independent and satisfy $\overset{(K)}{\tilde{\mathbf{l}}^T}[\overset{(K)}{\tilde{\lambda}}\mathbf{I} - n_j\tilde{\mathbf{A}}^j] = \mathbf{0}^T$.

Clearly, hyperbolicity of (1) means that m matrices $\tilde{\mathbf{A}}^j(\mathbf{w})$ as well as all linear combinations of those matrices $n_j\tilde{\mathbf{A}}^j(\mathbf{w})$ (with real coefficients n_j) are real simple (diagonable) matrices with real eigenvalues.

According to [1 - 5], the "balance of entropy"

$$\partial_t\tilde{h}^0(\mathbf{w}) + \partial_j\tilde{h}^j(\mathbf{w}) = \tilde{\sigma}(\mathbf{w},t,\mathbf{X}), \tag{4}$$

follows from (in other words, "is implied by" or "is a consequence of") the system (2), (3) if there are N functions $l_M(w^K)$, such that the equality

$$\begin{aligned}&\partial_t\tilde{h}^0(w^K) + \partial_j\tilde{h}^j(w^K) - \tilde{\sigma}(w^K,t,X^i) = \\ &l_M(w^K)[\partial_t w^M + \partial_j\tilde{f}^{jM}(w^K) - \tilde{d}^M(w^K,t,X^i)],\end{aligned} \tag{5}$$

holds for all functions $w^K(t,X^i)$. It follows from [4] (see, also comments in [3]) that the system (2), (3) implies (4) iff there exists the "entropy" function $\tilde{h}^0(\mathbf{w})$, such that its Hessian matrix is the common left symmetrizer of the matrices $\tilde{\mathbf{A}}^j(\mathbf{w})$, i.e.

$$\tilde{\mathbf{H}}(\mathbf{w})\mathbf{A}^j(\mathbf{w}) = [\tilde{\mathbf{H}}(\mathbf{w})\mathbf{A}^j(\mathbf{w})]^T = \tilde{\mathbf{A}}^{jT}(\mathbf{w})\tilde{\mathbf{H}}^T(\mathbf{w}), \tag{6}$$

where $H_{KL}(w^M) = \dfrac{\partial^2\tilde{h}^0(w^M)}{\partial w^K\partial w^L}$. As outlined in [3], the analysis of consistency of (2), (3) with (4), based on this result, can be accomplished in in the following steps: a) to determine the family of left symmetric symmetrizers for each $\tilde{\mathbf{A}}^j(\mathbf{w})$, b) to establish the conditions that $\tilde{\mathbf{A}}^j(\mathbf{w})$ have common left symmetric symmetrizers, c) to determine the family of common left symmetric symmetrizers, d) to impose the conditions that this family of symmetrizers contains at least one Hessian, e) to integrate the Hessian (or the family of Hessians) in order to determine the "entropies". If the introduced transformation of dependent variables $\mathbf{w} = \tilde{\mathbf{w}}(\mathbf{u})$ is substituted into the "entopy" $\tilde{h}^0(\mathbf{w})$ and into (4), (5), then the function $h^0(\mathbf{u}) = \tilde{h}^0(\tilde{\mathbf{w}}(\mathbf{u}))$ is the "entropy" function for the system (1) and satisfies the respective conservation equation.

3 Left Symmetric Symmetrizers and "Entropies"

For any simple $N \times N$ matrix $\mathbf{A}$, the family of all left symmetric symmetrizers $\mathbf{H}$ can be obtained from the results of [6] in the following form:

$$\mathbf{H} = \mathbf{S}^T\mathbf{Y}\mathbf{S}, \quad \mathbf{Y} = \mathbf{Y}^T, \tag{7}$$

where the N rows of $\mathbf{S}$ are formed of arbitrarily chosen set of N linearly independent left eigenvectors of $\mathbf{A}$, arranged in the order corresponding to the similarity transformation

$$\mathbf{A} = \mathbf{S}^{-1}\mathbf{D}\mathbf{S}, \quad \mathbf{D} = diag[\overset{(1)}{\lambda}, \ldots, \overset{(N)}{\lambda}]. \tag{8}$$

where $\overset{(K)}{\lambda}, K = 1, \ldots, N$ are the eigenvalues of $\mathbf{A}$ taken with multiplicities and $\mathbf{Y}$ are the symmetric matrices which commute with $\mathbf{D}$, i.e.

$$\mathbf{Y}\mathbf{D} = \mathbf{D}\mathbf{Y}, \quad \mathbf{Y} = \mathbf{Y}^T. \tag{9}$$

Suppose that $\mathbf{A}$ has $1 \leq k \leq N$ distinct eigenvalues $\overset{(\Gamma)}{\lambda}, \Gamma = 1, \ldots, k$ of the respective multiplicities l_Γ, $\sum_{\Gamma=1}^{k} l_\Gamma = N$. Then, the solutions of (9) take the block diagonal form

$$\mathbf{Y} = diag[\mathbf{Z}_1, \ldots, \mathbf{Z}_k], \quad \mathbf{Z}_\Gamma = {\mathbf{Z}_\Gamma}^T, \quad \Gamma = 1, \ldots, k, \tag{10}$$

with arbitrary entries $z_{\Gamma\hat{\Lambda}}, \hat{\Lambda} = 1, \ldots, \frac{1}{2}l_\Gamma(l_\Gamma + 1)$, in each symmetric $l_\Gamma \times l_\Gamma$ block $\mathbf{Z}_\Gamma$. Hence, the entries $z_{\Gamma\hat{\Lambda}}$ play the role of parameters and, according to (7), the matrices $\mathbf{H}$ depend linearly on those parameters. The trivial symmertizer $\mathbf{H} = \mathbf{O}$ corresponds to all $z_{\Gamma\hat{\Lambda}} = 0$ and any linear combination of symmstrizers $\mathbf{H}$ ia also a symmetrizer. Since $\mathbf{S}$ is fixed with the entries S_{KL} being given functions of the components A_{PQ} and of the eigenvalues $\overset{(\Gamma)}{\lambda}$ of $\mathbf{A}$, the number of linearly independent symmetric symmetrizers $\mathbf{H}$ (in the space of symmetric $N \times N$ matrices) corresponds to the number of linearly independent solutions of (9), that is to the number $s = \frac{1}{2}\sum_{\Gamma=1}^{k} l_\Gamma(l_\Gamma + 1)$, $N \leq s \leq \frac{1}{2}N(N+1)$, of parameters $z_{\Gamma\hat{\Lambda}}$.

In this way, for each real simple matrix $\tilde{\mathbf{A}}^j(\mathbf{w})$ with real distinct eigenvalues $\overset{(\Gamma)}{\lambda_j}(\mathbf{w})$, $\Gamma = 1, \ldots, k^j \leq N$, of the respective multiplicities ${l^j}_\Gamma$, $\sum_{\Gamma=1}^{k^j} {l^j}_\Gamma = N$, we obtain a family of all real left symmetric symmetrizers in a parametric form

$$\mathbf{H}^j(\mathbf{w}; z_{\Gamma\hat{\Lambda}}) = \mathbf{S}^{jT}(\tilde{A}^j_{PQ}(\mathbf{w}), \overset{(\Delta)}{\lambda_j}(\mathbf{w}))\mathbf{Y}^j({z^j}_{\Gamma\hat{\Lambda}})\mathbf{S}^j(\tilde{A}^j_{PQ}(\mathbf{w}), \overset{(\Delta)}{\lambda_j}(\mathbf{w})), \tag{11}$$

where the index j is not summed here and in all subsequent formulae. Performing the respective matrix multiplication in (11) and taking into account linear dependence on s^j parameters $z^j{}_{\Gamma\hat{\Lambda}}$,

$\hat{\Lambda} = 1, \ldots, \frac{1}{2} l^j{}_\Gamma (l^j{}_\Gamma + 1)$, $\quad s^j = \frac{1}{2} \sum_{\Gamma=1}^{k^j} l^j{}_\Gamma (l^j{}_\Gamma + 1)$, $\quad N \leq s^j \leq \frac{1}{2} N(N+1)$, we arrive at the folowing representation of the components of $\mathbf{H}^j$

$$H^j{}_{KL}(\mathbf{w}; z^j{}_{\Gamma\hat{\Lambda}}) = \sum_{\Gamma=1}^{k^j} \sum_{\hat{\Lambda}=1}^{\frac{1}{2} l^j{}_\Gamma (l^j{}_\Gamma+1)} a^j{}_{KL}{}^{\Gamma\hat{\Lambda}} (\tilde{A}^j_{PQ}(\mathbf{w}), \overset{(\Delta)}{\lambda_j} (\mathbf{w})) z^j{}_{\Gamma\hat{\Lambda}}. \tag{12}$$

The condition that the matrices $\tilde{\mathbf{A}}^j(\mathbf{w})$ have common nontrivial left symmetric symmetrizers $\tilde{\mathbf{H}}$ means that m families of the symmetrizers $\mathbf{H}^j(\mathbf{w}; z^j{}_{\Gamma\hat{\Lambda}})$ given by (11), (12), have nontrivial commom members. Therefore, the system of $M = \frac{1}{2} N(N+1) \frac{m!}{2(m-2)!}$ equalities

$$H^1{}_{KL}(\mathbf{w}; z^1{}_{\Gamma\hat{\Lambda}}) = H^2{}_{KL}(\mathbf{w}; z^2{}_{\Gamma\hat{\Lambda}}) = \ldots = H^m{}_{KL}(\mathbf{w}; z^m{}_{\Gamma\hat{\Lambda}}), \tag{13}$$

must be satisfied for $\sigma = \sum_{j=1}^{m} s^j = \frac{1}{2} \sum_{j=1}^{m} \sum_{\Gamma=1}^{k^j} l^j{}_\Gamma (l^j{}_\Gamma + 1)$,

$mN \leq \sigma \leq \frac{1}{2} mN(N+1)$, real parameters $z^j{}_{\Gamma\hat{\Lambda}}$, not all identically zero. In view of (12), the system (13) can be represented as a linear homogeneous sytem of M equations for σ parameters $z^j{}_{\Gamma\hat{\Lambda}}$

$$\begin{aligned} &\sum_{\Gamma=1}^{k^1} \sum_{\hat{\Lambda}=1}^{\frac{1}{2} l^1{}_\Gamma (l^1{}_\Gamma+1)} a^1{}_{KL}{}^{\Gamma\hat{\Lambda}} z^1{}_{\Gamma\hat{\Lambda}} = \sum_{\Gamma=1}^{k^2} \sum_{\hat{\Lambda}=1}^{\frac{1}{2} l^2{}_\Gamma (l^2{}_\Gamma+1)} a^2{}_{KL}{}^{\Gamma\hat{\Lambda}} = z^2{}_{\Gamma\hat{\Lambda}} = \ldots \\ &= \sum_{\Gamma=1}^{k^m} \sum_{\hat{\Lambda}=1}^{\frac{1}{2} l^m{}_\Gamma (l^m{}_\Gamma+1)} a^m{}_{KL}{}^{\Gamma\hat{\Lambda}} z^m{}_{\Gamma\hat{\Lambda}} , \end{aligned} \tag{14}$$

and rearranged to the matrix form

$$\mathcal{B}(a^j{}_{KL}{}^{\Gamma\hat{\Lambda}} (\tilde{A}^j_{PQ}(\mathbf{w}), \overset{(\Delta)}{\lambda_j} (\mathbf{w}))) \mathbf{z} = \mathbf{0}, \tag{15}$$

where $\mathcal{B}$ is the $M \times \sigma$ matrix with entries $a^j{}_{KL}{}^{\Gamma\hat{\Lambda}}$ and $\mathbf{z} = [z^1{}_{\Gamma\hat{\Lambda}}, \ldots, z^m{}_{\Gamma\hat{\Lambda}}]$ is the σ-dimensional vector. Hence, the necessary and sufficient condition for the matrices $\tilde{\mathbf{A}}^j(\mathbf{w})$ to have common nontrivial symmetric symmetrizers $\tilde{\mathbf{H}}$ is $\boldsymbol{rank\mathcal{B}} \leq \sigma - 1$. If it is satisfied, then the solution set of (15) depends on $q = rank\mathcal{B} - \sigma + 2$ parameters χ^α, which may depend on the unknowns

$\mathbf{w}$. Substituting $\mathbf{z}(\mathbf{w};\chi^\alpha(\mathbf{w}))$ calculated from (13), (14), (15) into (11) or (12) we arrive at the family of left symmetric symmetrizers $\tilde{\mathbf{H}}(\mathbf{w};\chi^\alpha(\mathbf{w}))$ of the matrices $\tilde{\mathbf{A}}^j(\mathbf{w})$.

The family of common left symmetric symmetrizers $\tilde{\mathbf{H}}(\mathbf{w};\chi^\alpha(\mathbf{w})) = [\tilde{H}_{KL}(\mathbf{w};\chi^\alpha(\mathbf{w}))] = [\tilde{H}_{LK}(\mathbf{w};\chi^\alpha(\mathbf{w}))]$ contains a Hessian iff the system of $\frac{1}{2}N(N+1)$ second-order p.d.e. for the "entropy" function $\tilde{h}^0(w^K)$

$$\frac{\partial^2\tilde{h}^0(w^K)}{\partial w^P\partial w^Q} = \tilde{H}_{PQ}(w^K;\chi^\alpha(w^K)), \quad K,P,Q=1,\ldots,N, \quad P\leq Q, \tag{16}$$

is integrable. According to the well known Frobenius theorem, the integrability conditon for (16) takes the form of the following system of equalities for the $\frac{1}{2}N(N+1)$ components of $\tilde{\mathbf{H}}(\mathbf{w};\chi^\alpha(\mathbf{w}))$

$$\frac{\partial\tilde{H}_{PQ}(w^K;\chi^\alpha(w^K))}{\partial w^R} = \frac{\partial\tilde{H}_{PR}(w^K;\chi^\alpha(w^K))}{\partial w^Q} = \frac{\partial\tilde{H}_{RQ}(w^K;\chi^\alpha(w^K))}{\partial w^P}, \tag{17}$$
$$P,Q,R=1,\ldots,N.$$

Integration of (16) gives the "entropies" for the system of conservation equations (2), (3).

4 "Entropies" for Isentropic Flow of Barotropic Fluid

The equations for isentropic flow of barotropic fluid can be written as a hyperbolic system of conservation equations in the normal form (2), (3) in the unknowns $\mathbf{w}^T = [\rho, \mathbf{m}^T]$, where ρ is the mass density, $\mathbf{m} = [m^j] = \rho\mathbf{v} = [\rho v^j]$ is the momentum density and $\mathbf{v} = [v^j]$ is the particle velocity. In this case, $\tilde{\mathbf{d}}(\mathbf{w},t,\mathbf{X}) = \mathbf{0}$ and the matrices $\tilde{\mathbf{A}}^j(\mathbf{w}), j=1,2,3$, take the form

$$\tilde{\mathbf{A}}^j =$$
$$\begin{bmatrix} 0 & \delta^{1j} & \delta^{2j} & \delta^{3j} \\ p'(\rho)\delta^{1j} - \dfrac{m^1m^j}{\rho^2} & \dfrac{m^1}{\rho}\delta^{1j} + \dfrac{m^j}{\rho} & \dfrac{m^1}{\rho}\delta^{2j} & \dfrac{m^1}{\rho}\delta^{3j} \\ p'(\rho)\delta^{2j} - \dfrac{m^2m^j}{\rho^2} & \dfrac{m^2}{\rho}\delta^{1j} & \dfrac{m^2}{\rho}\delta^{2j} + \dfrac{m^j}{\rho} & \dfrac{m^2}{\rho}\delta^{3j} \\ p'(\rho)\delta^{3j} - \dfrac{m^3m^j}{\rho^2} & \dfrac{m^3}{\rho}\delta^{1j} & \dfrac{m^3}{\rho}\delta^{2j} & \dfrac{m^3}{\rho}\delta^{3j} + \dfrac{m^j}{\rho} \end{bmatrix} \tag{18}$$

where $p'(\rho) = \dfrac{dp(\rho)}{d\rho}$ and $p(\rho)$ is a pressure. The matrices $\tilde{\mathbf{A}}^j$ have the following eigenvalues:

$$\overset{(1)}{\lambda_j} = \overset{(2)}{\lambda_j} = \rho^{-1}m^j, \quad \overset{(3)}{\lambda_j} = \rho^{-1}m^j + [p'(\rho)]^{\frac{1}{2}}, \quad \overset{(4)}{\lambda_j} = \rho^{-1}m^j - [p'(\rho)]^{\frac{1}{2}} \tag{19}$$

and the corresponding sets of N linearly independent left eigenvectors can be chosen as:

$$\overset{(1)}{\tilde{\mathbf{l}}_j^T} = [-\rho^{-1}m^k, \mathbf{e}_k], \qquad \overset{(2)}{\tilde{\mathbf{l}}_j^T} = [-\rho^{-1}m^l, \mathbf{e}_l],$$

$$\overset{(3)}{\tilde{\mathbf{l}}_j^T} = [[p'(\rho)]^{\frac{1}{2}} - \rho^{-1}m^j, \mathbf{e}_j], \overset{(4)}{\tilde{\mathbf{l}}_j^T} = [[p'(\rho)]^{\frac{1}{2}} + \rho^{-1}m^j, -\mathbf{e}_j], \tag{20}$$

$$j, k, l = 1, 2, 3, \qquad k \neq j, l \neq j, k < l,$$

where

$$\mathbf{e}_1 = [1, 0, 0], \quad \mathbf{e}_2 = [0, 1, 0], \quad \mathbf{e}_3 = [0, 0, 1]$$

denote the base vectors of the orthonormal Cartesian frame, in which the flow is considered. Substituting the matrices $\mathbf{S}^j$ formed of those left eigenvectors into (7), (8), (9), (10) together with the corresponding matrices $\mathbf{Y}^j$

$$\mathbf{Y}^j = diag\left(\begin{bmatrix} z^j{}_{11} & z^j{}_{13} \\ z^j{}_{13} & z^j{}_{12} \end{bmatrix}, z^j{}_{21}, z^j{}_{31}\right), \tag{21}$$

we obtain three families of all left symmetric symmetrizers $\mathbf{H}^j(\rho, m^k; z^j{}_{11}, z^j{}_{12}, z^j{}_{13}, z^j{}_{21}, z^j{}_{31})$ for the respective three matrices $\tilde{\mathbf{A}}^j$. In those families, the components $H^j{}_{11}$ take the form

$$H^j{}_{11} = [\rho^{-1}m^k]^2 z^j{}_{11} + [\rho^{-1}m^l]^2 z^j{}_{12} + 2\rho^{-2}m^k m^l z^j{}_{13} + [\overset{(3)}{\lambda_j}]^2 z^j{}_{21} + [\overset{(4)}{\lambda_j}]^2 z^j{}_{31},$$
$$j, k, l = 1, 2, 3, \quad k \neq j, \quad l \neq j, \quad k < l, \tag{22}$$

while the remaining components are given in the following Table:

	$j=1$	$j=2$	$j=3$
$H^j{}_{12}$	$-\overset{(3)}{\lambda_1} z^1{}_{21} - \overset{(4)}{\lambda_1} z^1{}_{31}$	$-\frac{m^1}{\rho} z^2{}_{11} - \frac{m^3}{\rho} z^2{}_{13}$	$-\frac{m^1}{\rho} z^3{}_{11} - \frac{m^2}{\rho} z^3{}_{12}$
$H^j{}_{13}$	$-\frac{m^2}{\rho} z^1{}_{11} - \frac{m^3}{\rho} z^1{}_{13}$	$-\overset{(3)}{\lambda_2} z^2{}_{21} - \overset{(4)}{\lambda_2} z^2{}_{31}$	$-\frac{m^2}{\rho} z^3{}_{12} - \frac{m^1}{\rho} z^3{}_{13}$
$H^j{}_{14}$	$-\frac{m^3}{\rho} z^1{}_{12} - \frac{m^2}{\rho} z^1{}_{13}$	$-\frac{m^3}{\rho} z^2{}_{12} - \frac{m^1}{\rho} z^2{}_{13}$	$-\overset{(3)}{\lambda_3} z^3{}_{21} - \overset{(4)}{\lambda_3} z^3{}_{31}$
$H^j{}_{22}$	$z^1{}_{21} + z^1{}_{31}$	$z^2{}_{11}$	$z^3{}_{11}$
$H^j{}_{23}$	0	0	$z^3{}_{13}$
$H^j{}_{24}$	0	$z^2{}_{13}$	0
$H^j{}_{33}$	$z^1{}_{11}$	$z^2{}_{21} + z^2{}_{31}$	$z^3{}_{12}$
$H^j{}_{34}$	$z^1{}_{13}$	0	0
$H^j{}_{44}$	$z^1{}_{12}$	$z^3{}_{12}$	$z^3{}_{21} + z^3{}_{31}$

In this case, the relations (13) take the form of the system of 30 linear equations

$$H^1{}_{KL} = H^2{}_{KL} = H^3{}_{KL}, \quad K, L = 1,2,3, \quad K \leq L \tag{23}$$

for 15 parameters $z^j{}_{\Gamma\hat{\Lambda}}$ and, according to (14), (15), its solution dependents on the single parameter χ and takes the following form:

$$\begin{aligned} &z^1{}_{11} = z^1{}_{12} = z^2{}_{11} = z^2{}_{12} = z^3{}_{11} = z^3{}_{12} \\ &= z^1{}_{21} + z^1{}_{31} = z^2{}_{21} + z^2{}_{31} = z^3{}_{21} + z^3{}_{31} = \chi \\ &z^j{}_{13} = 0, \quad z^j{}_{21} = z^j{}_{31}, \quad j = 1,2,3\ . \end{aligned} \tag{24}$$

Substitution of (24) into $H^j{}_{KL}$ given by (22) and listed in the Table yields the family of common left symmetric symmetrizers of the matrices (18)

$$\tilde{\mathbf{H}}(\rho, m^k) = \chi(\rho, m^k) \begin{bmatrix} p'(\rho) + \rho^{-2}\mathbf{m}^2 & -\rho^{-1}m^1 & -\rho^{-1}m^2 & -\rho^{-1}m^3 \\ -\rho^{-1}m^1 & 1 & 0 & 0 \\ -\rho^{-1}m^2 & 0 & 1 & 0 \\ -\rho^{-1}m^3 & 0 & 0 & 1 \end{bmatrix}, \tag{25}$$

where $\mathbf{m}^2 := (m^1)^2 + (m^2)^2 + (m^3)^2$. In this case, the system (16) consists of the following 10 equations for the "entropy" $\tilde{h}^0(\rho, m^k)$:

$$\begin{aligned} &\frac{\partial^2 \tilde{h}^0}{\partial \rho^2} = \chi(\rho, m^k)\left[p'(\rho) + \left(\frac{m^1}{\rho}\right)^2 + \left(\frac{m^2}{\rho}\right)^2 + \left(\frac{m^3}{\rho}\right)^2\right] \\ &\frac{\partial^2 \tilde{h}^0}{\partial \rho \partial m^i} = \chi(\rho, m^k)\rho^{-1}m_i, \quad i = 1,2,3, \\ &\frac{\partial^2 \tilde{h}^0}{\partial m^i \partial m^j} = \frac{\partial^2 \tilde{h}^0}{\partial m^j \partial m^i} = \chi(\rho, m^k)\delta_{ij} \quad i,j = 1,2,3. \end{aligned} \tag{26}$$

Then, the integrability conditions (17) take the form of the system of 39 equations which must to be satisfied by $p(\rho)$ and $\chi(\rho, m^k)$

$$\begin{aligned} &\left[p'(\rho) + \left(\frac{m^1}{\rho}\right)^2 + \left(\frac{m^2}{\rho}\right)^2 + \left(\frac{m^3}{\rho}\right)^2\right]\frac{\partial \chi}{\partial m^k} + \left[\frac{\chi}{\rho} + \frac{\partial \chi}{\partial \rho}\right]\frac{m^k}{\rho} = 0 \\ &\left[\frac{\chi}{\rho} + \frac{\partial \chi}{\partial \rho}\right]\delta_{kj} + \frac{m^k}{\rho}\frac{\partial \chi}{\partial m^j} = 0 \\ &\frac{\partial \chi}{\partial m^i}\delta_{jk} = \frac{\partial \chi}{\partial m^j}\delta_{ik} = \frac{\partial \chi}{\partial m^k}\delta_{ij}\ , \quad i,j,k = 1,2,3. \end{aligned} \tag{27}$$

It follows from (27) that $\chi = \tilde{\chi}(\rho)$ must satisfy the eqation

$$\frac{d\tilde{\chi}}{d\rho} = -\frac{\tilde{\chi}}{\rho} \, . \tag{28}$$

Integration of (26) with substituted solution of (28)

$$\chi = \tilde{\chi}(\rho) = A\rho^{-1} \tag{29}$$

gives the family of "entropies" for the system of conservation equations (2), (18)

$$\tilde{h}^0(\rho, m^j) = A\left[\phi(\rho) + \frac{\mathbf{m}^2}{2\rho}\right] + B\rho + C_k m^k + D \, , \tag{30}$$

where A, B, C_k, D are arbitrary real constsnts and

$$p(\rho) = \rho\frac{d\phi(\rho)}{d\rho} - \phi(\rho) \, . \tag{31}$$

Taking $A = 1, B = C_k = D = 0$ in (30), returning to original variables ρ, v^k and putting $\phi(\rho) = \rho e(\rho)$, we obtain the energy of the fluid in the isentropic motion as a particular "entropy" $h^0(\rho, v^k) = \rho e(\rho) + \frac{1}{2}\rho v_i v^i$, where $e(\rho)$ is the internal energy density. Hence, (31) is, in fact, the well-known constitutive relation for barotropic fluid $p(\rho) = \rho^2\frac{de(\rho)}{d\rho}$. This particular "entropy" is obtained if either Liu's method of Lagrange multipliers or the "main dependency relation" is applied to the system of conservation equations governing an isentropic flow of barotropic fluid since the balance of energy plays the role of the "additional conservation equation" in this case. It is seen from (30) that the "entropies" obtained here are the linear combinations of the energy density, i.e. density from the "additional conserwation equation", and of mass and and momentum densities, i.e. the densities from the basic system of conservation equations (2), (18).

Clearly, in our approach, the "entropies" $\tilde{h}^0(w^K)$ contain linear combination of the densities from the basic system of conservation equations (2), (3) since they are obtained as a result of integration of the system (16) of the second-order p.d.e., in which the densities from the basic system (2), (3) are the independent variables, and the boundary conditions are not imposed. If those linear terms are neglected then the "entropies" corresponding to the "additional conservation equation" are obtained. If the covex (concave) entropies are sought (which are required for symmetric hyperbolicity), then the family of common left symmetric symmetrizers $\tilde{\mathbf{H}}(\mathbf{w}; \chi^\alpha(\mathbf{w}))$ must contain positive definite (negative definite) matrices, what corresponds to positive definite (negative definite) matrices $\mathbf{Y}^j(z^j{}_{\Gamma\hat{\Lambda}})$ in (11), and the integrability conditions (17) should be imposed on the components of those matrices. In the case of physical problems, in which the "additional conservation equation" has the physical meaning of balance of entropy and the nonequlibrium phenomena are considered, the condition that entropy is maximal at equlibrium should be taken into account as the respective constraint imposed on the system (16).

References

1. I-S. Liu, Method of Lagrange multipliers for exploitation of the entropy principle, Arch. Rational Mech. Anal., **46**, 131-142, 1972
2. K.O. Friedrichs, On the laws of relativistic electro-magneto-fluid dynamics, Comm. Pure Appl. Math., **27**, 749-808, 1974
3. W. Larecki, Symmetrization of systems of conservation equations and the converse to the condition of Friedrichs and Lax, Arch. Mech., **49**, 865-876, 1997
4. G. Boillat, Sur l'existence et la recherche d'equations de conservation supplementaire pour les systemes hyperboliques, C.R. Acad. Sci. Paris, **278 A**, 909-914, 1974
5. K.O. Friedrichs, P.D. Lax, Systems of conservation equations with a convex extension, Proc. Acad. Sci. U.S.A., **68**, 8, 1686-1688, 1971
6. A. Jameson, E. Kreindler, P. Lancaster, Symmetric, positive semidefinite and positive definite real solutions of $\mathbf{AX} = \mathbf{XA}^{\mathbf{T}}$ and $\mathbf{AX} = \mathbf{YB}$, Linear Algebra and Its Applications, **160**, 189-215, 1992

TRANSFORMATIONS OF DEPENDENT VARIABLES AND ALTERNATIVE SYMMETRIC SYSTEMS CORRESPONDING TO CONSISTENT SYSTEM OF N+1 CONSERVATION EQUATIONS

W. LARECKI

Institute of Fundamental Technological Research, Polish Academy of Sciences, 00-049 Warszawa, Swietokrzyska 21, Poland

E-mail: wlarecki@ippt.gov.pl

In variuos problems of continuum physics, the "main dependency relation" (MDR) introduced by Friedrichs can be used for derivation of the restrictions on constitutive functions implied by the entropy inequality, and the MDR can be treated as a generalization of Liu's method of Lagrange multipliers of the Extended Thermodynamics. For the system of $N+1$ conservation equations for N dependent variables (unknowns) in $m+1$ independent variables which satisfies the MDR, an equivalent symmetric system of N first-order p.d.e. for original unknowns can be assigned to each solution of the MDR. In this paper, invertible transformations of dependent variables in the consistent (in the sense of the MDR) system of $N+1$ conservation equations are investigated. The respective transformations of the solutions of the MDR and the corresponding symmetric systems for new unknowns are analysed. It is shown that if the N independent components of an arbitrary solution of the MDR are taken as new unknowns then the corresponding symmetric system, except the right-hand side, is completely determined by $m+1$ potentials and by one additonal function which expresses the remaining dependent component of the solution of the MDR in terms of those independent components. In this way, we obtain a new class of symmetric systems which is a natural generalization of Godunov's class of symmertic conservative systems. It contains, as special cases, $N+1$ alternative symmetric conservative systems corresponding to those particular $N+1$ solutions of the MDR which have one (dependent) component equal to (-1).

1 Introduction

Various problems of continuum physics (in particular, the extended thermodynamics [1]) are governed by systems of N conservation equations for N unknowns $\mathbf{u}^T(\mathbf{X}) = [u^1(\mathbf{X}), \ldots, u^N(\mathbf{X})]$

$$\partial_\alpha \mathbf{f}^\alpha(\mathbf{u}) = \mathbf{d}(\mathbf{u}, \mathbf{X}) \ , \quad \alpha = 0, i, \quad i = 1, 2, \ldots, m \ , \tag{1}$$

while the thermodynamic restrictions take the form of unilateral differential constraints ("entropy inequality")

$$\partial_\alpha h^\alpha(\mathbf{u}) \geq \mu(\mathbf{u}, \mathbf{X}) \ , \tag{2}$$

where $\mathbf{X} := [X^\alpha]$ ($X^0 \in (0, I], I < \infty$ is the time and $X^i \in R^m$ are spatial coordinates), $\mathbf{u} \in \mathcal{D} \subset R^N, \mathcal{D}$ open bounded convex domain,

$\mathbf{f}^\alpha : \mathcal{D} \to R^N, h^\alpha : \mathcal{D} \to R$ are twice continously differentiable, $\partial_\alpha := \dfrac{\partial}{\partial X^\alpha}$, and the usual summation convention over repeated lower and upper indices is assumed.

It has been proved in [2] that: If the following condition (A.1):

$$\det\left[\nabla_{\mathbf{u}}\mathbf{f}^0(\mathbf{u})\right] \neq 0, \text{ for all } \quad \mathbf{u} \in \mathcal{D}\ , \tag{3}$$

holds for (1), then every Lipschitz continuous solution $\mathbf{u}(\mathbf{X})$ of (1) satisfies (2) iff the system of $N+1$ conservation equations

$$\mathbf{L}[\mathbf{u}] := \partial_\alpha \mathbf{g}^\alpha(\mathbf{u}) - \mathbf{b}(\mathbf{u}, \mathbf{X}) = \mathcal{A}^\alpha(\mathbf{u})\partial_\alpha \mathbf{u} - \mathbf{b}(\mathbf{u}, \mathbf{X}) = \mathbf{0},$$

$$\mathbf{g}^{\alpha T}(\mathbf{u}) = [\mathbf{f}^{\alpha T}(\mathbf{u}), h^\alpha(\mathbf{u})], \qquad \mathbf{b}^T(\mathbf{u}, \mathbf{X}) = [\mathbf{d}^T(\mathbf{u}, \mathbf{X}), \chi(\mathbf{u}, \mathbf{X})],$$

$$\chi(\mathbf{u}, \mathbf{X}) = \left[\nabla_{\mathbf{u}} h^0(\mathbf{u})\right]\left[\nabla_{\mathbf{u}}\mathbf{f}^0(\mathbf{u})\right]^{-1}\mathbf{d}(\mathbf{u}, \mathbf{X}), \quad \mathcal{A}^\alpha(\mathbf{u}) = \left[\nabla_{\mathbf{u}}\mathbf{g}^\alpha(\mathbf{u})\right], \tag{4}$$

satisfies the "main dependency relation" (MDR) and the inequality

$$\chi(\mathbf{u}, .) - \mu(\mathbf{u}, .) \geq 0. \tag{5}$$

Here, $\nabla_{\mathbf{w}}$ denotes gradient with respect to components of vector $\mathbf{w}$ and $(.)^T$ denotes the transpose. We recall that the MDR introduced in [3] requires the existence of $N+1$ functions $y_\Lambda : \mathcal{D} \to R$ (which nay be treated as components of $(N+1)$ - dimensional vector $\mathbf{y}(\mathbf{u})$), not all identically zero, such that (Property CI in [3])

$$\mathbf{y}^T(\mathbf{u})\mathbf{L}[\mathbf{u}] = \mathbf{y}^T(\mathbf{u})\left[\partial_\alpha \mathbf{g}^\alpha(\mathbf{u}) - \mathbf{b}(\mathbf{u}, \mathbf{X})\right] = 0 \tag{6}$$

for all functions $\mathbf{u} \in \mathcal{C}^1(R^{m+1})$, what is equivalent to the system of identities (Property CI' in [3])

$$\mathbf{y}^T(\mathbf{u})\mathcal{A}^\alpha(\mathbf{u}) = \mathbf{0}\ , \quad \mathbf{y}^T(\mathbf{u})\mathbf{b}(\mathbf{u}, \mathbf{X}) = 0. \tag{7}$$

Hence, the MDR can be directly applied for derivation of thermodynamic restrictions on constitutive functions $\mathbf{f}^\alpha, h^\alpha, \mathbf{d}$, provided that (A.1) holds. It follows form (6), (7) that the MDR coincides with Liu's method of Lagrange multipliers [1] if one component of $\mathbf{y}(\mathbf{u})$ is assumed to be (-1) and that conservation equation in (4) which is multiplied by (-1) in (6) is interpreted as the "additional conservation equation" implied by the remaining N equations of (4). From this point of wiev, the MDR can be understood as a generalization of Liu's method.

It is well known [1] that Liu's method leads to one symmetric system for the original unknowns and to one symmetric conservative system (introduced in [4]) for Lagrange multipliers treated as new unknowns. It has been demonstrated in [2] that if the conditions stated in this paper as (A.1), (A.2) are satisfied, then the MDR enables direct derivation of: a) a family of symmetric systems for the unknowns $\mathbf{u}(\mathbf{X})$ (parametrized by scalar functions of the unknowns), such that one symmetric system is assigned to each solution $\mathbf{y}(\mathbf{u})$ of (6), (7), b) a family of $N+1$ symmetric conservative systems assigned to those specific $N+1$ solutions $\mathbf{y}(\mathbf{u})$ for which one component is (-1), and in each such case the remaining N components of $\mathbf{y}(\mathbf{u})$ are taken as new unknowns in the corresponding symmetric conservative system.

The objective of this paper is to generalize and supplement the results of [2] by investigating transformations of dependent variables (unknowns) in the system (4) which satisfies the MDR and the conditions (A.1), (A.2). In Sec.3, it is demonstrated how symmetric systems for new dependent variables can be obtained from symmetric systems for $\mathbf{u}(\mathbf{X})$. Since (A.1), (A.2) imply that each solution $\mathbf{y}(\mathbf{u})$ has N components which are independent functions of $\mathbf{u}$, those indepnent components are taken as new unknowns and corresponding symmetric systems are derived in Sec.4. Except dissipative terms (right-hand side), such systems are completely determined by $m+1$ potentials ϕ^α (like symmstric conservative systems) and by one additional function which expresses the dependent component of $\mathbf{y}$ in terms of N independent components. A new class of symmetric systems obtained in this way is a natural generalization of the class of symmetric conservative systems, and it contains $N+1$ symmetric conservative systems as special cases. Various alternative symmetric systems offered by the MDR may be utilized in designing and analysing optimal numerical schemes for considered systems of conservation equations.

2 Symmetric Systems for Original Unknowns

According to [3], premultiplication of the system (4) by the $N\times(N+1)$ matrix $\mathcal{K}^T(\mathbf{u}) := [\nabla_{\mathbf{u}}\mathbf{y}(\mathbf{u})]^T$ gives the symmetric system of N quasilinear equations

$$\mathbf{M}[\mathbf{u}] := \mathcal{K}^T(\mathbf{u})\mathbf{L}[\mathbf{u}] = \mathbf{A}^\alpha(\mathbf{u})\partial_\alpha \mathbf{u} - \mathbf{c}(\mathbf{u},\mathbf{X}) = \mathbf{0} \ , \tag{8}$$

where

$$\mathbf{A}^\alpha(\mathbf{u}) = \mathbf{A}^{\alpha T}(\mathbf{u}) = \mathcal{K}^T(\mathbf{u})\mathcal{A}^\alpha(\mathbf{u}), \quad \mathbf{c}(\mathbf{u},\mathbf{X}) = \mathcal{K}^T(\mathbf{u})\mathbf{b}(\mathbf{u},\mathbf{X}), \tag{9}$$

and the following (A.2) condition of nonsingularity of the $(N+1)\times(N+1)$ matrix $\mathcal{M}(\mathbf{u})$, i.e.

$$det\mathcal{M}(\mathbf{u}) = det\begin{bmatrix}\mathbf{y}^T(\mathbf{u})\\ \mathcal{K}^T(\mathbf{u})\end{bmatrix} \neq 0 \text{ for all } \quad \mathbf{u}\in\mathcal{D} \tag{10}$$

ensures the equivalence of (8), (9) to (4) for differentiable solutions $\mathbf{u}(\mathbf{X})$. Clearly, (8), (9) assign a symmetric system to each solution $\mathbf{y}(\mathbf{u})$ of (7). Since (7) is an overdetermined linear system of $1 + N(m+1)$ equations for $N+1$ components of $\mathbf{y}$, its solution, if exists, is not unique. If the system (4) satisfies the MDR (6), (7) then (A.1) implies $Rank\mathcal{A}^0(\mathbf{u}) = N$ for all $\mathbf{u} \in \mathcal{D}$, so the solutions of (7) form a family of collinear vectors $\mathbf{y}(\mathbf{u})$. It has been shown in [2] that, in this case, symmetric systems corresponding to all $\mathbf{y}(\mathbf{u})$ which solve (7) form a family parametrized by real differentiable functions of the unknowns. It has been also proved in [2] that if the condition (A.2) holds for one particular solution of (7) then it holds for all $\mathbf{y}(\mathbf{u})$ that satisfy (1.6), provided that (A.1) is satisfied.

3 Transformations of Dependent Variables

An arbitrary invertible transformation of dependent variables $\mathbf{u} \to \mathbf{v}$ can be introduced in (4) by substitution $\mathbf{u} = \tilde{\mathbf{u}}(\mathbf{v})$ and the system $\tilde{\mathbf{L}}[\mathbf{v}] := \mathbf{L}[\tilde{\mathbf{u}}(\mathbf{v})] = \mathbf{0}$ of $N+1$ conservation equations for new unknowns $\mathbf{v}(\mathbf{X})$ is then obtained. It can be easily verified that the properties (6), (7) of the MDR are preserved under arbitrary invertible transformations of dependent variables, and the solutions of (6), (7) transform like $\mathbf{y} = \mathbf{y}(\mathbf{u}(\mathbf{v})) = \tilde{\mathbf{y}}(\mathbf{v})$, that is $\mathbf{y}^T(\mathbf{u})\mathbf{L}[\mathbf{u}] = \mathbf{0}$ for all $\mathbf{u}(\mathbf{X})$ iff $\tilde{\mathbf{y}}^T(\mathbf{v})\tilde{\mathbf{L}}[\mathbf{v}] = \mathbf{0}$ for all $\mathbf{v}(\mathbf{X})$. It follows from (8), (9) that substitution of $\tilde{\mathbf{u}}(\mathbf{v})$ and premultiplication by the nonsingular $N \times N$ matrix $[\nabla_{\mathbf{v}}\tilde{\mathbf{u}}(\mathbf{v})]^T$ directly leads to the symmetric system for $\mathbf{v}(\mathbf{X})$, corresponding to premultiplication of the system $\tilde{\mathbf{L}}[\mathbf{v}] = \mathbf{0}$ by the $N \times (N+1)$ matrix

$$\tilde{\mathcal{K}}^T(\mathbf{v}) := [\nabla_{\mathbf{v}}\tilde{\mathbf{y}}(\mathbf{v})]^T = [[\nabla_{\mathbf{u}}\mathbf{y}(\tilde{\mathbf{u}}(\mathbf{v}))][\nabla_{\mathbf{v}}\tilde{\mathbf{u}}(\mathbf{v})]]^T = [\nabla_{\mathbf{v}}\tilde{\mathbf{u}}(\mathbf{v})]^T \mathcal{K}^T(\tilde{\mathbf{u}}(\mathbf{v})) \ , \tag{11}$$

namely,

$$\begin{aligned} &\tilde{\mathbf{M}}[\mathbf{v}] := [\nabla_{\mathbf{v}}\tilde{\mathbf{u}}(\mathbf{v})]^T \mathbf{M}[\tilde{\mathbf{u}}(\mathbf{v})] = [\nabla_{\mathbf{v}}\tilde{\mathbf{u}}(\mathbf{v})]^T \mathcal{K}^T(\tilde{\mathbf{u}}(\mathbf{v}))\mathbf{L}[\tilde{\mathbf{u}}(\mathbf{v})] = \\ &\tilde{\mathcal{K}}^T(\mathbf{v})\tilde{\mathbf{L}}[\mathbf{v}] = \tilde{\mathbf{A}}^\alpha(\mathbf{v})\partial_\alpha \mathbf{v} - \tilde{\mathbf{c}}(\mathbf{v}, \mathbf{X}) = \\ &[\nabla_{\mathbf{v}}\tilde{\mathbf{u}}(\mathbf{v})]^T \mathbf{A}^\alpha(\tilde{\mathbf{u}}(\mathbf{v}))[\nabla_{\mathbf{v}}\tilde{\mathbf{u}}(\mathbf{v})]\partial_\alpha \mathbf{v} - [\nabla_{\mathbf{v}}\tilde{\mathbf{u}}(\mathbf{v})]^T \mathbf{c}(\tilde{\mathbf{u}}(\mathbf{v}), \mathbf{X}) = \mathbf{0} \ . \end{aligned} \tag{12}$$

Since the $N \times N$ matrices $\mathbf{A}^\alpha$ and $\tilde{\mathbf{A}}^\alpha$ in (12) are mutually related by congruent transformation by $[\nabla_{\mathbf{v}}\tilde{\mathbf{u}}(\mathbf{v})]^T$, symmetry of $\mathbf{A}^\alpha$ is equivalent to the symmetry of $\tilde{\mathbf{A}}^\alpha$ and the system (8) is symmetric hyperbolic for given $\mathbf{y}(\mathbf{u})$ ($\mathbf{A}^0$ positive definite) iff the system (12) is symmetric hyperbolic for corresponding $\tilde{\mathbf{y}}(\mathbf{v}) = \mathbf{y}(\tilde{\mathbf{u}}(\mathbf{v}))$ ($\tilde{\mathbf{A}}^0$ positive definite).

Applying the reasoning given in [3] to the symmetric system (12), we can see that nonsingularity of the $(N+1)\times(N+1)$ matrix

$$\tilde{\mathcal{M}}(\mathbf{v}) := \begin{bmatrix} \tilde{\mathbf{y}}^T(\mathbf{v}) \\ \tilde{\mathcal{K}}^T(\mathbf{v}) \end{bmatrix} = \begin{bmatrix} 1 & \mathbf{0}^T \\ \mathbf{0} & [\nabla_{\mathbf{v}}\tilde{\mathbf{u}}(\mathbf{v})]^T \end{bmatrix} \mathcal{M}(\tilde{\mathbf{u}}(\mathbf{v})) \ , \tag{13}$$

ensures the equivalence of (12) and $\tilde{\mathbf{L}}[\mathbf{v}] = \mathbf{0}$ for the differentiable solutions, so the condition (A.2) is preserved under arbitrary invertible transfornations of dependent variables.

4 Symmetric Systems for Independent Components of y

It has been proved in [2] that, for any solution $\mathbf{y}(\mathbf{u})$ of (7), the condition (A.2) is satisfied iff among $N+1$ components of $\mathbf{y}(\mathbf{u})$ there are N independent functions of $\mathbf{u}$ and the remaining component, say y_Γ, $1 \leq \Gamma \leq N+1$, wich is then expressible as $y_\Gamma(\mathbf{u}) = \tilde{y}_\Gamma(y_1(\mathbf{u}),\ldots,y_{\Gamma-1}(\mathbf{u}),y_{\Gamma+1}(\mathbf{u}),\ldots,y_{N+1}(\mathbf{u}))$, is not a homogeneous function of degree one of the arguments $y_1,\ldots,y_{\Gamma-1},y_{\Gamma+1},\ldots,y_{N+1}$.

Let (A.1), (A.2) be satisfied and let $\mathbf{y}(\mathbf{u})$ be an arbitrary solution of (7). Without loss of generality it can be assumed that the component y_Γ is expressible in terms of the remaining N independent components which, for notational convenience, will be denoted by $l_{\hat{\Delta}}(\mathbf{u}), \hat{\Delta} = 1,\ldots,\Gamma-1,\Gamma+1,\ldots,N+1$, and identified with the components of an N -dimensinal vector $\mathbf{l}$. Since $l_{\hat{\Delta}}(\mathbf{u})$ form a set of N independent functions of $\mathbf{u}$, there exist N functions $u^K = \tilde{u}^K(l_{\hat{\Delta}})$ and the function $y_\Gamma = \tilde{y}_\Gamma(l_{\hat{\Delta}})$, and $l_{\hat{\Delta}}(\mathbf{X})$ can be taken as new dependent variables. Then, choosen solution $\mathbf{y}(\mathbf{u})$ of (6),(7) assumes the form

$$\mathbf{y} = \mathbf{y}(\tilde{\mathbf{u}}(\mathbf{l})) = \tilde{\mathbf{y}}(\mathbf{l}) = [l_1,\ldots,l_{\Gamma-1},y_{\hat{\Gamma}}(l_{\hat{\Delta}}),l_{\Gamma+1},\ldots,l_{N+1}] \tag{14}$$

and, according to the results of Sect.3, the corresponding symmetric system (12) in new dependent variables $\mathbf{l}$ and reads

$$\tilde{\mathbf{M}}[\mathbf{l}] := \tilde{\mathcal{K}}^T(\mathbf{l})\tilde{\mathbf{L}}[\mathbf{l}] = \left[\nabla_{\mathbf{l}}\tilde{\mathbf{y}}(\mathbf{l})\right]^T \mathbf{L}[\tilde{\mathbf{u}}(\mathbf{l})] = \tilde{\mathbf{A}}^\alpha(\mathbf{l})\partial_\alpha \mathbf{l} - \tilde{\mathbf{c}}(\mathbf{l},\mathbf{X}) = \mathbf{0} \ , \tag{15}$$

where, according to (14), (11), (15), (12), the entries of the symmetric $N\times N$ matrices $\tilde{\mathbf{A}}^\alpha$ are of the following form:

$$\begin{gathered}\tilde{A}^{\alpha\hat{\Delta}\hat{\Sigma}}(l_{\hat{\Lambda}}) = \frac{\partial \tilde{g}^{\alpha\hat{\Delta}}(l_{\hat{\Lambda}})}{\partial l_{\hat{\Sigma}}} + \frac{\partial \tilde{y}_\Gamma(l_{\hat{\Lambda}})}{\partial l_{\hat{\Delta}}}\frac{\partial \tilde{g}^{\alpha\Gamma}(l_{\hat{\Lambda}})}{\partial l_{\hat{\Sigma}}} = \\ \frac{\partial}{\partial l_{\hat{\Sigma}}}[\tilde{g}^{\alpha\hat{\Delta}}(l^{\hat{\Lambda}}) + \tilde{g}^{\alpha\Gamma}(l_{\hat{\Lambda}})\frac{\partial \tilde{y}^\Gamma(l_{\hat{\Lambda}})}{\partial l_{\hat{\Delta}}}] - \tilde{g}^{\alpha\Gamma}(l_{\hat{\Lambda}})\frac{\partial^2 \tilde{y}_\Gamma(l_{\hat{\Lambda}})}{\partial l_{\hat{\Delta}}\partial l_{\hat{\Sigma}}} \ .\end{gathered} \tag{16}$$

In (16) and in all subsequent formulae the index Γ is not summed and $\tilde{g}^{\alpha\Delta}(l_{\hat{\Sigma}}) := g^{\alpha\Delta}(\tilde{u}^K(l_{\hat{\Sigma}}))$. Symmetry of $\tilde{\mathbf{A}}^\alpha(\mathbf{l})$ implies that there exist potentials $\phi^\alpha(\mathbf{l})$ such that

$$\frac{\partial\phi^\alpha(l_{\hat{\Sigma}})}{\partial l_{\hat{\Delta}}} = \tilde{g}^{\alpha\hat{\Delta}}(l_{\hat{\Sigma}})+\tilde{g}^{\alpha\Gamma}(l_{\hat{\Sigma}})\frac{\partial\tilde{y}^\Gamma(l_{\hat{\Sigma}})}{\partial l_{\hat{\Delta}}}\ , \quad \hat{\Delta},\hat{\Sigma} = 1,\ldots,\Gamma-1,\Gamma+1,\ldots,N+1\ , \tag{17}$$

and since the identities (7) in new dependent variables $l_{\hat{\Delta}}$ take the form

$$l_{\hat{\Delta}}\frac{\partial\tilde{g}^{\alpha\hat{\Delta}}(l_{\hat{\Sigma}})}{\partial l_{\hat{\Lambda}}} + \tilde{y}_\Gamma(l_{\hat{\Sigma}})\frac{\partial\tilde{g}^{\alpha\Gamma}(l_{\hat{\Sigma}})}{\partial l_{\hat{\Lambda}}} = 0\ , \tag{18}$$

integration of (17) yields

$$\phi^\alpha(l_{\hat{\Sigma}}) = l_{\hat{\Delta}}\tilde{g}^{\alpha\hat{\Delta}}(l_{\hat{\Sigma}}) + \tilde{y}_\Gamma(l_{\hat{\Sigma}})\tilde{g}^{\alpha\Gamma}(l_{\hat{\Sigma}})\ . \tag{19}$$

Hence, the components of $\tilde{\mathbf{A}}^\alpha(\mathbf{l})$ can be expressed in terms of $\phi^\alpha(\mathbf{l})$ and $\tilde{y}_\Gamma(\mathbf{l})$

$$\tilde{A}^{\alpha\hat{\Delta}\hat{\Lambda}}(l_{\hat{\Sigma}}) = \frac{\partial^2\phi^\alpha(l_{\hat{\Sigma}})}{\partial l_{\hat{\Delta}}\partial l_{\hat{\Lambda}}} - \tilde{g}^{\alpha\Gamma}(l_{\hat{\Sigma}})\frac{\partial^2\tilde{y}_\Gamma(l_{\hat{\Sigma}})}{\partial l_{\hat{\Delta}}\partial l_{\hat{\Lambda}}}, \tag{20}$$

since (16), (18) imply $m+1$ relations

$$l_{\hat{\Delta}}\frac{\partial\phi^\alpha(l_{\hat{\Sigma}})}{\partial l_{\hat{\Delta}}} - \phi^\alpha(l_{\hat{\Sigma}}) = \left[l_{\hat{\Delta}}\frac{\partial\tilde{y}_\Gamma(l_{\hat{\Sigma}})}{\partial l_{\hat{\Delta}}} - \tilde{y}_\Gamma(l_{\hat{\Sigma}})\right]\tilde{g}^{\alpha\Gamma}(l_{\hat{\Sigma}})\ , \tag{21}$$

wich enable elimination of $\tilde{g}^{\alpha\Gamma}(l_{\hat{\Sigma}})$ from (20). For assumed $\tilde{\mathbf{y}}(\mathbf{l})$ of the form (14), requirement (13) results in the following condition of equivalence of (15) and $\tilde{\mathbf{L}}[\mathbf{l}] = \mathbf{0}$:

$$det\tilde{\mathcal{M}}(\mathbf{l}) = det\left[\nabla_{\mathbf{l}}\tilde{\mathbf{u}}(\mathbf{l})\right] det\mathcal{M}(\tilde{\mathbf{u}}(\mathbf{l})) = \tilde{y}_\Gamma(l_{\hat{\Sigma}}) - l_{\hat{\Delta}}\frac{\partial\tilde{y}_\Gamma(l_{\hat{\Sigma}})}{\partial l_{\hat{\Delta}}} \neq 0\ , \tag{22}$$

which is satisfied on account of (A.2) since $\tilde{y}_\Gamma(l_{\hat{\Sigma}})$ cannot be a homogeneous function of degree one.

The conditions (A.1), (A.2) imply that the family of solutions of (7) contains $N+1$ vectors $\mathbf{y}(\mathbf{u})$, such that one (dependent) component is (-1). They will be denoted by

$$\overset{(\Sigma)}{\mathbf{y}}(\mathbf{u}) = [\overset{(\Sigma)}{y}_1(\mathbf{u}),\ldots,\overset{(\Sigma)}{y}_{\Sigma-1}(\mathbf{u}),-1,\overset{(\Sigma)}{y}_{\Sigma+1}(\mathbf{u}),\ldots,\overset{(\Sigma)}{y}_{N+1}(\mathbf{u})],\quad \Sigma = 1,2,\ldots,N+1. \tag{23}$$

For each $\overset{(\Sigma)}{\mathbf{y}}(\mathbf{u})$, the components $\overset{(\Sigma)}{y}_1, \ldots, \overset{(\Sigma)}{y}_{\Sigma-1}, \overset{(\Sigma)}{y}_{\Sigma+1}, \ldots, \overset{(\Sigma)}{y}_{N+1}$ form a set of N independent functions of $\mathbf{u}$ and therefore can be taken as new dependent variables. Identifying those independent components $\overset{(\Sigma)}{y}_\Delta$ with $l_{\hat{\Delta}}, \Delta, \hat{\Delta} = 1, \ldots, \Sigma-1, \Sigma+1, \ldots, N+1$, in (16), (17), (18), (19), (20), and observing that $\tilde{y}_\Sigma(\overset{(\Sigma)}{y}_\Delta) = \tilde{y}_\Sigma(l_{\hat{\Delta}}) = -1$ in this case, we conclude that the symmetric system (15), (20) for new unknowns $\overset{(\Sigma)}{y}_\Delta = l_{\hat{\Delta}}$ takes the symmetric conservative form [4]

$$\overset{(\Sigma)}{M}{}^{\Delta}[\overset{(\Sigma)}{y}_\Lambda] := \partial_\alpha \left(\frac{\partial \overset{(\Sigma)}{\phi}{}^{\alpha}(\overset{(\Sigma)}{y}_\Lambda)}{\partial \overset{(\Sigma)}{y}_\Delta} \right) - c^\Delta(u^K(\overset{(\Sigma)}{y}_\Lambda), X^\gamma) = 0 \,, \qquad \Delta, \Lambda = 1, 2, \ldots, \Sigma-1, \Sigma+1, \ldots, N+1 \,. \tag{24}$$

This symmetric conservative system corresponds to N equations

$$L^\Delta[u^K] := \partial_\alpha g^{\alpha\Delta}(u^K) - b^\Delta(u^K, X^\gamma) = 0 \,, \qquad \Delta = 1, 2, \ldots, \Sigma-1, \Sigma+1, \ldots, N+1 \tag{25}$$

of the system (4) with respectively transformed dependent variables, while the equation

$$L^\Sigma[u^K] := \partial_\alpha g^{\alpha\Sigma}(u^K) - b^\Sigma(u^K, X^\gamma) = 0 \tag{26}$$

corresponds to the additional conservation equation implied by the symmetric conservative system (24)

$$\overset{(\Sigma)}{L}{}^{\Sigma}[\overset{(\Sigma)}{y}_\Lambda] := \partial_\alpha \left(\sum_{\substack{\Delta=1 \\ \Delta\neq\Sigma}}^{N+1} \overset{(\Sigma)}{y}_\Delta \frac{\partial \overset{(\Sigma)}{\phi}{}^{\alpha}}{\partial \overset{(\Sigma)}{y}_\Delta} - \overset{(\Sigma)}{\phi}{}^{\alpha} \right) - \sum_{\substack{\Delta=1 \\ \Delta\neq\Sigma}}^{N+1} \overset{(\Sigma)}{y}_\Delta \, b^\Delta(u^K(\overset{(\Sigma)}{y}_\Lambda), X^\gamma) = 0. \tag{27}$$

It should be noted that symmetric conservative systems of the family (24), (27) are not equivalent for weak solutions (shock waves). For choosen Σ, N equations of the respective symmetric system (24) are satisfied on the shock, while (27) is replaced by the inequality (≥ 0) in weak sense. This means that the physical quantity corresponding to $g^{\alpha\Sigma}$ increases on the shock. The choice, which symmetric conservative systems from the family (24), (27) are adimissible for studying shock waves is therefore determined by physical meaning of involved quantities.

References

1. I. Müller, T. Ruggeri, "Extendend Thermodynamics", Springer Tracts in Natural Philosophy, Vol. 37, Springer 1993, (Chapter 6)

2. W. Larecki, Symmetric forms of the equations of heat transport in a rigid conductor of heat with internal state variables. I. Analysis of the model and thermodynamic restrictions via the "main dependency relation", II. Alternative symmetric systems, Arch.Mech., **50**, 143-205 (1998)
3. K.O.Friedrichs, On the laws of relativistic electro-magneto-fluid dynamics, Comm. Pure Appl. Math., **27**, 749-808 (1974).
4. S.K. Godunov, Interesting class of quasilinear systems, D.A.N, **139**, 521-523 (1961) (in Russian).

MECHANICS OF A VORTEX LATTICE IN SUPERCONDUCTORS: PHENOMENOLOGICAL APPROACH

B.T. MARUSZEWSKI
Poznań University of Technology, Institute of Applied Mechanics,
ul. Piotrowo 3, 60-965 Poznań, Poland
E-mail: maruszew@sol.put.poznan.pl

L. RESTUCCIA
Universita di Messina, Dipartimento di Matematica,
Contrada Papardo, Salita Sperone 31, 98166 Sant'Agata-Messina, Italy
E-mail: Lrest@dipmat.unime.it

The paper deals with a magnetomechanical description of the vortex lattice in the type-II superconductor. Both lattice and fluid properties of the vortex field have been taken into account. A specific stress tensor has been defined. Nonlinear field equations have been derived.

1. Introduction

The paper deals with the phenomenological aspect of superconductivity and, as such, it develops the mechanics of a vortex lattice as a new state and geometry in a medium. The superconductors belong generally to two classes of such materials. A type-I superconductor expels magnetic flux from the material and hence is in the Meissner state. That is possible only at the applied magnetic field strength less than its determined critical value. In contrast, a type-II superconductor, behave on the other way. For applied field strength less than the lower critical field that superconductor will exhibit the usual Meissner effect. Applied fields greater than the upper critical field strength destroy the superconductivity altogether. In between the lower H_{c1} and upper H_{c2} magnetic field strengths the superconductor is in the mixed or vortex state (Fig. 1) [1-6].

Magnetic flux can penetrate a type-II superconductor in the form of Abrikosov vortices (also called flux lines, flux tubes or fluxons) each carrying a quantum of magnetic flux. These tiny vortices of supercurrent tend to arrange themselves in a triangular flux-line lattice, which is more or less perturbed by material inhomogeneities that pin the flux lines. Pinning is caused by imperfections of the crystal lattice, such as dislocations, point defects, grain boundaries, etc. Hence, a honeycomb pattern of the vortex array presents some mechanical properties. The latter ones come mainly from the elastic properties of the superconducting body. However, since the vortices are formed from the applied magnetic field and around of each of them the supercurrent flows, there exist also some Lorentz force interactions among them. Those interactions form an origin of an additional mechanical (stress) field occurring in the type-II superconductor. That field near the lower critical magnetic intensity limit H_{c1} is also of the elastic character. However, if the density of the supercurrent is above its critical value and/or the temperature is sufficiently high, there occurs a flow of vortex lines in the superconduction body. Within such a situation vortices behave rather as a fluid than as an elastic lattice. The "fluidity" of the vortex array is also observed when the applied magnetic field tends to its upper critical limit H_{c2}. In this way we meet very interesting situation in a type-II superconductor. We can say, that there coexist two mechanical fields in the medium. One of them is of pure elastic character coming from the properties of crystal lattice of the superconductor. The second one comes from the vortex lattice, which, being of elastic character near the lower magnetic field strength limit H_{c1}, transfers smoothly into "fluid" form near the upper magnetic field strength limit H_{c2}.

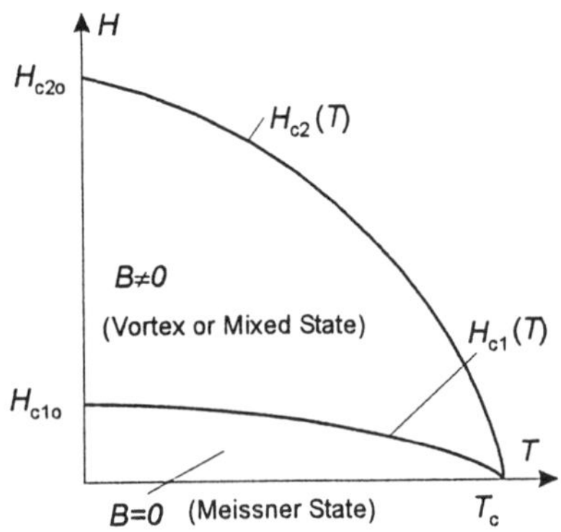

Fig. 1. Mean-field phase diagram of a type-II superconductor

The particular microscopic and mesoscopic description of the above phenomena within classical and quantum models are available in excellent monographs [1-6].

The paper aims at creation a magnetothermomechanical model of such defined type-II superconductor both for the "solid" and "fluid" states of the vortex field. The solid-fluid continuous description is based on a nonconventional irreversible thermodynamics (cf. [7] with many fundamental references). A specific definition of a tensor concerning the vortex field has been introduced. That field as well as the heterogeneity has been averaged within a mesoscale while the entire description is made within continuous phenomenological model in macroscale. The mathematical description expresses itself in the exploitation of the second law of thermodynamics (taken in the form of entropy inequality) with the help of Liu's theorem [8]. The main stress has just been put on the constitutive laws.

2. Thermodynamical model of a vortex field

As we have mentioned before vortices (hard vortices) are pinned, in general, on crystal lattice imperfections of the superconductor. However, if the Lorentz force of interactions between the vortices is much bigger that the pinning force the vortex lattice behaves elastically [4,5]. Such a situation occurs if an external current density $\mathbf{j}$ is applied to the vortex and is bigger than its critical value $\mathbf{j}_c$. So we assume following [4] that the pinning force is negligible in the sequel and we deal with soft vortices. The vortex motion (creep) is accompanied by an energy dissipation. That motion is damped by a force proportional to the vortex velocity. Hence, except for the elastic properties the vortex field is also of a viscous character. The resistivity in area of vortex creep is the same as the resistivity of the current which would flow inside the vortex core. Hence the viscosity coefficient reads [4]

$$\eta = \frac{\Phi_o \mu_o H_{c2}}{\rho_n} \tag{2.1}$$

where Φ_o is the magnetic flux, μ_o denotes the permeability of vacuum and ρ_n is the resistivity in the normal state.

In classical low-T_c superconductors the vortex lattice is mostly consisted of a parallel straight vortex lines arrangement which crossection forms a hexagonal symmetry (cf. [3,4]). However, recent research shows that the vortex lines can be curved or even tangled along the material [5,6,9]. Moreover, because the vortex lines can form, among others, sets called twisted triplets, twisted quadruplets, single loops or pair anihilation, the vortex field can be considered within three dimensions [9]. Because of the fact that each vortex line has sign (has definite vorticity), lines of the opposite signs anihilate. So the density of the vortex field can vary.

Taking into account all the above properties we create a phenomenological thermodynamical model of the vortex field in the superconductor assuming that:

- the field is consisted only of soft vortices,
- the characteristic volume of the body is sufficiently large for averaging all the physical quantities taken into considerations (its dimensions are much greater than the London penetration depth λ_o and the coherence length ξ),
- all the investigations are done within a continuous model,
- the mean vortex field comes from averaging curved vortex lines in three dimensions (in characteristic volume),
- any creation or anihilation of the vortices is omitted,
- the mass density ρ of the vortex field concerns the density of the material in the normal state as the counterpart in the mixed type-II superconductor

(i.e. it is mass of the normal part of the body related to the total volume of the material),

- the only interaction between vortices is due to the Lorentz force,
- the inclusion of thermal fluctuations in the phenomenology of type-II superconductor is here neglected since we consider only depinned (soft) vortices [5],
- the energy dissipation occurs only because of the viscosity of the vortex field caused by the ohmic-like resistivity (normal-state resistivity) inside the vortex core [5],
- because of very low temperatures of the considered material, the relaxation feature of the thermal field is taken into account,
- only small deformations of the vortex field are considered.

Following the above assumptions we create now the extended-like thermodynamic model for the viscoelastic field of vortices in the type-II superconductor. The general form of the vector of state (the set of independent variables) reads (cf. [7,10]):

$$C = \left\{\varepsilon_{ij}, \dot{\varepsilon}_{ij}, \varphi, A_i, T, T_{,i}, \psi, \psi^*, \psi_{'i}, \psi^*_{'i}, q_i, j_i^S\right\}, \tag{2.2}$$

where ε_{ij} denotes the strain tensor, its time derivative in (2.2) indicates the viscoelasticity of the vortex field, φ and A_i are the scalar and vector potentials, respectively, T is the absolute temperature, ψ is the order parameter (the wave function of a Cooper pair) and ψ^* is its compex conjugate, q_i is the heat flux and j_i^S is the supercurrent density. The fundamental laws that govern the set (2.2) are as follows:

$$\dot{\rho} + \rho v_{k,k} = 0, \tag{2.3}$$

$$\rho \dot{v}_k - \sigma_{ik,i} - \epsilon_{ijk}\left(j_i^N + j_i^S\right) B_j - f_i = 0, \tag{2.4}$$

$$\rho \dot{e} - \sigma_{ik} v_{k,i} - q_{k,k} - \left(j_i^N + j_i^S\right) \mathcal{E}_j - \rho r = 0, \tag{2.5}$$

$$\frac{1}{\mu_o} A_{i,kk} - j_i^N - j_i^S = 0, \tag{2.6}$$

$$\dot{q}_k - Q_k(C) = 0, \tag{2.7}$$

$$\dot{\psi} - \Psi(C) = 0, \tag{2.8}$$

$$\dot{\psi}^* - \Psi^*(C) = 0, \tag{2.9}$$

$$\dot{j}_k^S - J_k^S(C) = 0, \tag{2.10}$$

v_k denotes the velocity of the vortex field point, σ_{ik} is the viscoelastic symmetric stress tensor, j_i^N is the normal current, B_j is the magnetic induction, f_k is the body force, e is the internal energy density, $\mathcal{E}_i$ is the electromotive intensity, r is the heat source distribution. Moreover, the electromagnetic field quantities satisfy the Maxwell's equations and the following relations hold true

$$\mathcal{E}_i = E_i + \epsilon_{ijk} v_j B_k, \tag{2.11}$$

$$E_i = -\varphi_{,i} - \frac{\partial A_i}{\partial t}, \tag{2.12}$$

$$B_i = \epsilon_{ijk} A_{k,j}, \tag{2.13}$$

$$B_i = \mu_o H_i \tag{2.14}$$

where H_k is the magnetic field strength. We follow in the sequel the assumption that φ vanishes by gauging [3].

The use of the second law of thermodynamics in the form of the entropy inequality

$$\rho \dot{s} + \Phi_{k,k} - \frac{\rho r}{T} \geq 0 \tag{2.15}$$

gives us a possibility to determine all the constitutive functions that in our case form the set

$$\mathcal{Z} = \left\{ \sigma_{ik}, e, s, \Phi_k, j_k^N, Q_k, \Psi, \Psi^*, J_k^S \right\} \tag{2.16}$$

The most effective way to do that is the use of Liu's theorem [8]. For details see [7,10]. For the purpose of the paper we are only interested in determination of mechanical phenomenological properties of the vortex lattice. Therefore, we confine here to description of the magnetomechanical coupling omitting even any thermal influences on the processes occurring in the body.

3. Constitutive and field theory

As we have mentioned before we are mostly interested in mechanical properties of the vortex field vs. magnetic field $H_{c1} < H < H_{c2}$ as it was shown in Fig. 1. Having several possibilities to create proper constitutive law for the stress tensor σ_{ij}, the main quantity of our interest, we choose it as based on isotropic polynomial representations of tensor function of tensor, vector and scalar variables [11, 12]. Since our description is phenomenological and the vortex continuous field results from proper averaging with respect to the characteristic volume in macroscale we assume that the considered lattice is isotropic. This is, of course, rough approximation (the vortex lattice is of the hexagonal symmetry) but sufficient to catch the transfer from the lattice to fluid up and to catch the viscoelastic and then viscous properties of the vortex field up.

Reducing now for our purpose the set (2.2) to

$$C_1 = \{\varepsilon_{ij}, \dot{\varepsilon}_{ij}, B_i\} \tag{3.1}$$

we can create the required representation for σ_{ik} [11,12] as follows:

$$\sigma_{ij} = \beta^1_\sigma \delta_{ij} + \beta^2_\sigma \varepsilon_{ij} + \beta^3_\sigma \varepsilon_{ik}\varepsilon_{kj} + \beta^4_\sigma \left(B_i B_j - B_s B_s \delta_{ij}\right) + 2\beta^5_\sigma \epsilon_{kjl}\varepsilon_{ik} B_l +$$

$$+ \left(\beta^6_\sigma \epsilon_{iks}\epsilon_{ljp} B_s + 2\beta^7_\sigma \epsilon_{ljp}\varepsilon_{ik} + 2\beta^8_\sigma \epsilon_{ikp}\left(B_j B_l - B_s B_s \delta_{lj}\right)\right)\varepsilon_{kl} B_p +$$

$$+\beta^9_\sigma \delta_{ij} + \beta^{10}_\sigma \dot{\varepsilon}_{ij} + \beta^{11}_\sigma \dot{\varepsilon}_{ik}\dot{\varepsilon}_{kj} + 2\beta^{12}_\sigma \epsilon_{kjl}\dot{\varepsilon}_{ik} B_l +$$

$$+ \left(\beta^{13}_\sigma \epsilon_{iks}\epsilon_{ljp} B_s + 2\beta^{14}_\sigma \epsilon_{ljp}\varepsilon_{ik} + 2\beta^{15}_\sigma \epsilon_{ikp}\left(B_j B_l - B_s B_s \delta_{lj}\right)\right)\dot{\varepsilon}_{kl} B_p +$$

$$+2\beta^{16}_\sigma \epsilon_{ljp}\dot{\varepsilon}_{ik}\dot{\varepsilon}_{kl} B_p + \beta^{17}_\sigma \left(\varepsilon_{ik}\dot{\varepsilon}_{kj} + \dot{\varepsilon}_{ik}\varepsilon_{kj}\right) +$$

$$\beta^{18}_\sigma \left(\varepsilon_{ik}\varepsilon_{ks}\dot{\varepsilon}_{sj} + \dot{\varepsilon}_{ik}\varepsilon_{ks}\varepsilon_{sj}\right)\beta^{19}_\sigma \left(\varepsilon_{ik}\dot{\varepsilon}_{ks}\dot{\varepsilon}_{sj} + \dot{\varepsilon}_{ik}\dot{\varepsilon}_{ks}\varepsilon_{sj}\right), \tag{3.2}$$

where the coefficients β^n_σ can be functions of the invariants

$$\varepsilon_{kk},\quad \varepsilon_{ij}\varepsilon_{ij},\quad \varepsilon_{ij}\varepsilon_{jk}\varepsilon_{ki},\quad -2B_j B_j,\quad \varepsilon_{ij} B_i B_j,\quad -\varepsilon_{kk} B_s B_s,$$

$$\varepsilon_{ij}\varepsilon_{jk} B_i B_k - \varepsilon_{ij}\varepsilon_{ij} B_s B_s,\quad \left(\varepsilon_{ij}\varepsilon_{jk} B_k B_m - \varepsilon_{ij}\varepsilon_{jm} B_s B_s\right)\epsilon_{nir}\varepsilon_{mn} B_r,$$

$$\dot{\varepsilon}_{kk},\quad \dot{\varepsilon}_{ij}\dot{\varepsilon}_{ij},\quad \dot{\varepsilon}_{ij}\dot{\varepsilon}_{jk}\dot{\varepsilon}_{ki},\quad \dot{\varepsilon}_{ij} B_i B_j, -\dot{\varepsilon}_{kk} B_s B_s, \tag{3.3}$$

$$\dot{\varepsilon}_{ij}\dot{\varepsilon}_{jk} B_i B_k - \dot{\varepsilon}_{ij}\dot{\varepsilon}_{ij} B_s B_s,\quad \left(\dot{\varepsilon}_{ij}\dot{\varepsilon}_{jk} B_k B_m - \dot{\varepsilon}_{ij}\dot{\varepsilon}_{jm} B_s B_s\right)\epsilon_{nir}\dot{\varepsilon}_{mn} B_r,$$

$$\varepsilon_{ij}\dot{\varepsilon}_{ji},\quad \varepsilon_{ij}\varepsilon_{jk}\dot{\varepsilon}_{ki},\quad \varepsilon_{ij}\dot{\varepsilon}_{jk}\dot{\varepsilon}_{ki},\quad \varepsilon_{ij}\varepsilon_{jk}\dot{\varepsilon}_{ks}\dot{\varepsilon}_{si},\quad \varepsilon_{ij}\dot{\varepsilon}_{jk}\epsilon_{iks}B_s,$$

$$\varepsilon_{ij}\dot{\varepsilon}_{jk}\epsilon_{irs}\epsilon_{krp}B_sB_p,\quad \varepsilon_{ij}\dot{\varepsilon}_{ls}\epsilon_{jkr}\epsilon_{klp}\epsilon_{sim}B_rB_pB_m,$$

$$\varepsilon_{ij}\dot{\varepsilon}_{jk}\dot{\varepsilon}_{ks}\epsilon_{sir}B_r,\quad \varepsilon_{ij}\varepsilon_{jk}\dot{\varepsilon}_{ks}\epsilon_{sir}B_r.$$

Since we have confined in the paper to the linear theory of the vortex lattice, i.e. its deformation is small and is described by the linear strain tensor

$$\varepsilon_{ij} = \frac{1}{2}\left(u_{i,j} + u_{j,i}\right), \tag{3.4}$$

where u_i denotes the displacement vector, and the velocity of the lattice point $v_i = \dot{u}_i$ is also small, the representation (3.2) is reduced to a linear form

$$\sigma_{ij} = \left(\beta_\sigma^1\left(\varepsilon_{kk}, B_kB_k\right) + \beta_\sigma^9\left(\dot{\varepsilon}_{kk}, B_kB_k\right)\right)\delta_{ij} + \beta_\sigma^2\left(B_kB_k\right)\varepsilon_{ij} + \beta_\sigma^{10}\left(B_kB_k\right)\dot{\varepsilon}_{ij} \tag{3.5}$$

where we have used (3.3). To determine the coefficients β_σ^1, β_σ^2, β_σ^9, and β_σ^{10} we assume that the expression (3.5) should be of the similar form as for a viscoelastic isotropic body [13]. Hence, we propose

$$\sigma_{ij} = 2M\varepsilon_{ij} + L\varepsilon_{kk}\delta_{ij} + 2\mathcal{M}\dot{\varepsilon}_{ij} + \mathcal{L}\dot{\varepsilon}_{kk}\delta_{ij} \tag{3.6}$$

where M, L, $\mathcal{M}$, $\mathcal{L}$ are the material coefficients all of them concerning both the lattice and fluid state. They will be determined in the sequel. Relations between coefficients in (3.5) and (3.6) are as follows

$$\beta_\sigma^1\left(\varepsilon_{kk}, B_kB_k\right) = L\left(B_kB_k\right)\varepsilon_{kk}, \tag{3.7}$$

$$\beta_\sigma^9\left(\dot{\varepsilon}_{kk}, B_kB_k\right) = \mathcal{L}\left(B_kB_k\right)\dot{\varepsilon}_{kk}, \tag{3.8}$$

$$\beta_\sigma^2\left(B_kB_k\right) = 2M\left(B_kB_k\right), \tag{3.9}$$

$$\beta_\sigma^{10}\left(B_kB_k\right) = 2\mathcal{M}\left(B_kB_k\right). \tag{3.10}$$

Since one of the characteristic properties of σ_{ik} is the fact, that its definition should concern both the viscoelastic lattice and the viscous fluid, we split (3.6) into the trace and deviatoric parts:

$$\sigma_o = \sigma_{kk}, \qquad \tau_{ij} = \sigma_{ij} - \frac{1}{3}\sigma_o\delta_{ij}. \tag{3.11}$$

So,

$$\sigma_o = (2M + 3L)\,\varepsilon_{kk} + (2\mathcal{M} + 3\mathcal{L})\,\dot{\varepsilon}_{kk} \tag{3.12}$$

$$\tau_{ij} = 2M\varepsilon_{ij} - \frac{2}{3}M\varepsilon_{kk}\delta_{ij} + 2\mathcal{M}\dot{\varepsilon}_{ij} - \frac{2}{3}\mathcal{M}\dot{\varepsilon}_{kk}\delta_{ij}. \tag{3.13}$$

Now, we assume that the pure lattice exists only for H_{c1} and the pure fluid exists only for H_{c2}. For magnetc field intensities $H_{c1} < H < H_{c2}$ there occurs coexistence of the lattice and fluid. This way the following superpositions hold true:

$$\sigma_o = \sigma_o^{lattice} + \sigma_o^{fluid}, \tag{3.14}$$

$$\tau_{ij} = \tau_{ij}^{lattice} + \tau_{ij}^{fluid}. \tag{3.15}$$

Hence, introducing the following parameters dependent on the first magnetic invariant $-2B_sB_s$ in (3.3) with the use of (2.14) as

$$\alpha = \left(\frac{H_{c2} - H}{H_{c2} - H_{c1}}\right)^2, \qquad \alpha = \begin{cases} 0 & if\ H = H_{c2} \\ 1 & if\ H = H_{c1} \end{cases} \tag{3.16}$$

$$\beta = \left(\frac{H - H_{c1}}{H_{c2} - H_{c1}}\right)^2, \qquad \beta = \begin{cases} 0 & if\ H = H_{c1} \\ 1 & if\ H = H_{c2} \end{cases} \tag{3.17}$$

we can rewrite (3.12), (3.13) with (3.14), (3.15) as

$$\sigma_o^{lattice} = \alpha\,(2\mu + 3\lambda)\,\varepsilon_{kk} + \alpha\,(2\mu_L + 3\lambda_L)\,\dot{\varepsilon}_{kk}, \tag{3.18}$$

$$\sigma_o^{fluid} = -\beta p, \tag{3.19}$$

$$\tau_{ij}^{lattice} = 2\mu\alpha\varepsilon_{ij} - \frac{2}{3}\mu\alpha\varepsilon_{kk}\delta_{ij} + 2\mu_L\alpha\dot{\varepsilon}_{ij} - \frac{2}{3}\mu_L\dot{\varepsilon}_{kk}\delta_{ij} \tag{3.20}$$

$$\tau_{ij}^{fluid} = \beta D\dot{\varepsilon}_{ij}, \tag{3.21}$$

where λ, μ are the Lame constants of the lattice, λ_L, μ_L are the viscoelastic constants of the lattice, p is the pressure of the fluid and D is the viscosity of the fluid. We have assumed in (3.21) that the viscosity during compression is negligible so the viscous coefficient by $\dot{\varepsilon}_{kk}$ vanishes. Introducing now tilt modulus K and shear modulus G both for elastic and viscous states of the vortex lattice as

$$3K = 2\mu + 3\lambda, \qquad 3K_L = 2\mu_L + 3\lambda_L, \tag{3.22}$$

$$G = \mu, \qquad G_L = \mu_L, \tag{3.23}$$

and following [4,16] that

$$K_L = 3D = 3\eta, \tag{3.24}$$

$$G_L = D = \eta, \tag{3.25}$$

where η is defined by (2.1), from (3.6)-(3.25) we finally obtain the required form of the stress tensor σ_{ij}:

$$\sigma_{ij} = \left[\left(\frac{1}{3}\alpha K - \frac{2}{3}\alpha G\right)\varepsilon_{kk} - \frac{2}{3}\alpha\eta\dot{\varepsilon}_{kk} - \beta p\right]\delta_{ij} + 2\alpha G\varepsilon_{ij} + 2(\alpha+\beta)\eta\dot{\varepsilon}_{ij} \tag{3.26}$$

where

$$\alpha + \beta = \begin{cases} = 0 & if\ H = H_{c1}\ or\ H = H_{c2} \\ = f(H) & if\ H_{c1} < H < H_{c2} \end{cases}. \tag{3.27}$$

We have assumed in (3.26) that for K_L there is no mechanical relaxation observed if a medium is in the form of crystal lattice [17]; the viscosity of the vortex lattice deals only with shear [4]. Therefore, we have put $K_L = 0$ in (3.26). The final form of σ_{ij} for $H = H_{c1}$ concerns only the vortex lattice and for $H = H_{c2}$ only the pure fluid. The transfer from the lattice state to the fluid one is continuous, not in the form of a phase transition. That problem is still open because there has not been any convincing proof whether the transfer is smooth or singular [5].

We apply now the proposed tensor (3.26) to the phenomenological field theory for the vortex field resulting from the thermodynamical model presented in the previous section. We confine here only to the magnetomechanical features of the considered field. On using now (2.4), (2.6), (2.10)-(2.14) and the Maxwell equations we obtain (cf. [14])

- *the momentum balance*

$$\sigma_{ik,k} + \epsilon_{ikr}\epsilon_{kls}H_{s,l}B_r = \rho\dot{v}_i, \tag{3.28}$$

- *the first London equation*

$$\frac{\partial j_k^S}{\partial t} = \frac{1}{\Lambda}\left(E_k + \epsilon_{kls} v_l B_s\right), \qquad \Lambda = \mu_o \lambda_o^2 \tag{3.29}$$

- *the Maxwell equations*

$$\epsilon_{ijk} E_{k,j} = -\frac{\partial B_i}{\partial t}, \qquad B_{k,k} = 0, \qquad \epsilon_{ijk} H_{k,j} = j_i^S, \qquad E_{k,k} = 0, \tag{3.30}$$

where λ_o is the London penetration depth. Eliminating from (3.28)-(3.30) E_k, j_k^S, B_k, and σ_{ik} with the help of (3.26) and utilizing (2.14) and (3.4) we arrive at the following set of the looked for field equations:

$$\left[\alpha G + (\alpha+\beta)\,\eta\frac{\partial}{\partial t}\right] u_{i,jj} + \frac{1}{3}\left[(K+G)\,\alpha + (\alpha+3\beta)\,\eta\frac{\partial}{\partial t}\right] u_{j,ij} +$$

$$+ \left[\alpha_{,j} G + (\alpha+\beta)_{,j}\,\eta\frac{\partial}{\partial t}\right](u_{i,j} + u_{j,i}) + \frac{1}{3}\alpha_{,i}\left(K - 2G - 2\eta\frac{\partial}{\partial t}\right) u_{k,k} -$$

$$-\beta p_{,i} - \beta_{,i} p - \mu_o\left(H_{r,i} - H_{i,r}\right) H_r = \rho \ddot{u}_i \tag{3.31}$$

$$\lambda_o^2 H_{i,kk} - H_i + u_{i,k} H_k - u_{k,k} H_i = 0. \tag{3.32}$$

The equations (3.31), (3.32) describe dynamics of the vortex field in the type-II superconductor. Although the elastic properties of that field were modeled within linear theory, the field equations are strongly nonlinear. Basing on the fundamental properties of the magnetic field distribution in a superconductor we suggest that the influence of displacements of vortices on the value and form of the magnetic field is negligible. Hence the "mechanomagnetic" coupling in (3.32) can be omitted. That suggestion shows that the only reason to influence the "mechanics" of vortices is obviously the magnetic field. Such a situation occurs just in the case when pinning is excluded from the considerations.

Acknowledgement
The paper has been supported by DS 21-843/98 grant.

References

1. D.R. Tilley, J. Tilley, *Superfluidity and superconductivity* (Van Nostrand Reinhold Co., New York, 1974).

2. M. Tinkham, *Introduction to superconductivity* (McGraw-Hill Book Co., 1975).

3. T.P. Orlando, K.A. Delin, *Foundations of applied superconductivity* (Addison-Wesley Publ. Co., Reading, 1991).

4. M. Cyrot, D. Pavuna, *Introduction to superconductivity and high-T_c materials* (World Scientific, Singapore 1992).

5. G. Blatter, M.V. Feigel'man, V.B. Geshkenbein, A.I. Larkin,. V.M. Vinokur, *Rev. Mod. Phys.*, **66**, 1125-1388 (1994).

6. E.H. Brandt, *Rep. Prog. Phys.*, **58**,1465-1594 (1995).

7. B. Maruszewski, sl Arch. Mech. **50**, 497-508 (1998).

8. I-Shih Liu, *Arch. Rat. Mech. Anal.*, **46**, 131-148 (1972).

9. A. Schönenberger, A. Larkin, E. Heeb, V. Geshkenbein, G. Blatter, *Phys. Rev. Lett.* (to appear in 1998).

10. B. Maruszewski, *Proceedings of the Second International Symposium on Thermal Stresses and Related Topics* (Rochester Institute of Technology, Rochester, 1997), 691-694.

11. G.F. Smith, *Constitutive equations for anisotropic and isotropic materials* (North-Holland, Amsterdam, 1994).

12. G.F. Smith, *Int.J.Engng.Sci.*, **9**, 899-916 (1971).

13. C. Rymarz, *Continuum mechanics* (PWN, Warszawa), 1993 (in Polish).

14. L. Restuccia, B. Maruszewski,*Proceedings of 9th International Conference on Waves and Stability in Continuous Media, INCOWASCOM'97* (Bari, 1998).

15. B. Maruszewski, *Int.J.Engng.Sci.*, **26**, 1217-1230 (1988).

16. A.H. Cotrell, *The mechanical properties of matter* (Wiley, New York, 1964).

17. J.D. Ferry, *Viscoelastic propertiers of polymers* (Wiley, New York, 1961).

MECHANISM OF INTERACTIONS AND DIFFUSION IN THE FOOD MICROSTRUCTURE SYSTEMS

TADEUSZ MATUSZEK
Technical University of Gdansk
Faculty of Mechanical Engineering
11/12 G. Narutowicza St., Gdansk, Poland

In this paper some aspects of interactions and diffusion in the food microstructure systems, within the processes involved in changing the structure of raw food materials have been analysed. All food systems are composed of ingredients that are physically and chemically balanced to give finally a harmonised product. Water and fat with their possible interaction as well as proteins and lipids, as the most reactive of the major food components for the microstructure objectives have been taken into consideration. Individual cells in the dispersed whole food systems are separated from one another and their environment by a complex membrane structure that serves as a diffusion barrier with all dynamic process aspects. Based on the structure of the membrane and its permeability which differs with respect to the different ions and molecules, a mechanism of the food ingredient's interaction similar to the Drefus' theory have been accounted for. It has been concluded that the water activity that among other things strongly depends on the behaviour of electrolytes as well as on hysteresis of adsorption isotherm, critically determines the keeping properties of food microstructure. It has been suggested that defined well thickness of the interfacial layer and adequate location of the equimolecular surface will better characterise structure formation and its stability process in food microstructure.

1 Introduction

Food belongs to the very complex and dissipative systems. Regarding the mechanism of interaction and diffusion within such a system, it is necessary to say that the properties of food microstructure are not determined completely by their chemical composition, but also to a great extend by the physical structure of various constituents and interaction between them as well.

In general, the overall mechanisms governing all changes at the food processing plant, among other things can be considered through the following aspects:

structural related to the position and orientation of water molecules in relation to each other and to macromolecules, together with their activities within the spatial geometry and energy relationships of this space arrangements under food engineering process operations parameters;

dynamic related to details of molecular motion of water and their contribution to the hydrodynamic properties changes of food systems during the processing time and after it;

thermodynamic related to water in equilibrium with its surroundings at certain values of the relative humidity, pressure and temperature during processing and storage data.

The food technologist and food engineers would like to be able to predict the desire properties of any admixture of food components over a wide range of processing and distribution conditions. In order to make this happen it is necessary to understand more about the factors that affect:

- all levels of complex food structure spatial organisation;
- kinetics competitive of the food microstructure components adsorption at fluid interfaces and their "attachments" for time being to the skeleton of the solid structure, and to be sure finely how do they fit so much in? but also can single chain of proteins (for instance) alone form a simple or other organised supra-molecular systems?
- developing the time-dependent properties of semi-product and final product in the varieties of mixture raw materials components with their sites reaction or at the structure interfaces, and with drastically changes the physical properties of either phase when dispersing at enormous surface area per unit volume (for instance 1 dm^3 could contain between 1 and $3000m^2$);
- fouling problems at various part of the food plant equipment applied that cause a lot of changes of processing performance in many types of membranes because of their barriers resistance requirements as far as expected food quality factors are concerned.

2 Some cases of the food components interaction mechanism

In food structure water and fat are the two main solvents for the food constituents and they determine to a large degree the extent of the possible interactions and resultant diffusion mechanisms. Proteins and lipids are the most reactive of major food components. When we look at it through the microstructure objectives and food product criteria, it can be stated that most of processing operations are strongly related to the above mentioned thermal properties of foods with their structural, dynamics, and thermodynamics aspects. The expected tendency to arrive at any internal equilibrium for the behaviour of food microstructure can be performed on the basis of the two processes, i.e., evaporation with solute effect and condensation with sedimentation effect along with all factors governing the equilibrium in the system comprising them [6]. Based on the distortion mechanism of the food microstructure, the interfacial region between its components is often regarded as a discontinuity surface. In case of evaporation when a rarefaction wave occurs in the system, the bubble flux density distribution on the heating surface is an indirect source of information of the bubble formation processes behaviour and their influence on the mechanism responsible for the surrounding nucleation sites activation within the system. In case of condensation a structural breakdown always presupposes some molecular or microscopic processes for changing the consistency like an effect of Brownian motion and microscopic forces such as disperse, steric, and electrostatic forces. It is presented in Figure 1, and on corresponding to it in Figure 2 and 3 with description of the interaction mechanism at the surface of interface energy equilibrium.

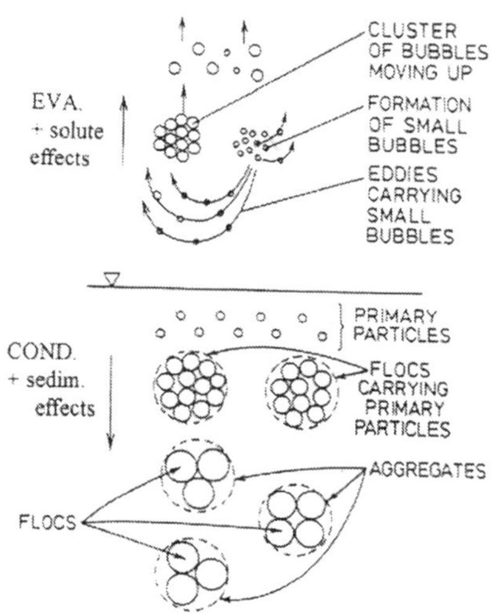

Figure 1. Schematic diagram of the evaporation and condensation phenomena[5].

When dispersing two phases the mechanism of interfaces interaction can drastically change physical properties of either phase. It is known that the interfaces of pure phases usually offer a poor functionality. Therefore, the interface requires suitable physical structure to be involved in any mechanism of its processing changes. It is often said that the surfaces or interfacial active

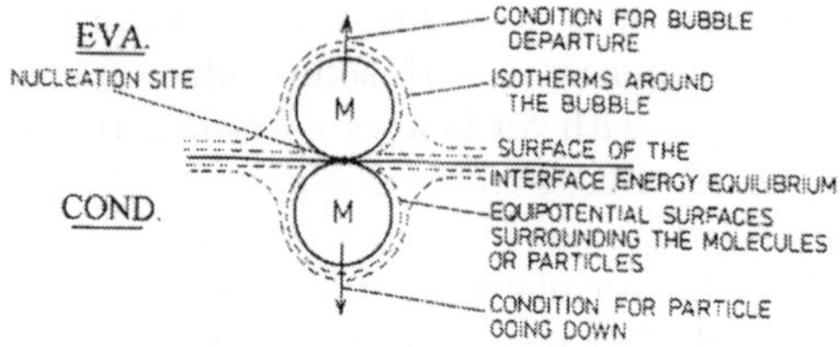

Figure 2. Mechanism of interaction at the interface energy equilibrium surface as regards the molecular transportation phenomena [6].

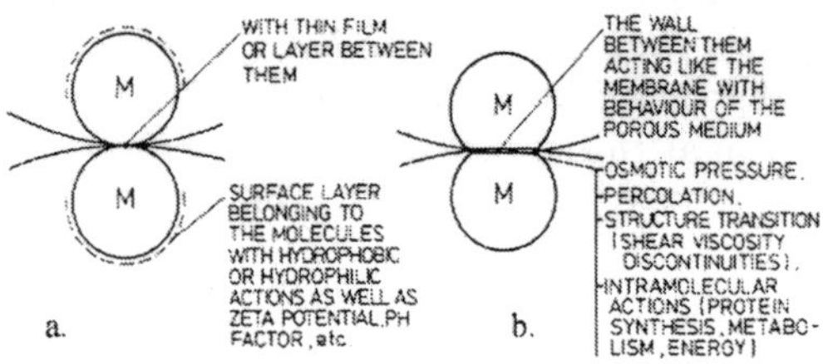

Figure 3. Interaction between molecules; a)in close contact; b)in collision contact with or without damaging effects [6].

molecules are usually considered as an amphipathic cases, i.e., they contain some distinct hydrophobic and hydrophilic regions. Those regions are acting against natural contracting forces. The process of amphipathic molecules adsorption is causing the interfacial packing and next interfacial expansion, and as resultant force of it is known as interfacial pressure. That pressure, under equilibrium is balanced by normal contracting forces of the pure substances interfacial tension. In many cases the value of interfacial tension is mainly responsible for interaction mechanism within food microstructure components, and subsequently strongly influencing the efficiency of food structure formation process. For example, structure stability of dispersions such as foams, emulsions, aerosols, powders and sols required low interfacial tension for efficient engineering process formation. On the other hand, the stability against coalescence and flocculation may require different interfacial properties. In this case the droplets surface potential may sufficiently prevent aggregation and coalescence by charge repulsion. Some steric effects for large surfactant head-groups or proteins may physically prevent coalescence, but may also through bridging effects encourage flocculation. Furthermore, the rheological properties like interfacial visco-elasticity of droplets surface, can determine whether coalescence occurs, particularly in shear flow.

When dilute dispersions are considered, the interaction mechanism and formation of some bi-layers structure from single chain components of amphotropic behaviour requires an element of order to promote the alignment of the molecules in bi-layers and to maintain molecular packing [10]. The order of those dilute dispersion components can also be brought about intermolecular association including those resulting from electrostatic forces, hydrogen bonding, steric interaction, or polymerisation.

In the mixture of water and protein molecules, and with water and surface active material components in both separate cases some mechanisms of film stabilisation can be observed. Based on protein interactions and rapid diffusion of surfactant, the film streching deformation regarding the Gibbs and Marangoni effects is taking place. In mixed system of that both there is no interaction and reduced mobility exists only. There is also lack of streching and the mechanism, as regards protein interaction, is expressed by slow diffusion process only. This mechanism is very useful and important in creating some stability parameters of variety food systems. Among those the fermentation processes are the most known ones.

One possible explanation of the above cases in the interaction mechanism is like that. The component of mixture elements moves along a way from its initial site, in which the stored energy of primary structure is being converted into kinetic energy of motion and also into oscillations about either the displacement centre of motion or average cluster site. This in turn is based on the molecular dynamic forces. The driving force for cluster expansion during any state of equilibrium is represented by the osmotic pressure, which is derived from a molecular theory of water activity. It is known that the different solute molecules cause extensive changes in the orientation of local water molecules and thus affect the water activity. And it is also accepted that water activity and not water content in the food components structure reflects the ability of adsorbed water to support solute mobility [1]. It is assumed that the rates of interaction mechanism changes in food microstructure, and as the resultant of subsequently ongoing intramolecular activities are directly dependent on the molecular mobility of partially hydrated and soluble solute molecules. The water activity, which among other things strongly depends on the behaviour of electrolytes, as well as hysteresis of the adsorption isotherm, critically determined the keeping properties of food microstructure. In case of clusters they may differ substantially in the character of the bonds associated with the complex food system. Cluster provides the ability to attach one or more solvent molecules to the reacting species. The reactions of clusters and the Van der Waals molecules are usually divided into intra- and intercluster processes, and further into neutral and ionic cluster reactions, with such effects as kinematic, dynamic, steric and electronic, and energetic as well [9].

3 Mechanism of diffusion in the food system

The assumption of basic structural elements in molecules themselves and in the molecular interaction contact between them depends on the current state of food microstructure characterisation. The same is related to the molecular energetic level, and to the resultant accuracy of mechanism description. The common expectation within this general area of a diffusion membrane technology, was to predict the selectivity and steady state transport properties of the food microstructure constituents. Moreover, also to establish molecular architecture and membrane chemical composition within this spatial microstructure arrangement. This information would hopefully present in a better understanding of the role of processing energy applied to achieve structure stability. It is also belief that it will determine the size of food structure in many states, and various compositions and concentrations too, as results of adequately recognised mechanism of interactions and diffusion in this system.

In order to explain the changes of the structural food microstructure components, the cell wall of these elements and the surface layer belonging to them have been compared to the geometric structure as a very similar to the porous medium [6], and is presented in Figs 4–6.

In all cases every structure equilibrium assessment was based on the assumption that structures were spherical. For this shape of molecule, the forces at equilibrium are equal in all directions. The interfacial regions are characterised by the presence of density gradients with adequately value related to the diffusion mechanism process. Regarding the mechanism of molecular transportation phenomena based on the interaction and diffusion structure transition effects, it is a matter of importance to pointed out the differences between the active and passive flow through the shunt system existing at the surface and in the wall of the molecules. As it can be seen in Figure 7 the shunt system looks like a set of membranes located inside the cross-section area of the cell wall, and its geometric structure is very similar to the porous medium.

In a whole food microstructure system individual cells are separated from one to another and their environment by a complex membrane structure composed mainly of lipids and proteins. These membranes serve a diffusion barrier

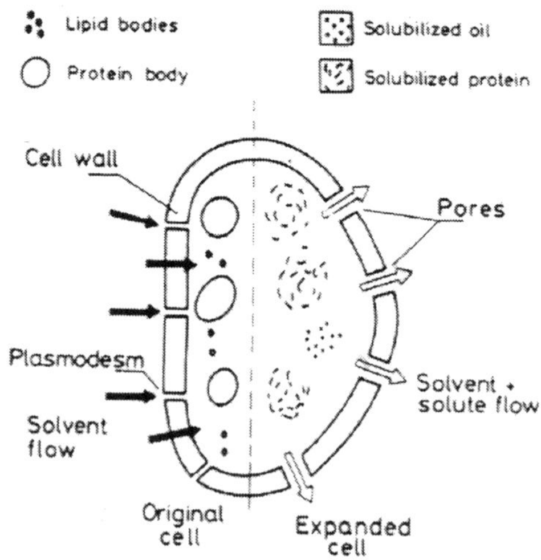

Figure 4. Model for solvent and solute transport from intact cells [10].

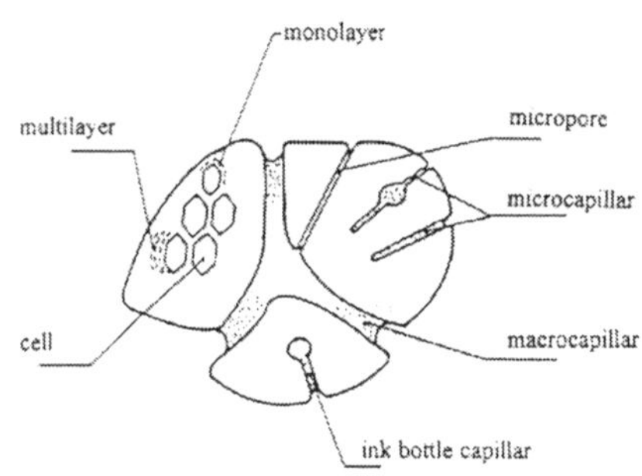

Figure 5. Schematic diagram representing the basic water activity depressing mechanism acting at the microstructure level [11].

with all dynamic process aspects, and contain transport system and receptors that allow a selective interaction of the cell with the environment, both natural and processing one. The structure of the membrane is such that its permeability differs with respect to different ions and molecules. For any given membrane there are substances that do not pass through at all, whereas others

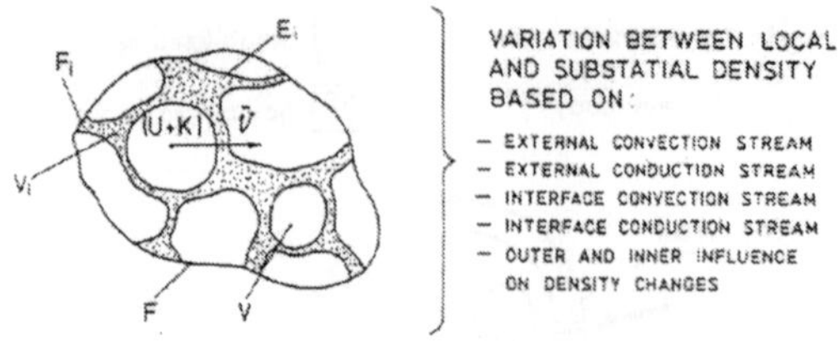

Figure 6. Macroscopic balance and free surface energy of moving system in a porous structure (adopted from Bucewicz [3] with permission and modified).

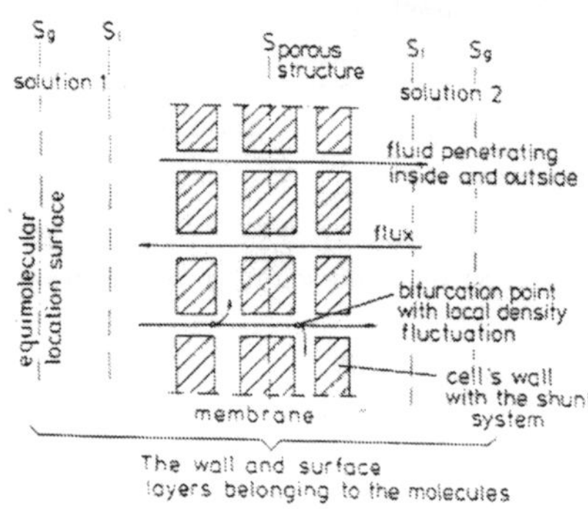

Figure 7. Segment of the porous structure: g-gas, l-liquid, S-equimolecular surface with expected mixture of the homogeneous components.

pass trough at rates that may vary from barely detectable to those comparable with the rate of diffusion in water. Factors affecting the passage of a molecule or ion through a membrane are its size, charge, oil-water partition coefficient, and concentration gradient of the food microstructure components.

Dynamic diffusion mechanism and properties are often related to the exchange processes of molecular constituents between fat, water, and other ingredients interacting at the interfacial regions. But water and fat are again dependent on the interfacial layer fluidity. Therefore, the membrane in this case behaves as if it is pierced by small pores that allow the free passage of water-soluble molecules and of water themselves. Among food microstructure components there are some substances that move in by passive diffusion. However, certain substances are insoluble in the membrane and pass into or out of cells at the rate faster than can be accounted for by passive diffusion. Substances that enter cells in these ways make use of specific membrane diffusion and transport mechanisms.

A number of a molecular-based theory of the micro-phase membrane diffusion separation process mechanism, including also so call excitable membranes, have been recently evolved. The most important of them have been reviewed by Mauritz [7]. The Dreyfus' theory was chosen because it seems to be the most appropriate for description of food ingredients interaction in the dispersed food system. The theory is based on the following assumptions:

- entropy enters the formalism only through the inclusion of a rubber-elastic 'entropy-energy' in the amount of kT,
- specific nature of the dipole ordering process and its mathematical representation,
- volume density of charge in the radial direction from the centre of a cluster,
- principle of calculation is based on maximum monomer packing density.

The Dreyfus' model and the resultant expression for the volume charge density is presented in Figure 8.

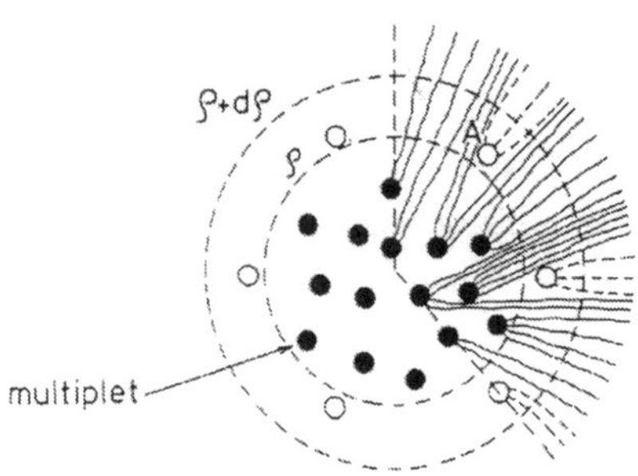

$$\nu g(\rho) = C + \frac{1}{\rho} - \frac{l^2}{2\rho^2}(1+\alpha)\left(1-\exp\left[-\left\{\frac{2\rho}{l(1+\alpha)}\right\}\right]\right)$$

Figure 8. Dreyfus' model for the condensation of multiplets and resultant streching of polymers chains in cluster formation; ν-volume density, g(ρ)-charge density, C-mole fraction of ion pairs, l-length of monomer, ρ-radius, dρ-tickness of newly-added spherical shell, and α-dimensional parameter [7].

The Dreyfus' model was taken because of its strong relation and correspondence to the radial distribution of interaction and diffusion transport concept. The dimensional parameter α could represent the effective fractional increase in the volume of charged groups of the food microstructure components due to tightly bound to them water molecules. The equation is independent of the multiplet order, and the local high density of charge is depicted that gradually decays to the average bulk value.

4 Concluding remarks

As it has been difficult to estimate the irregular surface of the food microstructure components, and too in a porous structure, therefore the quality of food as a whole regarding the relation to interaction and diffusion mechanisms is not yet entirely understood. For the same reason it cannot be presented in the way as expected by those who are involved in the food process engineering plant.

It has been suggested that structural evaporation and condensation effects and their stability under the process conditions are influenced by many factors. Namely, such as molecule shapes, its sizes and frequency, volumes, surfaces, fractional and proportion between the element of the porous structure and interfacial processes there, moisture content, and biochemical, microbiological, enzymatic and water activities.

A workable theory still needs to be developed to establish a quantitative relationships between interaction and diffusion mechanisms related to the molecule transformation phenomena, and to predict the behaviour of time dependence substances under working conditions. In this way, some hope can be expressed to be somewhat closer to an understanding of those effects that are vital to rational food processes design.

References

1. Birch G.G., Lindley M.G. (1986). Interactions of Food Components. Elsevier Applied Sc., London.
2. Brauer H. (1985). Fundamentals of Biochemical Engineering, Vol.2., VCH, Weiheim.
3. Bucewicz A. (1990). Termodynamiczna teoria procesów transportowych w porowatych materiałach budowlanych. Wyd. Pol. Wrocławskiej, 40. 18 (in polish).
4. Gatinol R., Seppecher P. (1986). Modelisation of fluid-fluid interfaces with material properties. J.of Theoretical and Appl. Mech., 225.

5. Matuszek T. (1987). A mathematical approach to the molecular vibration and its structural condensation effects. Proc., 7th World Congress of Food Sc. & Tech., Singapore.
6. Matuszek T. (1990). Dynamic diffusion processes in the porous structure. Proc., 14th Int. Conf. On Vibration in Physical Systems, Poznan, Poland.
7. Mauritz K.A. (1988). JMS-REV., Macromolecular Chem. Phys., C. 28(1), 65–98.
8. Prigogine J., Stuart A.R. (1988). Evolution in Size Effects in Chemical Dynamics, Part 2., Elsevier Applied Sc., London.
9. Ringsdorf H., Schlarb B., Venzmer J. (1988). Angew. Chem. Int. Ed. 27, 113–158.
10. Schneider F.H. (1980). Zur extraktiven Lipid-Freisetzung aus planzlichen Zellen. Fette, Seifer, Anstrichmittel, 82, 16–23.
11. Schneider F.H. (1981). Zur Wasserbestimung in vegetabilen Olrohstoffen. Fette, Seifer, Anstrichmittel, 83, 329–37.
12. Scott S. (1990). Shaping future flavours. Int. Food Ingredients Magazine, 3, 15.

MODELLING OF MOLECULE VIBRATION IN THE DISPERSIVE SYSTEMS

TADEUSZ MATUSZEK, HENRYK OLSZEWSKI
Technical University of Gdansk,
Faculty of Mechanical Engineering

In this work the modeling of molecule vibration and structure network transformation for its stability and possible deformation have been analysed. Consolidation and rarefaction phenomena have been taken into consideration with regard to the interaction between molecules and, during transition effects, some cases of existing processes in the clay, soil, sea floor sand, sandy soil and rocks. The changes of internal friction forces generate changes in volume and shape of structural elements as well as in the stability of intermolecular contact in such dispersive systems. The two-phase soil model, and the demo version of a Working Model 3.0 program were used to build up the soil particles motion equations. In this system for calculation of molecule vibration effects the fourth-order Runge-Kutta algorithm and some numerical techniques for soil particles dynamic problem simulation have been taken into account. Based on data taken from the AP Institute, and modelling of pile-soil interaction, some practical results of the structure deflections have been presented.

1 Introduction

Soil, as an example of the dispersive system, was treated like a set of separate, small, rocky elements together with small amounts of clay particle minerals with weakly connections between them. The vibration of particle-molecules have been analysed. Empty spaces in between porous structure contains partially or is completely filled by water. The shape and the spacing of solid particles are the case of random characteristic. Soil as a porous, multi-phase, grain structure with its stochastic characteristic can be also characterised as a material with large participation of "empty" spaces in which the "empty" volume is comparable with the volume of solid particles of soil frame. The particles of soil frame form the solid phase. Soil contains liquid and gas phase in "empty" spaces as well. In this case it is a three-phase soil. Gravel, sands in aquiferous layers, and in most cases clay and foam are two-phase soils. Two-phase soils are usually saturated with water. Grain structure of soil has a small connections between solid particles of the frame. These connections are made by intermolecular, Van der Waals forces and electrostatic forces which are incomparably small when compared with chemical and ionic bonds of solid bodies.

2 An attempt to soil simulation problem

The models of soil currently applied to the theory of classical mechanics are made in the macro-scale. The same scale of soil models are commonly used for analysis of the laboratory experiments results. The microscopic features of soils well described in the work [1] like: porosity, multiphasity and granularity are used to justify the macroscopic phenomenon in soils, and to characterise the features of soils at macro-scale. The following macroscopic characteristic features of soils can be distinguished [1]:

1. dominant role of plasticity regardless of stress in soil (elastic deformations are smaller then 10^{-5}Pa),
2. relationship between plasticity of soil and changes of soil volume,
3. dilatation,
4. frictional character of resistance of soil.

Stresses in soil are divided and placed into individual phases. In case of two-phase soil with saturated structure the Terzaghi's principle of effective stress is used [4]. According to this principle normal stress σ is an algebraic sum of effective stress σ' and pressure of water in pores $u : \sigma = \sigma' + u$. The effective stress σ' is a ratio of force applied to soil frame and its complete surface. The effective stress is responsible for deformation and boundary conditions of two-phase soil with saturated structure. The normal stress σ determines soil equilibrium. In order to determine the soil equilibrium the pressure of water in the pores must be calculated.

The most characteristic aspect of modelling phenomena in soil is a description of scale used. The following ratio is an indicator of scale problem:

$$\frac{l_0}{\overline{d}} \tag{1}$$

where:

l_0 — characteristic dimension of structure laying on the soil (i.e. width of foundation footing). This structure is usually seen as a boundary value problem of soil calculations.

$\overline{d}$ — average, substitute diameter of soil particles.

The estimation of this ratio is very difficult, and in many cases is not possible. The ratio of $l_0/\overline{d}$ is in the range of $10^5 \div 10^8$ for clay and loam (Fig. 1). The value of this ratio is comparable to the scale value which takes place in the steel constructions. For example between thickness construction elements, and average substitute diameter of metallic crystal. In case of sands the ratio of $l_0/\bar{d}$ is in the range of $10^3 \div 10^5$, and is comparable with the relation between dimensions of concrete particles and dimensions of reinforced concrete construction. The scale effect in the micro-scale modelling of soil causes sometimes nonsensical results, similar to those observed in constructional materials which have homogeneous properties.

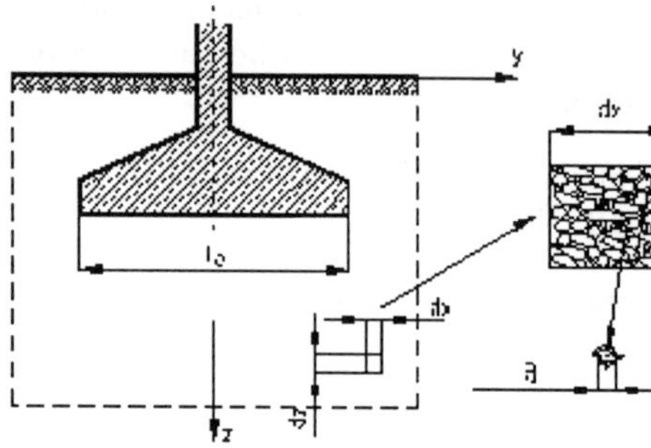

Figure 1. Modelling of soil-scale problem [2]

Exceptions to the above mentioned rule and comments on it, are some analysis with the scale ratio in the range of $1 \div 10^2$ (for example results in the research of coarse-grained soils [11, 13]). In this case the probabilistic models are much more effective when compared to microscopic models.

3 Microscopic model of soil

The solid phase from the particles of clay and rocks has been modelled as a two-phase soil. The model of soil is a porous, multi-phase structure with homogeneous and mixed characteristic, placed under the gravity forces (Fig. 2). Particles were connected by linear and nonlinear spring-damping elements with random characteristic of:

- spring force:

$$F = kx^3, \quad F = kx^2, \quad F = kx, \quad F = k, \quad F = \frac{k}{x}, \quad F = \frac{k}{x^2} \tag{2}$$

- damping force:

$$F = cv^3, \quad F = cv^2, \quad F = cv, \quad F = c, \quad F = \frac{c}{v}, \quad F = \frac{c}{v^2} \tag{3}$$

where:

k — stiffness coefficient
c — damping coefficient
x — deformation
v — velocity of deformation

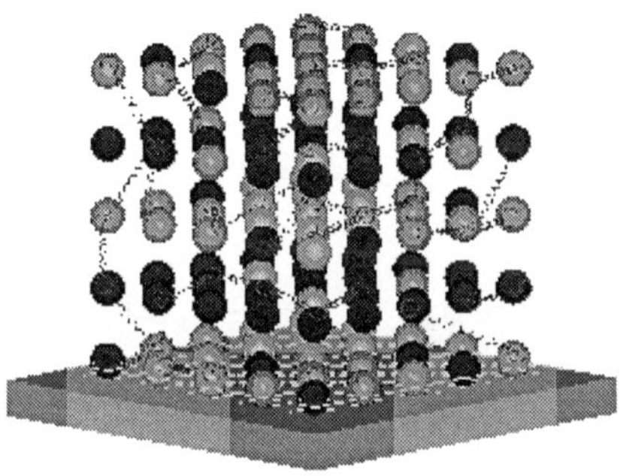

Figure 2. The mixed various layers of soil with particles connected by linear and nonlinear spring-damping elements (isometric view)

Based on the model, first described by Buczkowski and Kleiber [4], and later on modified by Matuszek [9], the contact phenomena between soil particles for their motion and deformation changes were considered, regarding the interaction of the spring and damping forces within this system. Relevant internal friction forces inside the soil will generate changes in volume and shape of structural elements as well as in the stability of intermolecular contact in such dispersive systems.

This two-phase soil model was further taken as an example of the spatial mechanical system arrangements, with solid particles connected with spring-damping elements. The dynamic analysis of the mechanical system requires solution of the motion equations. The equations of molecules motion were solved numerically. The fourth-order Runge-Kutta algorithm was used for solving ordinary differential equations with larger step size and greater accuracy. Since the algorithm is fourth-order one, the truncation error remains relatively small even for a relatively large step size. The major disadvantage of this algorithm is that the function $f(y,t)$ must be evaluated four times at each time step. In addition, the values of the function are not used in any subsequent computation.

The demo version of Working Model 3.0 program was used to build the equations of motion of the soil model and some dynamic problem of particles-molecules using numerical techniques have been solved. The box of soil with dimensions 1mm*1mm*1mm of five soil mixed layers with clay and rock particles was lowered from 0.1mm height. The time dependent deformation of structure during simulation was calculated. The deformation of soil was measured during simulation after 0.1s, 0.24s, 0.4s, 0.56s (Fig 3).

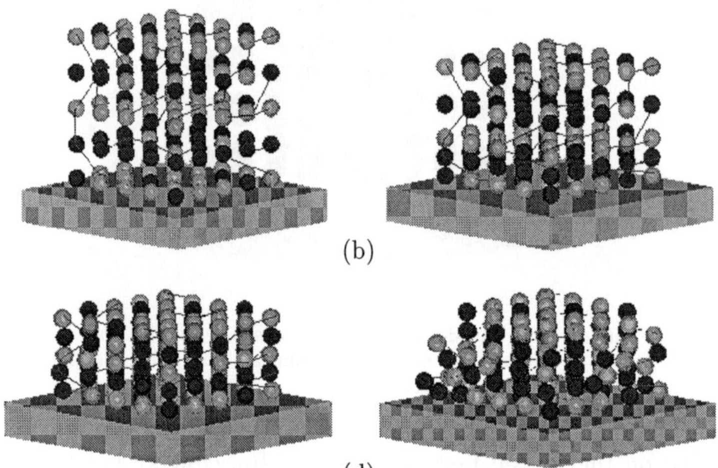

(a) (b) (c) (d)

Figure 3. The deformation of soil after: a) 0.1s, b) 0.24s, c) 0.4s, d) 0.56s

4 Case study of pile-soil interaction modelling

The behavior of a pile supported structure was analysed for several static load conditions. Finite deflection of the pile was accounted for. The soil exhibits non-linear force-deformation behavior of both along and transverse to the pile axis. Because of the non-linear behavior of the pile-soil combination, the structure-foundation system stiffness is a function of displacement. In a linear analysis theory the structural stiffness matrix is formed based on undeformed structure, and it does not change even if the structure deforms. In case of significant non-linearity, the stiffness matrix of the deformed shape cannot be known until the deformed shape is known as well, but again the deformed shape cannot be found until the stiffness matrix is found.

Iterative methods were used for solving this type of problems. The method starts with an initial assumption for the displacement. Based on these displacement the stiffness matrix was found. After that the new displacement was also found using the same stiffness matrix, and then another updated stiffness matrix is found again, and the process is repeated until the calculated displacements in an iteration step are the same (within a tolerance), as those from the previous iteration.

The technique described above is not a very practical one for the structure with many degrees of freedom. The case study structure was divided into two parts that had their interface at the pile-head joints. The piles down from the pile-head joints were the non-linear elements, and the structure above of the pile-head joints was linear. The basic role of the structure elements above the pile-head joints was to connect every single part of the piles to each other, and to transfer loads to the pile-heads structure soil system components.

The characteristics of the soil was specified either in terms of basic soil properties (unit weight, shear strength, etc.) or in terms of displacement data (P-Y force-displacement curves and T-Z moment-rotation curves). The axial resistance was also specified in terms of adhesion capabilities, and axial bearing capacities at arbitrary points along the pile.

The P-Y force-displacement curves for lateral soil resistance and T-Z moment-rotation curves for axial resistance were taken as an input for many soil strata. The soil properties related to resistance to axial, lateral and torsional loads applied at the pile-head have been specified. Longitudinal force distributed along the length of the pile and bearing forces both at the end of pile and at intermediate points, have been considered as the forces acting against the axial loads. The resistance was specified in terms of basic soil mechanical properties from which the friction and end bearing capacities were evaluated.

Pile-head lateral loads were seen as a resistance to the normal forces distributed transversely to the pile axis along its length. These resistance were specified in terms of the input P-Y curves. Torsional loads were balanced by shear capacities distributed along the length of the pile.

4.1 Example problem solution

The structure shown in Figure 4 is 50ft. height. The first level of floor is on 20ft. The "legs" of the structure are the tubes with 30" outside diameter and 1" wall thickness. The rest of them are the tubular elements with 16" outside diameter and 0.5" wall thickness. The plates at both level have 0.5" thickness.

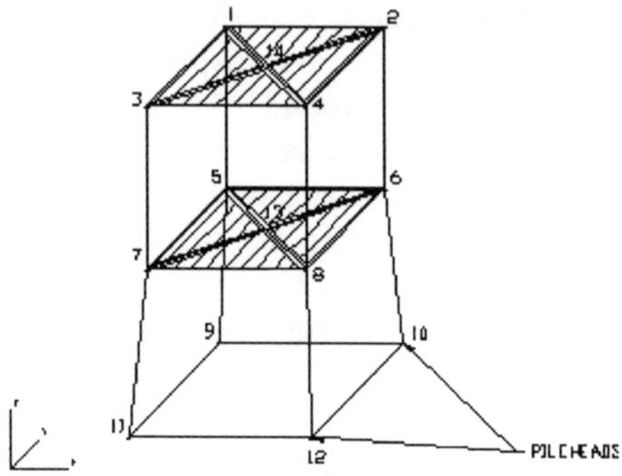

Figure 4. Example of pile structure

The properties of soil were selected from the AP Institute [1]. Two different layers of soil were defined. The first one (with ID SOL1) is between 0 and 50ft. depth. The second one (with ID SOL2) is between 50ft. and 132ft. The calculations and analysis were made for different load conditions. Based on that the pile head forces, and internal forces on structure both in structural coordinates, and force comparison together with deflections of pile heads both in structural coordinates have been achieved and some of them are presented in the Table 1.

Besides, the deflection as well as internal loads and stress of piles with respect to the distance from pile head were analysed. The final results of the structure deflection, including pile-soil interaction for two load cases, are shown in the Figure 5.

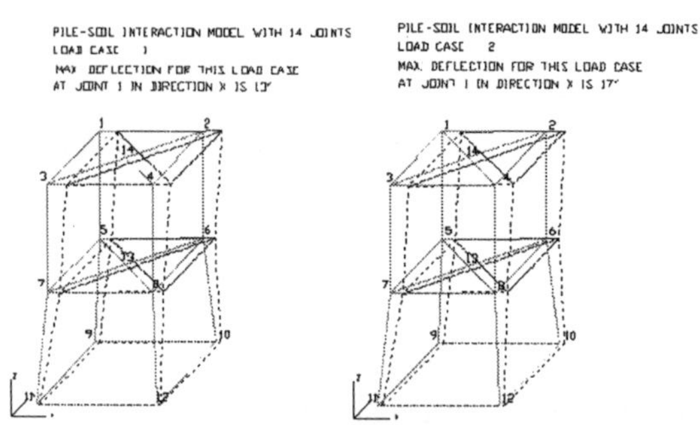

Figure 5. Deflections of the pile structure for two load cases

5 Concluding remarks

Based on the results presented it can be stated that, for the initial pile head loads and their displacement together with intermediate calculation both in pile head coordinates, the progress of iterations for lateral pile solution have been shown.

Next, the final pile head forces within the structure and moments both in pile head and structural coordinates together with comparison the forces in the piles with those in the structure have also been computed. Subsequently, the final pile head deflection as the results from iterative analysis were printed in both pile head and structure coordinates.

It has been assumed that proper modelling of the molecule vibration within the pile head soil layers can strongly influence the results of structure deflection as well as make them more reliable as far as its stability is concerned.

References

1. American Petroleum Institute 17th Edition.
2. Recommended practice 2A-WSD (RP 2A-WSD), 20th Edition, 1993.
3. Bishop A.W.: The principle of effective stress, Teknisk Ukeblad, 39, 1955.
4. Buczkowski R., Kleiber M., Finite element analysis of elastic-plastic-plane contact problem with nonlinear interface compliance. J. of Theoret. and Appl.Mech.,1992,4,30,pp.855–883.
5. Gryczmanski M.: Wprowadzenie do opisu sprężysto-plastycznych modeli gruntów, Polska Akademia Nauk, Studia z zakresu inżynierii nr 40, Warszawa 1995 (in polish).
6. Galka A., Telega J.J., Tokarzewski S.: Nonlinear transport equation and macroscopic properties of microheterogeneous media. Arch. Mech., 49, 2, pp. 293–319, Warszawa 1997.
7. Jarzebowski A.: Constitute modelling of deformation response of unsaturated soils. Compt. Rend. 6-me Colloque Franco-Polonais de Mecanique des Sols Applique, Ses. I.3, pp. 130–137, Douai 1993.
8. Kempa W.: On the description of the consodilation phenomenon by means of a two-component continuum. Arch. Mech., 49, 5, pp.893–917, Warszawa 1997.
9. Matuszek T., Food system network transformation based on the contact phenomena. 1st International Symposium on Food Rheology and Structure, Zurich, 1997,pp.369–373.
10. Piechor K.: Kinetic theory and thermocapillarity equations, Arch. Mech., 46, 6, pp. 937–951, Warszawa 1994.
11. Topolnicki M.: Shear band formation and associated changes of clay structure during isochoric plane strain deformation. Technical University of Gdansk, Gdansk.

12. Topolnicki M.: Nowe aspekty niektorych podstawowych wlaociwooci mechanicznych gruntow spoistych. IX Krajowa Konferencja Mechaniki Gruntow i Fundamentowania, Krakow 1993 (in polish).
13. Topolnicki M., Niemunis A.: Numerical simulation of stress-strain behavior of peat for biaxial deformation at plane strain. Advances in Understanding and Modelling the Mechanical Behavior of Peat, den Haan, Termaat & Edil (eds), pp. 253–266, Rotterdam 1994.

Table 1. * * P i l e G r o u p S u m m a r y * *
group ID = PL1

distance from pile head [ft]	DEFLECTIONS			INTERNAL LOADS			STRESSES				pile head ID	cri- tical load case	maxi- mum unity check
	lateral [in]	axial [in]	rot. [rad]	bending moment [in-kip]	shear [kips]	axial load [kips]	bending stress [ksi]	axial stress [ksi]	shear stress [ksi]	comb. stress [ksi]			
0.0	2.672	-0.043	0.01590	7643.6	73.0	183.9	11.96	2.02	1.60	13.98	9	1	0.536
4.0	1.949	-0.040	0.01429	11118.6	72.7	177.2	17.39	1.94	1.60	19.34	9	1	0.734
7.9	2.547	-0.052	0.01939	15972.8	140.4	214.3	24.99	2.35	3.08	27.34	9	2	1.034
11.9	1.694	-0.048	0.01618	20910.1	60.5	205.0	32.71	2.25	1.33	34.96	9	2	1.316
15.9	1.010	-0.044	0.01247	21804.1	19.6	195.5	34.11	2.15	0.43	36.25	9	2	1.363
19.9	0.503	-0.041	0.00886	19996.5	56.5	186.1	31.28	2.04	1.24	33.32	9	2	1.253
23.8	0.158	-0.038	0.00572	16541.6	82.6	176.8	25.88	1.94	1.81	27.82	9	2	1.048
27.8	0.052	-0.034	0.00323	12440.3	85.8	167.6	19.46	1.84	1.88	21.30	9	2	0.806
31.8	0.160	-0.032	0.00144	8533.3	76.2	158.6	13.35	1.74	1.67	15.09	9	2	0.575
35.8	0.197	-0.029	0.00026	5254.5	60.7	149.7	8.22	1.64	1.33	9.86	9	2	0.380

NONLINEAR THERMOMECHANICS OF ELECTROELASTICITY USEFUL IN SMART MATERIALS AND STRUCTURES

Gérard A. MAUGIN
Université Pierre et Marie Curie,
Laboratoire de Modélisation en Mécanique
(UMR 7607 CNRS)
Tour 66, 4 Place Jussieu
75252 Paris Cédex 05, FRANCE
E-mail : gam@ccr.jussieu.fr

We review the useful notions of forces and generalized forces (stresses, couples) both in the physical framework and in the thermodynamical one (so-called configurational forces) which offer interesting prospects of exploitation in presently developed smart materials and structures. Of necessity this is achieved in a finite-strain, form-invariant formalism that helps conceptually in the approach, especially in so far as the second type of forces is concerned. Applications in the electroelastic framework involve both classical notions of processors and actuators using coupled electromechanical effects as also local computations relevant to the fracture of, and the propagation of phase-transition fronts in, nonlinear electroelastic materials and shape-memory alloys.

1 Introduction

In studying electromagnetomechanical interactions the new basic ingredient introduced is that of *force* exerted by an applied *electric* field on an element of electrically polarized matter (we ignore here magnetization effects). Reciprocally, in certain circumstances the deformation or motion of ponderable or continuous matter causes the production of an electric field (a potential difference) **E** which may serve some practical purpose. Modern adaptive structures and rapidly developing "smart" materials exploit some of these electromechanical interactions in an efficient manner. More generally, we present here the various notions of *forces* which are acting on or can be duly exploited in such materials. The *thermomechanical* background mentioned in the title indicates that we are interested not only in the forces per se, but also in the *mechanical power* these can expend, as well as ultimately in the *dissipation* they produce in an *irreversible evolution*.

The notion of force is multifaceted. It must be understood in the large including:

(i) *classical physical forces*, e.g., forces acting per unit "quantity" - mass, volume or charge – of the material;

(ii) *generalized physical forces*, e.g., stresses, couples, hyperstresses, etc;

(iii) *thermodynamical* or *configurational* forces, which are forces acting on defects or singularities of the electromechanical field, and which acquire a physical meaning through the *power* they expend and the *dissipation* they produce.

All these may be exploited for all practical purposes to a larger or lesser degree (signal processing, sensors, actuators, functional and gradient materials, phase transformations, etc.)

While applications and engineering approaches consider only but the simplest electromechanical couplings such as *linear piezoelectricity*, establishing correctly invariant representations in the case where fields of sufficiently large intensity are present and geometrical nonlinearities must be envisaged, is a necessity. This is the case not only in circumstances which, by definition, involve *nonlinearities* (e.g., electroelasticity of higher order than piezoelectricity, phase transitions, shape-memory alloys) but also whenever small fields are superimposed on bias fields – the latter playing an important role of control in some drastic effects (e.g., stability) – and where geometrical nonlinearities are involved (e.g., in shell-like bodies). Such a general rotationally invariant framework was firmly established in the 1960s-1970s. The interested reader will find the corresponding developments in several treatises [1]-[5]. These require a good understanding of nonlinear continuum mechanics and of basic electromagnetism in the form given by theoretical physicists. The thermodynamical framework then allows one to build a *constitutive theory* according to needs.

Here we shall briefly review the above mentioned three types of forces of interest in the present context. The corresponding detailed analysis is to be found in the cited technical works.

2 Physical forces at regular points

At the most fundamental level we have the *Lorentz force* acting on the elementary point charge α:

$$\mathbf{f}^{\alpha} = q^{\alpha}\mathbf{e}^{\alpha}\left(\mathbf{x}^{\alpha}, t\right),$$
$$\textit{together with Maxwell's "micro" equations},$$
$$\nabla \times \mathbf{e}(\mathbf{x}, t) = \mathbf{0}, \nabla.\mathbf{e} = \sum_{\alpha \in \Delta V} q^{\alpha} ,$$

where the electric field $\mathbf{e}^{\alpha}$ is evaluated at the position of the point charge, and ΔV is *representative volume element*. This force, like Newton's force of gravitation, is acting at the placement point of the charge in *physical space* at actual time instant t, and is thus a *physical* force in its own right. The Lorentz theory of "electrons" either in its naïve volume-average form or in its more sophisticated phase-space average form, is one of the first successful homogenization theory (i.e., passing from a mesoscopic to a macroscopic scale) that provides averaged continuum concepts such as that of *electric polarization* $\mathbf{P}$ per unit volume :

$$\mathbf{P}(\mathbf{x}, t) = \frac{1}{\Delta V}\sum_{\alpha} q^{\alpha}\left(\mathbf{x}^{\alpha} - \mathbf{x}\right) \textit{ such that}$$
$$\mathbf{e}\left(\mathbf{x}^{\alpha}, t\right) = \mathbf{E}(\mathbf{x}, t) + \left(\mathbf{x}^{\alpha} - \mathbf{x}\right).\nabla\mathbf{e}(\mathbf{x}, t) + \ldots,$$
$$\mathbf{E}(\mathbf{x}, t) \equiv \mathbf{e}(\mathbf{x}, t),$$
$$\nabla \times \mathbf{E} = \mathbf{0} \text{ , } \nabla.\mathbf{E} = q^{P} \equiv -\nabla.\mathbf{P}$$

In the electrically neutral bodies that we consider in the *quasi-electrostatic* framework this is the basic material electric property since there is no free-charge density and magnetic phenomena are ignored. The averaging procedure simultaneously provides sensible expressions for macroscopic (physical) *ponderomotive force and couple* (for general nonlinear media where electric field and polarization are not necessarily aligned) and expended *interaction power* per unit volume :

$$\mathbf{f}^{em} = (\mathbf{P}.\nabla)\mathbf{E} = q^{P}\mathbf{E} + div\,\mathbf{t}^{P} \text{ , } with \ \mathbf{t}^{P} = \mathbf{P} \otimes \mathbf{E},$$
$$\mathbf{c}^{em} = \mathbf{P} \times \mathbf{E} = skew\,\mathbf{t}^{P} ,$$
$$w^{em} = \mathbf{f}^{em}.\mathbf{v} + \mathbf{c}^{em}.\Omega + \rho\, h^{em} \text{ , } with$$
$$\Omega = \frac{1}{2}\nabla \times \mathbf{v} \text{ , } \rho h^{em} = \mathbf{E}.\hat{\mathbf{P}} + tr\left(\mathbf{t}^{P}.\mathbf{D}\right), \ \mathbf{D} = (\nabla\mathbf{v})_{S} .$$

The first three quantities, a vector, an axial vector or skewsymmetric tensor, and a scalar, are but source terms in the general balance laws of continuum thermomechanics (balances of physical linear and angular momenta, first law of thermodynamics) which are complemented by a statement of the second law of thermodynamics. Here $\mathbf{v}$ is the matter velocity, ρ is the density, $\mathbf{D}$ is the rate of strain, Ω is the vorticity and the superimposed caret denotes a *convective (Lie) time derivative.* From the above one can introduce the *electromagnetic stress stensor* $\mathbf{t}^{em}$, the *free field* (symmetric) Maxwell stress tensor $\mathbf{t}^{em.f}$, the electromagnetic traction $\mathbf{T}^{em}$ at an oriented surface of unit normal **n,** another form of the electromagnetic bulk energy source, the *quasi-electrostatic Poynting vector* $\mathbf{S}^{e}$, and the Eulerian electrostatic potential $\varphi(\mathbf{x},t)$ by

$$\mathbf{f}^{em} = div\,\mathbf{t}^{em}\;,\mathbf{t}^{em} = \mathbf{t}^{em.f} + \mathbf{t}^{P},\; \mathbf{t}^{em.f} \equiv \mathbf{E}\otimes\mathbf{E} - \frac{1}{2}E^{2}\,\mathbf{1},$$

$$\mathbf{T}^{em} = \mathbf{n}.\left[\mathbf{t}^{em}\right],$$

$$w^{em} = -\frac{\partial}{\partial t}\left(\frac{1}{2}E^{2}\right)_{x} - \nabla.\left(\mathbf{S}^{e} - \mathbf{v}\left(tr\,\mathbf{t}^{P}\right)\right),$$

$$\mathbf{S}^{e} = -\varphi\frac{\partial(\mathbf{E}+\mathbf{P})}{\partial t}\;,\;\mathbf{E} = -\nabla\varphi\;.$$

Like in all continuum theories the final outcome of this type of approach is the formulation of a local statement of the second law known as the *Clausius-Duhem inequality.* This is a constraint imposed on any thermo-*dynamical* evolution with possible large excursions outside equilibrium.

$$-\rho\left(\dot{\psi} + \eta\dot{\theta}\right) + tr\left(\mathbf{t}(\nabla\mathbf{v})^{T}\right) + \rho\mathbf{E}.\dot{\pi} + \theta\,\mathbf{q}.\nabla\theta^{-1} \geq 0\,,$$

where ψ, η and θ are respectively the free (Helmholtz) energy , the entropy - both per unit deformed mass -, and the thermodynamical temperature, $\mathbf{q}$ is the heat flux vector, π is the electric polarization per unit mass in the actual configuration, and a superimposed dot denotes material time differentiation.

Typical reversible (except for heat conduction) behaviors from this type of approach are those of coupled *thermo-electroelasticity* in the finite-strain framework, that are appropriately described by a free energy density (per unit reference volume), material density at the reference configuration, and *material* heat-flux vector of the form

$$
\begin{aligned}
W &= \overline{W}\left(\mathbf{C}, \overline{\mathbf{E}}, \theta; \mathbf{X}\right), \\
\rho_0 &= \overline{\rho}_0(\mathbf{X}), \\
\mathbf{Q} &= \overline{\mathbf{Q}}\left(\nabla_R \theta, \mathbf{C}, \overline{\mathbf{E}}, \theta; \mathbf{X}\right),
\end{aligned}
$$

where the four arguments in W are, respectively, the Green-Cauchy finite-strain tensor, the *material* electric field, the temperature, and material coordinates -whenever the material is inhomogeneous. $\nabla_R\theta$ denotes the material gradient of temperature. Such forms are sufficiently general to account for the *nonlinear electroelasticity* of many materials, including *shape-memory alloys with electric properties and nonconvex potential*, nonlinear electroelasticity useful in *nonlinear* signal processing (convolution and correlation operations), and *electroelastic polymers.* A comprehensive theory of electromechanical *phase transitions* in solids must be based on these expressions. Corresponding nonlinear irreversible processes may include viscosity, plasticity, damage, electric relaxation and electric hysteresis in addition to heat conduction. While the phenomena of viscosity, plasticity and damage are now well apprehended on a sound thermodynamical basis thanks to the theory of so-called *internal variables of state* [6] and appropriate mathematical arguments of convexity, the same tools are now being successfully applied to *electric relaxation* (of interest in the dynamics of electrically polarized ceramics) and *electric hysteresis* (of interest in ferroelectric bodies); cf.[6]-[7].

What must be emphasized in the above is that next to the notions of ponderomotive force and couple – which are at least quadratic in the fields and may be equivalently represented by non-necessarily symmetric stresses of electric origin (akin to Maxwell stresses) – the thermodynamic approach also introduces electromechanical couplings via internal- or free-energy considerations. The latter are reflected in the presence of stress contributions involving the electric field as well. It is only in *linear piezoelectricity* - in some sense a rather elementary theory having the same status as linear elasticity - involving small fields but no bias fields (so-called linearized theory about a virgin state) that the former quantities are neglected – being at least second order – and there remains only the ubiquitous piezoelectric effect, crystal symmetry permitting. In all other cases the more general theory involving nonlinear terms – both physical and geometrical – have to be considered, a fact often overlooked by careless authors. In particular, this applies to applications in nonlinear signal processing [3], in problems of electro-acoustoelasticity, in matters of stability, when strong waves propagate, in phase-transition problems and obviously while examining "piezoelectric" polymers subjected to large deformations. It is also the case in the electro-rheology of fluids [2]. For shock wave studies a detailed presentation of basic equations in the material framework of finite strains and fields is given by Ani and Maugin [8].

3 Configurational forces at singular sets

Typically, configurational forces [9]-[11] are "forces" defined as the thermodynamical conjugates of the displacements (or displacement velocities) of defects or field singularities, such displacements taking place not in physical space but on the material manifold – i.e., the set of material points constituting the body itself. The most illustrative example of such defects or singularities is the *tip of a crack* in extension in a body. While the crack extends irreversibly (we cannot solder back its faces), a thermodynamical force is associated to this extension via the duality inherent in mechanics. As this extension is irreversible, the configurational force acting on the crack – "sucking" the crack in the material – dissipates the body energy. That there is indeed dissipation is shown by infrared thermography [6] which exhibits a localized heat source at the tip of the crack. The accompanying dissipation inequality – due solely to topological changes and not to any intrinsic dissipation of the material in the bulk – then serves as the basis for the construction of a *criterion of progress* of the defect in question. This is the case with the energy-release rate in fracture. In such a process the global dissipation can be expressed as (cf. Dascalu and Maugin in pure elasticity [12] and in electroelasticity [13] -[14]:

$$\Phi = \mathbf{F}.\overline{\mathbf{V}} \geq 0,$$

where **V** with a superimposed bar denotes the material velocity of the crack tip, and **F** is the *global material force* given by (here the integration is performed over a cylindrical surface around the crack tip line),

$$\mathbf{F} = \int_{\partial D_r} \mathbf{N}.\mathbf{b}\, dS \ ,$$

wherein **b** is the *electroelastic Eshelby material stress* defined by

$$\mathbf{b} = \left(W - \frac{1}{2}\rho_0 \mathbf{v}^2\right)\mathbf{1}_R - \mathbf{T}.\frac{\partial \chi}{\partial \mathbf{X}} - \left(\overline{\mathbf{E}} + \overline{\mathbf{P}}\right).\frac{\partial \phi}{\partial \mathbf{X}},$$

in which **T** is the first Piola-Kirchhoff stress, $\chi(\mathbf{X},t)$ is the general motion of the continuum, and $\phi(\mathbf{X},t)$ is the material electrostatic potential. The very structure of **b** shows that it automatically accounts for the material *gradients* of the basic potentials (χ,ϕ) of electroelasticity. The component of **F** along the propagation velocity of the crack tip is the quantity to be compared to a critical value (thoughness or critical energy-release rate) in the thermodynamical formulation of a criterion of progress of the crack. In the case where thermal equilibrium has not been reached yet, an additional (volume-dependent) contribution involving entropy and the material gradient of temperature must be added to the above-given expression of **F**. Such an expression was introduced by Epstein and Maugin [16] under the name of *material thermal force of quasi-inhomogeneity*.

Similar notions basing on a general theory known as the *theory of material inhomogeneities* and *forces* [10]-[11] allow one to find the expression of the configurational forces and their associated dissipation for other singularity sets such as those created by a dislocation, an expanding cavity, and progressing singular fronts of which shock waves and phase-transition fronts are but examples. The essential ingredients there are the balance of so-called *material momentum* (also called *pseudo-momentum* or canonical momentum) and the notion of Eshelby energy-momentum tensor. Like the concept of energy, these two notions capture all fields, whether mechanical or electric, present in the theory. They were duly exploited in electroelasticity, providing thus the expression of useful *driving forces*. The electroelastic fracture problem is dealt with in this manner by Dascalu and Maugin [13]-[14] – see also [15]. The electric Peach-Koehler force on a dislocation is evaluated by Maugin and Epstein [17]. The elementary electroelastic Peach-Koehler force dF^{PK} is found to be given by

$$dF^{PK} = \left(\vec{b}.(\sigma - \mathbf{E}\otimes\mathbf{P})\times\vec{\tau}\right)dl\ ,$$

where $\boldsymbol{b}$ is Burgers's vector, σ is Cauchy's stress, τ is a unit tangent to the dislocation line , and dl an increment along this line . Following the same methodology, the *driving force* on *thermo-electroelastic phase-transition fronts* (e.g., in polarized ferroelastic bodies or shape-memory alloys) is given by Maugin and Trimarco [18]. In the absence of dislocations at the front - so-called *coherent front* -, the front progress is shown to be strictly normal and the driving force, called *Hugoniot-Gibbs force* H_{PT}, is none other than the jump of the double normal component of the « static » part of the electroelastic Eshelby stress tensor expressed on the basis of the *free* energy W - because temperature is continuous at the front - i.e., symbolically,

$$H_{PT} = \mathbf{N}.\left[\mathbf{b}_S(W)\right].\mathbf{N}.$$

The dissipation per unit surface at the front Σ is given by

$$\Phi_\Sigma = -H_{PT}V_N \geq 0 \ ,$$

where V_N is the normal velocity of the front. This corresponds to the presence of a generally nonvanishing localized *hot heat source* at Σ. Whenever we impose that H_{PT} vanishes identically although there is indeed progress of the front, we are in the situation of the *nondissipative* Landau's theory of phase transitions, and the vanishing of the driving force is shown to correspond exactly to the *Maxwell rule of equal areas* (or construction of the *Maxwell line*). This is a *singular* and somewhat irrealistic case in phase-transition theory. Another such *singular* case is found in the study of the propagation of *electroelastic shock waves*, where, in the absence of a true shock structure, the celebrated *Hugoniot relation* is given by the identical vanishing of the driving force H_S where this *Hugoniot functional* is given by (cf.Maugin [19]-[20]) :

$$H_S = \mathbf{N}.\left[\mathbf{b}_S(E)\right].\mathbf{N} \ ,$$

where it is the « static » part of the Eshelby stress based on the electroelastic *internal* energy E that is logically involved. This vanishing condition shows the artificiality of the « classical » shock-wave theory since, to be consistent, as there is indeed *dissipation* across a shock wave (that is, entropy growth), it can be shown exactly that this dissipation is none other than the *power expended* by the said driving force [20].

The driving force on a ferroelectric domain wall or phase boundaries is to be found in like manner, whether these "singular sets" are considered as discontinuity surfaces of zero thickness or they are localized, more or less *solitonic,* structures of non-vanishing thickness (cf. [21]).

4 Comments

In all we have evoked all types of electromechanical 'forces', whether of a physical or configurational nature, respectively at regular material points (per unit volume of surface) and singular sets of material points. For some other electroelastic materials such as ionic crystals (e.g., alkali halides), ferroelectrics , and electroelastic semi-conductors, we need consider more complex phenomenological descriptions (examined in [1] and/or given in works by the author and co-workers, e.g., with J.Pouget [22] and N.Daher [23]). In the case of both ionic crystals and ferroelectrics of the molecular-group type, the *gradient of electric polarization* needs to be taken into account as an additional thermodynamical variable of state. This allows one to exhibit linear electroelastic couplings of a new type in centro-symmetric ionic crystals, to favor the phenomenon of *field localization*, and :o account for ferroelectric ordering in the latter case. In both cases the corresponding configurational forces are modified as they have for essential property to capture strong field gradients (e.g., in transition zones, boundary layers, etc.) -see an above made remark on the structure of the Eshelby stress tensor.

The magnetic analog of what has been exposed herein above has been worked out to a large extent with special attention paid to magnetoelastic couplings (magnetostriction is the most popular one and even takes greater importance with the develoment of materials exhibiting 'giant" magnetostriction) and hysteresis effects (works by G.A.M. and co-workers). In this respect see Sabir and Maugin [24] for the magnetoelastic fracture of para- or soft magnets, and Fomethe and Maugin for the magnetoelastic fracture of hard ferromagnets [25] and the propagation of phase-transition fronts in such materials [26]. The full electrodynamics that couples polarization effects and magnetization ones outside quasi-electro- or magneto-statics (full electromagnetic regime) is accessible but obviously cumbersome and perhaps unnecessarily complicated. Far more interesting for further developments of efficient adaptive materials is the possibility to design, and to evaluate the effective properties of, *composite materials* including electroelastic and/or magnetoelastic components through various techniques of homogenization – in linear electroelasticity see the works of Parton et *al* (1984), Telega (1990), Zhou-Hsieh-Maugin (1986), Benveniste (1992), Bio Wang (1990), T.Chen (1993), Dunn and Taya (1993), and Turbe and Maugin (1991) depending on the technique used ; cf.[7] for references. Artificial *magneto-electric* components (Harshe and Newnham) using electroelastic and magnetoelastic forces as intermediate agents are certainly among the most spectacular and surprising achievements in this line of thought.

Acknowledgments

This contribution was prepared while the author was a visiting professor at Kyoto University, Department of Energy-conversion Science, Graduate School of Energy Science, Sakyo-ku, Kyoto 606-8501 (1997-98). The hospitality offered by Prof. Tatsuo Inoue is gratefully acknowledged.

References

1. G.A.Maugin, *Continuum Mechanics of Electromagnetic solids* (Elsevier, Amsterdam, 1988).
2. A.C.Eringen and G.A.Maugin, *Electrodynamics of Continua*, Vol. I.and II (Springer-Verlag, New York , 1990).
3. G.A.Maugin, J.Pouget. R.Drouot and B.Collet, *Nonlinear Electromechanical Couplings* (J.Wiley and Sons, New York, 1992).
4. D.F.Nelson, *Electric, Optic and Acoustic Interactions in Dielectrics* (J. Wiley and Sons, New York, 1979).
5. H.F.Tiersten, *A Development of the Equations of Electromagnetism in Material Continua*, Springer Tracts in Natural Philosophy, Vol.No.36 (Springer-Verlag, New York, 1990).
6. G.A.Maugin, *Thermomechanics of Plasticity and Fracture* (Cambridge University Press, U.K., 1992).
7. G.A.Maugin, *New Materials in Electromagnetism* (Quaderni del C.N.R., Vol.49, Firenze, Italy, 1994).
8. W.Ani and G.A.Maugin, , *J.Acoust.Soc.Amer.*, **85,** 599 (1989).
9. G.A.Maugin, *Thermomechanics of Nonlinear (complex) Irreversible Behaviours* (World Scientific, Singapore, 1998, in press).
10. G.A.Maugin, *Material Inhomogeneities in Elasticity* (Chapman and Hall, London, 1993).
11. G.A.Maugin, *Appl.Mech.Rev., (A.S.M.E.),* **48**, 213 (1995).
12. C.Dascalu and G.A.Maugin, *C.R.Acad.Sci.Paris*, **II-317**, 1135 (1993).
13. C.Dascalu and G.A.Maugin, *Int.J.Engng.Sci.,* **32,** 755 (1994).
14. C.Dascalu and G.A.Maugin, *Quart.J.Mech.Appl.Math.*, 48, 237 (1995).
15. J.S.Lee, G.A.Maugin and Y.Shindo (editors), *Electromagneto-elastic Materials and Structures*, Vol.AMD-61 (A.S.M.E., New York, 1993).
16. M.Epstein and G.A.Maugin, *C.R.Acad.Sci.Paris*, **II-320**, 63 (1995).
17. G.A.Maugin and M.Epstein, *Proc.Roy.Soc.Lond.*, **A433**, 299 (1991).
18. G.A.Maugin and C.Trimarco, *Math.Mech.Solids*, **2,** 199 (1997).

19. G.A.Maugin, *ARI (Spinger)*, **50,** 41 (1997).
20. G.A.Maugin, *ARI (Springer)*, **50,** 141 (1998).
21. G.A.Maugin and T.Inoue, in *Proceedings of the 3rd EUROMECH-MECAMAT Symposium*, Oxford, UK, November 1998 (to be published by Editions de Physique, Paris, 1999).
22. G.A.Maugin and J.Pouget, *J.Acoust.Soc.Amer.*, **68**, 575 (1980)
23. N.Daher and G.A.Maugin, *Int.J.Engng.Sci.*, **25**, 1093 (1987).
24. M.Sabir and G.A.Maugin, *Int.J.Non-linear Mechanics*, **31**, 425 (1996).
25. A.Fomethe and G.A.Maugin, *Int.J.Non-linear Mechanics*, **33**, 85 (1998).
26. A.Fomethe and G.A.Maugin, *Int.J.Appl.Electrom.Mech.*, **8,** 143 (1997).

THERMODYNAMICAL MODELS OF CONTINUAL DESCRIPTION OF THE COUPLED PROCESSES IN THIN-FILM SYSTEMS

T.NAGIRNY

Centre of mathematical modelling IAPMM NAS of Ukraine,
15, Dudaev's str., Lviv, 290005, Ukraine, phone (0322) 74 11 68
E-mail burak@cmm.lviv.ua

Y.BURAK

Centre of mathematical modelling IAPMM NAS of Ukraine and
Rzeshuwe Politechnical Institute, Department of constraction and ingeneering,
6, Powstatsow Warshawy Str., 35-959 Rzeshuwe, Poland

The basic equations of local gradient thermomechanics are reduced. It shows, that account of gradient of chemical potential enables to describe interface phenomena in three-dimensional approach as well as possibility of investigation the size effect of strength and plasticity points. The environment of body and admixtures influence of nearsurface inhomogenety together with surface tension and breaking load of layer are under consideration.

1 Introduction

It is important and interesting subject to analyze thermomechanical problems in thin films and thin-films systems. It's well known that such elements of the fields gradientality in the equilibrium state are characterized. We use the methods of non-equilibrium thermodynamics and solid mechanics to formulate the set of equations for describing physic-mechanical processes in mechanical structure with such elements. We shall not attempt to compare the approach considered in this paper to different approaches in nonlocal theories of solids which have been developing recently and have studied interface phenomena. We shall formulate the constitutive equations for geometrical linearized approach by using the values per unit volume.

2 Thermoelastic solids

Let us consider a deformable solids where the mechanical and thermal processes are basic.

2.1 Balance equations

If the convective component of time derivative is neglected then the balance equations for these fields are following [1-5]

$$\frac{\partial E}{\partial \tau} = \vec{\nabla} \cdot \left(\hat{\sigma} \cdot \vec{v} - T\vec{J_s} - H\vec{J_m} \right),$$

$$\frac{\partial \vec{k}}{\partial \tau} = \vec{\nabla} \cdot \hat{\sigma},$$

$$\frac{\partial S}{\partial \tau} = -\vec{\nabla} \cdot \vec{J_s} + \sigma_s,$$

$$\frac{\partial \varrho}{\partial \tau} = -\vec{\nabla} \cdot \vec{J_m}. \tag{1}$$

The first one of these equations represents the total energy E conservation law for open thermodynamical systems. E is superposition of internal U and kinetic K energies

$$E = U + K \tag{2}$$

and $\hat{\sigma}$ is the Cauchy stress tensor; $\vec{v}$ is the velocity vector; S, T, H are entropy, absolute temperature and chemical potential; $\vec{J_s}$, $\vec{J_m}$ are the vectors of entropy and mass fluxes; $\vec{\nabla}$ is the Hamilton vector-operator; τ is time; "$\cdot$" denotes the scalar product. The trace of (1-2) represents the conservation law for mechanical motion momentum $\vec{k}$. The Eqs.(1-3),(1-4) represent the conservation law for entropy and mass corresponding; σ_s is entropy production for which we have according to the second law of thermodynamics [6-9]

$$\sigma_s \geq 0. \tag{3}$$

For the classical model of thermoelastic body $\vec{J_s}$ is the flux of continuum points relative to the mass centre. Therefore

$$\vec{J_m} = 0 \tag{4}$$

and

$$\frac{\partial \varrho}{\partial \tau} = 0 \qquad \Longrightarrow \qquad \varrho = \varrho_* = const. \tag{5}$$

In this case the kinetic K and internal U energies are defined in a space of parameters

$$K = K\left(\vec{k}\right), \qquad U = U\left(S, \hat{e}\right), \tag{6}$$

where $\hat{e}$ is the strain tensor

$$\hat{e} = \frac{1}{2}\left[\vec{\nabla} \otimes \vec{u} + \left(\vec{\nabla} \otimes \vec{u}\right)^{T}\right] \tag{7}$$

and $\vec{u}$ is the displacement vector; $\otimes$ is the tensor product; superscript "T" denotes transposition.

The changes of dU, dK are expressed as

$$dU = TdS + \hat{\sigma} : d\hat{e}, \qquad dK = \vec{v} \cdot d\vec{k}. \tag{8}$$

Unlike the classical formulation [6-9] for the local gradient model of thermoelastic body we suppose [1-5]

$$\vec{J}_s = -\frac{\partial \vec{\Pi}_s}{\partial \tau} + \vec{j}_s, \qquad \vec{J}_m = -\frac{\partial \vec{\Pi}_m}{\partial \tau}, \tag{9}$$

where $\vec{\Pi}_s, \vec{\Pi}_m$ are the vectors of entropy and mass displacements; $\vec{j}_s$ is the irreversible component of $\vec{J}_s$. Vector $\vec{J}_m$ does not contain irreversible component $\vec{j}_m$ because we consider one component solids.

Substituting (9) into (1) gives the equations of mass and entropy conservation

$$\frac{\partial}{\partial \tau}\left(\varrho - \vec{\nabla} \cdot \vec{\Pi}_m\right) = 0,$$

$$\frac{\partial}{\partial \tau}\left(S - \vec{\nabla} \cdot \vec{\Pi}_s\right) + \vec{\nabla} \cdot \vec{j}_s = \sigma_s. \tag{10}$$

This formulae distinguish from ones for classical mathematical model of thermoelastic body by the second terms in brackets.

2.2 Constitutive equations

With allowance for (2) and (9) we can write (1) as

$$\frac{\partial K}{\partial \tau} + \frac{\partial U}{\partial \tau} = \vec{v} \cdot \frac{\partial \vec{k}}{\partial \tau} + T\frac{\partial S}{\partial \tau} + H\frac{\partial \varrho}{\partial \tau} + \vec{\nabla} H \cdot \frac{\partial \vec{\Pi}_m}{\partial \tau} + \vec{\nabla} T \cdot \frac{\partial \vec{\Pi}_s}{\partial \tau} + \hat{\sigma} : \frac{\partial \hat{e}}{\partial \tau} -$$

$$-T\sigma_s - \vec{\nabla} T \cdot \vec{j}_s. \tag{11}$$

We define the change dK of kinetic energy K by expression (8) and we postulate the classical form for the entropy production σ_s

$$\sigma_s = -\vec{j}_s \cdot \frac{\vec{\nabla} T}{T}. \tag{12}$$

From (11) according to (12) and to the potential description we get

$$dU = TdS + Hd\varrho + \vec{\nabla} H \cdot d\vec{\Pi}_m + \vec{\nabla} T \cdot d\vec{\Pi}_s + \hat{\sigma} : d\hat{e}, \tag{13}$$

which is generalization of Gibbs' relation for local gradient approach. Therefore we get the set of basic state parameters

$$S, \varrho, \vec{\Pi}_m, \vec{\Pi}_s, \hat{e}. \tag{14}$$

In this space (space of basic state variables) the internal energy U is the potential function

$$U = U\left(S, \varrho, \vec{\Pi}_m, \vec{\Pi}_s, \hat{e}\right). \tag{15}$$

According to (13) the conjugate for the basic parameters are

$$T, H, \vec{\nabla} H, \vec{\nabla} T, \hat{\sigma}. \tag{16}$$

Base parameters determine the conjugate parameters by such state equations

$$T = \frac{\partial U}{\partial S}, \quad H = \frac{\partial U}{\partial \varrho}, \quad \vec{\nabla} H = \frac{\partial U}{\partial \vec{\Pi}_m}, \quad \vec{\nabla} T = \frac{\partial U}{\partial \vec{\Pi}_s}, \quad \hat{\sigma} = \frac{\partial U}{\partial \hat{e}}. \tag{17}$$

Using Legandre transformation

$$F = U - TS - H\varrho - \vec{\Pi}_m \cdot \vec{\nabla} H - \vec{\Pi}_s \cdot \vec{\nabla} T, \tag{18}$$

instead the chosen set of parameters $S, \varrho, \vec{\Pi}_m, \vec{\Pi}_s, \hat{e}$ we have the set $T, H, \vec{\nabla} H$, $\vec{\nabla} T, \hat{e}$. In the new variables the constitutive relations can be written in the form

$$S = -\frac{\partial F}{\partial T}, \quad \varrho = -\frac{\partial F}{\partial H}, \quad \hat{\sigma} = \frac{\partial F}{\partial \hat{e}},$$

$$\vec{\Pi}_s = -\frac{\partial F}{\partial (\vec{\nabla} T)}, \quad \vec{\Pi}_m = -\frac{\partial F}{\partial (\vec{\nabla} H)}. \tag{19}$$

When the displacements of mass and entropy are absent then

$$\frac{\partial F}{\partial(\vec{\nabla} H)} = 0, \qquad \frac{\partial F}{\partial(\vec{\nabla} T)} = 0$$

and free energy F can't depend on temperature and chemical potential gradients.

Let's notice, that the new parameters have a clear physical interpretation and the old ones do not; however they occur explicitly in the balance laws and that makes them mathematically attractive. In order to develop of the theoretical analysis we shall assume that the temperature field is homogeneous in body under homogeneous action. Therefore we must put the coefficients equal zero that characterize the influence $\vec{\Pi}_s$ on F.

The expression for entropy production (12) is a product of thermodynamical force $\left(-\vec{\nabla}T/T\right)$ and flux $\vec{j}_s$. In non-equilibrium thermodynamics it establishes relation between this flux and force. In a common case we can write

$$\vec{j}_s = \vec{j}_s\left(-\frac{\vec{\nabla}T}{T}; U\right), \qquad \vec{j}_s\,(0; U) = 0. \tag{20}$$

The exploitation of the constitutive equations (19),(20) in the general case of nonisotropic material is difficult and results are of a little restrictive. Therefore we shall consider these equations only for isotropic materials.

In the space of deflections of state parameters

$$\eta = H - H_*, \quad \theta = T - T_*, \quad \hat{e} = \hat{e} - 0 \tag{21}$$

by using a quadratic approximation for F

$$F = F_* - S_*\theta - \varrho_*\eta + \frac{1}{2}\lambda e^2 + \mu\hat{e} : \hat{e} - (3\lambda + 2\mu)\,(\alpha_t\theta + \alpha_m\eta)\, e +$$
$$+\frac{1}{2}\beta_t\theta^2 + \beta_{tm}\eta\theta + \frac{1}{2}\beta_m\eta^2 + \frac{1}{2}\chi_m\vec{\nabla}\eta \cdot \vec{\nabla}\eta \tag{22}$$

we get from (19)

$$\hat{\sigma} = 2\mu\hat{e} + \left[\lambda e - (3\lambda + 2\mu)\,(\alpha_t\theta + \alpha_m\eta)\right]\hat{I},$$
$$s \equiv S - S_* = -\frac{\beta_t}{T_*}\theta - \beta_{tm}\eta + (3\lambda + 2\mu)\,\alpha_t e,$$
$$\varrho - \varrho_* = -\beta_m\eta - \beta_{tm}\theta + (3\lambda + 2\mu)\,\alpha_m e,$$
$$\vec{\Pi}_m = -\chi_m\vec{\nabla}\eta, \tag{23}$$

where S_*, T_*, ϱ_*, H_* are the initial values of S, T, ϱ, H; $\lambda, \mu, \alpha_t, \alpha_m, \beta_t, \beta_m, \chi_m$ are the material constants; $\hat{I}$ is the identity tensor; $e \equiv \hat{e} : \hat{I}$.

For the linear approach kinetic equation (20) has the form

$$\vec{j}_s = -\lambda_s \frac{\vec{\nabla} T}{T}. \tag{24}$$

where λ_s is the thermal conductivity.

The set of equations (1),(7),(9),(23),(24) together with the expression

$$\vec{k} = \varrho_* \vec{v} \tag{25}$$

for mechanical impulse are the full set of linearized equations for isotropic thermoelastic body model at local gradient approach. Such set of equations on $\vec{u}, \theta, \eta$ has the form

$$\mu \nabla^2 \vec{u} + (\lambda + \mu) \vec{\nabla} \left(\vec{\nabla} \cdot \vec{u} \right) - (3\lambda + 2\mu) (\alpha_t \theta + \alpha_m \eta) = \varrho_* \frac{\partial^2 \vec{u}}{\partial \tau^2},$$

$$\frac{\partial}{\partial \tau} \left(-\beta_t \theta + (3\lambda + 2\mu) \alpha_m T_* \vec{\nabla} \cdot \vec{u} - \beta_{tm} T_* \eta \right) - \lambda_s \nabla^2 \theta = 0,$$

$$\chi_m \nabla^2 \eta - \beta_m \eta - \beta_{tm} \theta + (3\lambda + 2\mu) \alpha_m \vec{\nabla} \cdot \vec{u} = 0. \tag{26}$$

As an example of the application Eqs.(26), consider some particular cases for isothermal approach ($T = T_* = const$, $\theta = 0$).

2.3 Interface phenomena. Size effects

Let us consider equilibrium state an isotropic elastic layer (domain $\mid x \mid \leq l$ in the Cartesian coordinates (x, y, z)). In this case linearized set of model equations can be written in the form

$$\frac{d^2 \eta}{dx^2} - \alpha^2 \eta - \alpha_1^2 \sigma = 0,$$

$$\frac{d\sigma_x}{dx} = 0, \qquad \frac{d^2 \sigma_y}{dx^2} = \frac{d^2 \sigma_z}{dx^2} = A \frac{d^2 \eta}{dx^2}, \tag{27}$$

where $\sigma \equiv \hat{\sigma} : \hat{I}$; $\sigma_x, \sigma_y, \sigma_z$ are the components of stress tensor; α^2, α_1^2, A are the constants

$$A = -2\mu\alpha_m \frac{3\lambda + 2\mu}{\lambda + \mu}, \quad \alpha_1^2 = -\frac{\alpha_m}{\chi_m}, \quad \alpha^2 = \frac{\beta_m - 3(3\lambda + 2\mu)\alpha_m^2}{\chi_m}.$$

Eqs.(27) along with the conditions on the layer boundary ($x = \pm l$)

$$\sigma_x = 0, \qquad \eta = \eta_a = H_a - H_* \tag{28}$$

and

$$\int_{-l}^{l} \sigma_y \, dx = 0 \tag{29}$$

discribe mechanical equilibrium of layer in local gradient approach. We suppose that H_* is the characteristic of medium and H_a ($H_a \neq H_*$) is the chemical potential on the body boundary. Condition ($H_a \neq H_*$) reflects interaction of material point and environment which differs at internal body region and at boundary (near surface region of body).

We shall process investigation of this problem neglecting stress (strain) influence on the chemical potential. Howerver if we carried out all studied accurately the regularity of consider fields would be the same. This results are not presented for on other reason than immense size of solution expressions. For this approach from (27)-(29) we get

$$\eta(x) = \eta_a \frac{ch(\alpha x)}{ch(\alpha l)},$$

$$\sigma_x(x) = 0, \quad \sigma_y(x) = \sigma_z(x) = A\eta_a \left(\frac{ch(\alpha x)}{ch(\alpha l)} - \frac{th(\alpha l)}{\alpha l} \right). \tag{30}$$

It is clear from (30) that chemical potential η/η_a equals 1 on the body boundary and decreases at $\mid x \mid \to 0$. In the middle of layer

$$\eta(0) = \eta_a \; \frac{1}{ch(\alpha l)} \tag{31}$$

and, therefore, for thick layers chemical potential approximately equals one for media

$$H \approx H_*. \tag{32}$$

Inhomogenety in distribution of η is of nearsurface kind for such bodies unlike the thin films where

$$H \neq H_* \tag{33}$$

in all region of body.

Hight stress is on the body boundary $x = \pm l$

$$\sigma_y(\pm l) = \sigma_z(\pm l) = A\eta_a \left(1 - \frac{th(\alpha l)}{\alpha l}\right). \tag{34}$$

Surface stress in layer depends on thickness. $\sigma_y(\pm l)$ increase at $l \to \infty$ to the quantity of surface stress for semispace. Therefore the local gradient approach in thermomechanics describes the interface phenomena as well as size effects in elastic body.

2.4 Surface tension

Surface tension f_p acording to [10] is definded as

$$f_p = \int_{x_a}^{l} \sigma_y(x)\, dx, \tag{35}$$

where the point x_a satisfies equation

$$\sigma_y(x_a) = 0. \tag{36}$$

By neglecting the influence of stress on the chemical potential from (30), (36) we get

$$x_a = l\left(1 - \frac{ln(\alpha l)}{\alpha l}\right),$$

$$f_p = \frac{A\eta_a}{\alpha}\left[1 - \frac{1}{\alpha l}\left(1 + ln(\alpha l)\right)\right]. \tag{37}$$

At $l \to \infty$ surface tension f_p increases to one for semispace f_p^{∞}

$$f_p^{\infty} = \frac{A\eta_a}{\alpha}. \tag{38}$$

It is interesting to denote that surface tension as well as surface stress in layer ($\mid x \mid \leq l$) is greiter comparing to ones in cylinder ($\mid r \mid \leq R = l$) and sphere ($\mid r \mid \leq R = l$). In cylinder they are greiter ones in sphere. In terms of surface tensions in semispace f_p^{∞}, layer f_p, cylinder f_p^c and sphere f_p^s we can write

$$f_p^{\infty} > f_p > f_p^c > f_p^s. \tag{39}$$

2.5 Size effects of strength

If the layer is under the force loaded by efforts $\vec{F} = \{0, 2\sigma_y^0 l, 0\}$ then surface stress are describe by the formulae

$$\sigma_y(\pm l) = \sigma_y^0 + A\eta_a \left(1 - \frac{th(\alpha l)}{\alpha l}\right),$$

$$\sigma_z(\pm l) = A\eta_a \left(1 - \frac{th(\alpha l)}{\alpha l}\right). \tag{40}$$

We suppose that theoretical breaking stress (the first classical theory in fracture mechanics) is σ^p. Maximum extending stress is surface stress σ_y at $x = \pm l$. Then according to (40) we can write

$$\sigma^k = \sigma^p - A\eta_a \left(1 - \frac{th(\alpha l)}{\alpha l}\right). \tag{41}$$

For the thick layers

$$\sigma^k = \sigma^p - A\eta_a \equiv \sigma_+. \tag{42}$$

The quantity σ_+ can be interpreted as strength point for load of the thick bodies. Using (42), from (41) we obtain

$$\sigma^k = \sigma_+ + A\eta_a \frac{th(\alpha l)}{\alpha l}, \tag{43}$$

which describes the size effect of breaking load. As follows from equations (42), (43) magnitude of σ^k change between the limits σ^p and σ_+, i.e.

$$\sigma^p > \sigma^k \geq \sigma_+. \tag{44}$$

It is interesting to denote that the same results for the plasticity point are presented.

It can be shown using of the mass conservation for all region of a body that the magnitude of H_a is determined by physical and geometrical body characteristics and the ones of body environment. On this point of view we can tell about the influence of environment on the body fracture.

Now we study the change the exploitation of classical deformation criteria in fracture mechanics yields to which. Constitutive equation (22) for strain tensor $\hat{e}$ can be writed in the form

$$\hat{e} = \frac{1}{2\mu}\hat{\sigma} + \left(\alpha_m \eta - \frac{\lambda}{3\lambda\,(3\lambda + 2\mu)}\sigma\right)\hat{I}, \tag{45}$$

where $\sigma \equiv \hat{\sigma} : \hat{I}$ is the bulk component of stress tensor.

Taking into account that

$$\sigma_x = 0, \qquad \sigma_y = \sigma_y^0 + \sigma_z \tag{46}$$

from (45) we have

$$e_y = \frac{\lambda + \mu}{\mu(3\lambda + 2\mu)}\sigma_y^0 + \frac{\lambda + 2\mu}{2\mu(3\lambda + 2\mu)}\sigma_z + \alpha_m \eta. \tag{47}$$

If the load is absent $\left(\vec{F} = 0\right)$ then

$$e_y = e_y^* = \frac{\lambda + 2\mu}{2\mu(3\lambda + 2\mu)}\sigma_z + \alpha_m \eta.$$

Loading of layer yield to change of deformation on magnitude

$$e_y - e_y^* = \frac{\lambda + \mu}{\mu(3\lambda + 2\mu)}\sigma_y^0. \tag{48}$$

Therefore if the strain criteria of fracture mechanics are base and the stress-strain state of layer without the load we consider as the initial state then size effects are absent. This results correspond to experimental investigations of fracture size effects [11]. Corelation between the stress and strain criteria arrive if in the second case the initial stress-strain state of layer call forth the boundary of layer take into account.

3 Solid solutions

In this case the equation of balance energy E is modified to the form

$$\frac{\partial E}{\partial \tau} = \vec{\nabla} \cdot \left(\hat{\sigma} \cdot \vec{v} - T\vec{J}_s - \sum_i H^{(i)} \vec{J}_m^{(i)}\right), \tag{49}$$

where $i = \overline{0, N}$ ($i = 0$ corresponds the skeleton and $i = \overline{1, N}$ - admixtures); $N + 1$ is a namber of solid solutions components.

For two components solids ($N = 1$) analogously to (9) we suppose

$$\vec{J}_s = -\frac{\partial \vec{\Pi}_s}{\partial \tau} + \vec{j}_s, \qquad \vec{J}_m^{(0)} = -\frac{\partial \vec{\Pi}_m^{(0)}}{\partial \tau} + \vec{j}_m^{(0)}, \qquad \vec{J}_m^{(1)} = \vec{j}_m^{(1)} \tag{50}$$

and

$$\vec{j}_m^{(0)} + \vec{j}_m^{(1)} = 0. \tag{51}$$

In relations (49) vector of admixture mass displacement is absent because we consider admixtures for ideal gas approach.

Acording to (50), (51) the equations of mass conservation can be transformed into the form

$$\frac{\partial}{\partial \tau}\left(\varrho^{(0)}+\varrho^{(1)}-\vec{\nabla}\cdot\vec{\Pi}_m^{(0)}\right)=0,$$

$$\frac{\partial \varrho^{(1)}}{\partial \tau}+\vec{\nabla}\cdot\vec{j}_m^{(1)}=0. \tag{52}$$

If

$$\vec{\Pi}_m^{(0)}=0$$

we would obtain the diffusive equations of classical model of solid solutions.

3.1 Constitutive equations

Taking into account the results of second section together with supposition

$$\sigma_s=-\vec{j}_s\cdot\frac{\vec{\nabla}T}{T}-\sum_i\vec{j}_m^{(i)}\cdot\frac{\vec{\nabla}H^{(i)}}{T} \tag{53}$$

from balance equation of internal energy we get Gibb's relation in the form

$$dU=TdS+\sum_i H^{(i)}d\varrho^{(i)}+\vec{\nabla}H^{(0)}\cdot d\vec{\Pi}_m^{(0)}+\hat{\sigma}:d\hat{e}. \tag{54}$$

In this case free energy

$$F=U-TS-\sum_i H^{(i)}\varrho^{(i)}-\vec{\Pi}_m^{(0)}\cdot\vec{\nabla}H^{(0)} \tag{55}$$

is the potential function in space of the state parameters

$$F=F\left(T,\{H^{(i)}\},\vec{\nabla}H^{(0)},\hat{e}\right), \tag{56}$$

where $\{\cdot\}$ denotes the set of corresponding values, i.e. $H^{(0)}, H^{(1)}$ in (56). For change of F we have 1-differential form

$$dF=SdT+\sum_i\varrho^{(i)}dH^{(i)}+\vec{\Pi}_m^{(0)}\cdot d\left(\vec{\nabla}H^{(0)}\right)+\hat{\sigma}:d\hat{e}, \tag{57}$$

and corresponding state equations

$$S = -\frac{\partial F}{\partial T}, \quad \varrho^{(i)} = -\frac{\partial F}{\partial H^{(i)}}, \quad \hat{\sigma} = \frac{\partial F}{\partial \hat{e}},$$

$$\vec{\Pi}_m^{(0)} = -\frac{\partial F}{\partial(\vec{\nabla} H^{(0)})}, \quad i = 0, 1. \tag{58}$$

The expression (53) for entropy production σ_s is basis for formulating the kinetic equations. Thermodynamical fluxes $\vec{j}_s, \{\vec{j}_m^{(i)}\}$ are functions of thermodynamical forces $(-\vec{\nabla} T)/T, \{-(\vec{\nabla} H^{(i)})/T\}$

$$\vec{j}_m^{(i)} = \vec{j}_m^{(i)} \left(-\frac{\vec{\nabla} T}{T}, -\frac{\vec{\nabla} H^{(0)}}{T}, -\frac{\vec{\nabla} H^{(1)}}{T}; U \right), \quad i = 0, 1$$

$$\vec{j}_s = \vec{j}_s \left(-\frac{\vec{\nabla} T}{T}, -\frac{\vec{\nabla} H^{(0)}}{T}, -\frac{\vec{\nabla} H^{(1)}}{T}; U \right),$$

$$\vec{j}_m^{(i)}(0,0,0;U) = 0, \quad \vec{j}_s(0,0,0;U) = 0. \tag{59}$$

Comparing these results with classical models we can see that in our case the gradient of skeleton chemical potential is present like a thermodynamical force. Denote that constitutive equations of thermoelastic bodies as well as solid solutions can be obtained using the functional methods (method of reological deformable systems) [12, 13].

The set of equations obtained above are the full set of equations for thermoelastic solid solutions model for local gradient approach. Linearized set of equations for isothermal quasistatic approach on $\vec{u}, \eta^{(0)}, \eta^{(l)}$ has the form

$$\mu \nabla^2 \vec{u} + (\lambda + \mu) \vec{\nabla} \left(\vec{\nabla} \cdot \vec{u} \right) - (3\lambda + 2\mu) \left(\alpha_m^{(0)} \eta^{(0)} + \alpha_m^{(1)} \eta \right) = 0,$$

$$\frac{\partial}{\partial \tau} \left(-\beta_m^{(0)} \eta^{(0)} - \beta_m^{(1)} \eta^{(l)} + (3\lambda + 2\mu) \alpha_m^* \vec{\nabla} \cdot \vec{u} + \chi_m \nabla^2 \eta^{(0)} \right) -$$

$$-\nabla^2 \left(\lambda_m^{(0)} \eta^{(0)} + \lambda_m^{(1)} \eta \right) = 0,$$

$$\chi_m \nabla^2 \eta^{(0)} - a_0 \eta^{(0)} - a_1 \eta^{(l)} - a_u \vec{\nabla} \cdot \vec{u} = 0, \tag{60}$$

where $\alpha_m^{(0)}, \alpha_m^{(1)}, \beta_m^{(0)}, \beta_m^{(1)}, \alpha_m^*, \lambda_m^{(0)}, \lambda_m^{(1)}, a_0, a_1, a_u$ are constants.

3.2 Interface phenomena. Size effects

We use obtained relations to study the equilibrium interface phenomena in isotropic elastic layer (domain $|x| \leq l$) under the force loaded by efforts $\vec{F} = \{0, 2\sigma_y^0 l, 0\}$.

For layer we have such boundary conditions

$$\vec{n} \cdot \hat{\sigma} = 0, \quad \eta^{(i)} = \eta_a^{(i)} = H_a^{(i)} - H_*^{(i)}, \quad i = 0, 1 \qquad \text{at} \quad x = \pm l \tag{61}$$

and equilibrium condition

$$\int_{-l}^{l} \sigma_y \, dx = 2\sigma_y^0 l \quad \text{at} \quad y \to \pm\infty, \tag{62}$$

where $\vec{n}$ is the external normal.

Solution of this problem can be written in the form

$$\eta^{(0)}(x) = \eta_a^{(0)} + \frac{\alpha_0^2 \eta_a^{(0)} + \alpha_1^2 \eta_a^{(l)} + \alpha_\sigma^2 \sigma_y^0}{\nu^2 + 2MN\alpha_\sigma^2} \left(\frac{ch(\nu x)}{ch(\nu l)} - 1 \right),$$

$$\eta^{(l)}(x) = \eta_a^{(l)} - a \frac{\alpha_0^2 \eta_a^{(0)} + \alpha_1^2 \eta_a^{(l)} + \alpha_\sigma^2 \sigma_y^0}{\nu^2 + 2MN\alpha_\sigma^2} \left(\frac{ch(\nu x)}{ch(\nu l)} - 1 \right),$$

$$\sigma_y(x) = \sigma_y^0 + M \frac{\alpha_0^2 \eta_a^{(0)} + \alpha_1^2 \eta_a^{(l)} + \alpha_\sigma^2 \sigma_y^0}{\nu^2 + 2MN\alpha_\sigma^2} \left(\frac{ch(\nu x)}{ch(\nu l)} - \frac{th(\nu l)}{\nu l} \right),$$

$$\sigma_z(x) = \sigma_y(x) - \sigma_y^0, \tag{63}$$

where $M, \alpha_0^2, \alpha_1^2, \alpha_\sigma^2, \nu$ are constants and

$$N = \frac{th(\nu l)}{\nu l} - 1.$$

Let notice that ν^{-1} is the characteristic of the nearsurface nonohomogenety region and ν^{-1} amounts to $10^{-2}m$ for body with admixture while such characteristic for body without admixture is α^{-1} and $\alpha^{-1} \sim 10^{-7}m$.

Stress is highest at the layer surface $x = \pm l$. From (50) for the surface stress $\sigma_y(\pm l)$ we obtain

$$\sigma_y(x) = \sigma_y^0 - \left(\alpha_0^2 \eta_a^{(0)} + \alpha_1^2 \eta_a^{(l)} + \alpha_\sigma^2 \sigma_y^0 \right) \frac{MN}{\alpha^2 + 2MN\alpha_\sigma^2}. \tag{64}$$

If σ^p is theoretical breaking stress of material without admixtures (the first classical theory) then according to (64) we can write

$$\sigma_{ad}^{k} = \frac{\sigma^{p}\left(\alpha^{2}+2MN\alpha_{\sigma}^{2}\right)+MN\left(\alpha_{0}^{2}\eta_{a}^{(0)}+\alpha_{1}^{2}\eta_{a}^{(l)}\right)}{\alpha^{2}+2MN\alpha_{\sigma}^{2}}, \tag{65}$$

where σ_{ad}^{k} is the magnitude of breaking load for layer with admixture.

Therefore using formulae (33), (65) we can describe the influence of admixture on breaking point of load. If the influence of stress on the chemical potentials $\{\eta^{(i)}\}$ is neglected then for the thick layers we get

$$\sigma_{ad}^{k} = \sigma^{k} - A\eta_{a}^{(0)} + M\frac{\alpha_{0}^{2}\eta_{a}^{(0)}+\alpha_{1}^{2}\eta_{a}^{(l)}}{\alpha_{0}^{2}-a\alpha_{1}^{2}} \tag{66}$$

where σ^{k} is defined by (21) and is the breaking point of load for layer without the admixture.

References

1. Ya.Y.Burak, T.S.Nagirny, *Appl.mech.* **12**, 12 (1992). (in rus.)
2. Y.Y.Burak, T.S.Nagirny *Physico-chemical Mechanisc of Material*, 4 (1993) (in ukr.)
3. Y.Y.Burak, T.S.Nagirny, O.R.Hrytsyna *Reports of NASU*, 9 (1991) (in ukr).
4. Ya.Burak, T.Nagirny, *Zeszyty naukowe politechniki Rzeszowskiej. Mechanika* **48**, 151 (1996).
5. Ya.Burak, Yu.Govda, T.Nagirny in *Vibrations in Physical systems* (Poznan, Poland, 1996).
6. A.Munster *Thermodunamique de processes irreversible.*- New York: McGraw-Hill. 1969.
7. J.W.Gibbs *On the equilibrium of Heterogeneous substances. Collected Works.*- Vol.1, New Haven. 1948.
8. I.Prigogine *Introduction to Thermodynamics of Irreversible Processes.*- Springfield. 1955.
9. S.R.De Groot *Thermodynamics of Irreversible Processes.*- Amsterdam. 1952.
10. Y.Y.Burak, B.P.Galapats, B.M.Hnidets *Physikomechanical Prosses in Electroconductive Bodies.*- Kyiv: Naukova Dumka. 1978 (in ukr).
11. M.Jono, A.Sigeta., T.Kimura., S.Konzaki *J. Soc. Mater. Sci. Jap.* **41**, 470 (1992).
12. T.S.Nagirny *Reports of NASU*, 2 (1992) (in ukr).
13. T.S.Nagirny *Reports of NASU*, 2 (1997) (in ukr).

LIQUID CRYSTAL THEORY AS AN EXAMPLE OF MESOSCOPIC CONTINUUM MECHANICS

C. PAPENFUSS-, W. MUSCHIK*
INSTITUT FÜR THEORETISCHE PHYSIK
TECHNISCHE UNIVERSITÄT BERLIN
HARDENBERGSTR. 36
D-10623 BERLIN GERMANY

Abstract
The mesoscopic concept in continuum mechanics consists of extending the domain of the balance equations by the set of mesoscopic variables and of introducing a local distribution function of these variables as a statistical element. The mesoscopic balance equations defined on this extended domain are considered. The liquid crystals theory of uniaxial molecules represents one example of mesoscopic continuum mechanics.

1 Introduction: Complex materials

Let us call substances consisting of sperical particles with interactions depending only on the inter-particle distance *simple materials*. However most of the interesting and practically important materials have a more complicated internal structure. We will call these materials with internal degrees of freedom *complex materials*. Some examples of such complex materials are presented in the following.

1.1 Steel

Steel consists of microcrystallites of different orientations of the crystal axes (see figure 1)

Under the action of a mechanical load the microcrystallites can be reoriented. The resulting macroscopic properties are history dependent material properties and after effects.

1.2 Shape memory alloys

These materials show a very spectacular hysteresis effect: If the original shape is deformed at low temperature, and the material is heated up then above a critical temperature the original undeformed sheape is recovered. After cooling down below the critical temperature the material "remembers " the defor-

*E-MAIL: MUSCHIK@PHYSIK.TU-BERLIN.DE

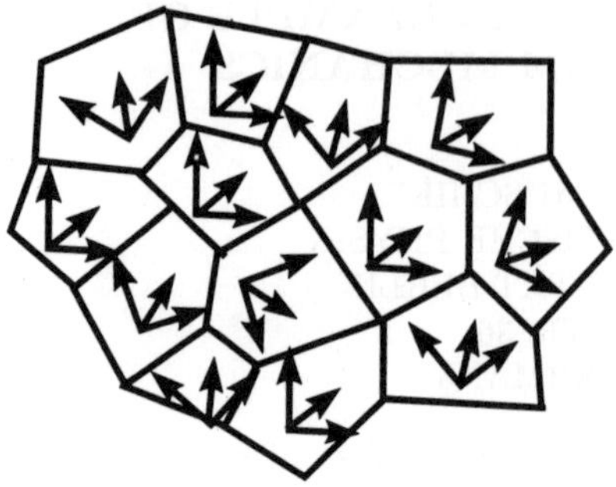

Figure 1. Microcrystallites of different orientations of the crystal axes

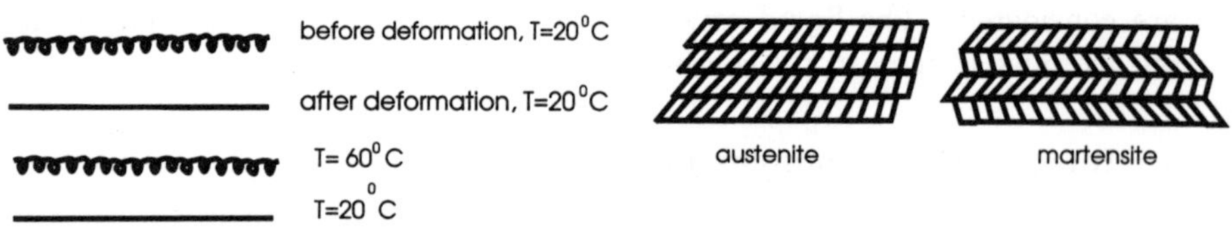

Figure 2. The memory effect observed in certain alloys. The underlaying stress-strain diagrams can be explained by a phase transition between martensite and austenite at a temperature between 20 and 60C.

mation and recovers the deformed shape again. There are several applications of these shape memory alloys, like switches controlled by temperature, and even medical applications [1]. A schematic picture of the hysteresis effect is shown in figure 2. The underlaying internal structure is a microcrystalline material with different orientations of the microcrystallites. In addition the material can undergo a phase transition between martensite and austenite. The phase transition at the critical temperature and reorientation of crystal axes under mechanical load can explain the different stress-strain diagrams at different temperatures resulting in the hysteresis effect.

1.3 Polymer melts and solutions

Polymer molecules are long chains of organic subunits. Polymer melts or polymer solutions with a low molecular weight solvent show non-Newtonian flow behaviour [2,3]. Let us consider as an example a plane Couette flow (see figure 3) with the velocity in the x-direction, the gradient in the y-direction, and the shear rate defined by $\gamma = \frac{\partial v_x}{\partial y}$. In a Newtonian fluid the stress tensor

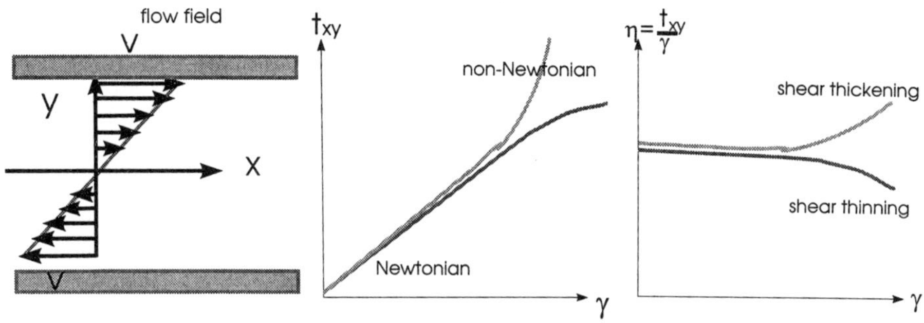

Figure 3. A plane Couette flow. For polymer melts and solutions non-Newtonian behaviour is observed.

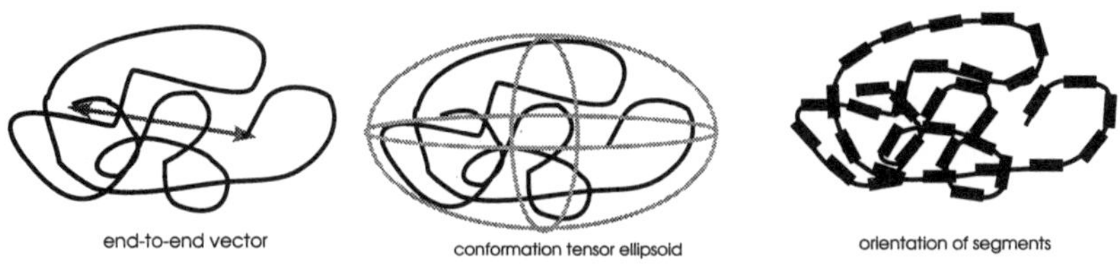

Figure 4. Different levels of description of the conformation of the polymer chain

component t_{xy} depends linearly on the shear rate. In polymer melts and solutions however we observe a nonlinear dependence, i.e. shear thinning or shear thickening, respectively (see figure 3).

This non-Newtonian behaviour can be explained assuming stretching and reorientation of the polymer chains in the flow field. The theoretical description of the iternal degrees of freedom connected with the conformation of the polymer chain is possible on different levels of accuracy (see figure 4). The simplest description is in terms of an end-to-end vector connecting the two ends of the molecule. More information is included in the conformation tensor. The tensor ellipsoid of the conformation tensor is the smallest ellipsoid including the polymer chain. The most elevated description is in terms of the positions and orientations of the N subunits of the polymer molecule.

1.4 Liquid crystals

The liquid crystalline state shows properties of the liquid phase as well as those of the solid phase. For instance liquid crystals show fluid-like non-

Newtonian flow behaviour. On the other hand they are anisotropic like solids, showing birefringence, anisotropic electric and heat conductivity and so on [4,5]. Thermotropic liquid crystals consist of rigid non-spherical particles. In the uniaxial case the particles are rotation symmetric. The axis of rotation symmetry is called microscopic director $\boldsymbol{n}$. The molecules can be rod-like or disc-like. In all liquid crystallie phases there exists an orientational order of the microscopic directors. Liquid crystals consisting of rigid particles without rotation symmetry are called liquid crystals of biaxial molecules. Then the orientation of at least one of the molecular axes shows a long range order in the liquid crystalline phase.

2 Mesoscopic concept

There are two principally different possibilities to deal with complex materials within continuum thermodynamics: The first way is to introduce additional fields depending on position and time. These fields can be internal variables [6,7], order or damage parameters [4,8], Cosserat triads [9], directors [10,11], alignment and conformation tensors [12,13]. The other way is a so called mesoscopic theory. The idea is to enlarge the domain of the field quantities. The new mesoscopic fields are defined on the space $\mathbb{R}_x^3 \times \mathbb{R}_t \times \mathcal{M}$. The manifold $\mathcal{M}$ is given by the set of values the internal degree of freedom can take. Therefore the choice of $\mathcal{M}$ depends on the complex material under consideration (examples see below), but if the internal structure is known, the nature of $\mathcal{M}$ is clear. We will see later that the manifold $\mathcal{M}$ should be such that differentiation and integration is possible on it.

We call this way of dealing with the internal structure of complex materials a mesoscopic concept, because it includes more information than a macroscopic theory on $\mathbb{R}_x^3 \times \mathbb{R}_t$, but no interactions on the molecular level are assumed like in a microscopic approach. The mesoscopic level is between the microscopic and the macroscopic level. The domain of the mesoscopic field quantities $\mathbb{R}_x^3 \times \mathbb{R}_t \times \mathcal{M}$ is called mesoscopic space. In table 1 the physical meaning of the additional mesoscopic variable and of the manifold Mare given for the examples considered in the previous section. The orientation of a triad of crystal axes or molecular axes is determined by a rotation matrix $\in SO(3)$ mapping the actual triad to a reference system.

Beyond the use of additional variables $\boldsymbol{m}$ the mesoscopic concept introduces a statistical element, the so-called *mesoscopic distribution function* (MDF) $f(\boldsymbol{m}, \boldsymbol{x}, t)$ generated by the different values of the mesoscopic variable

steel	orientation of crystal axes of microcrystallites	$SO(3)$
shape memory alloys	fraction of different phases and orientation of crystal axes	$[0,1] \times SO(3)$
polymer solutions	orientation and length of end-to-end vector	$\mathbb{R}^3$ (or $S^2 \times [0,r]$)
liquid crystals of uniaxial molecules	orientation of particle = microscopic director	S^2
liquid crystals of biaxial molecules	orientation of three molecular axes	$SO(3)$

Table 1. The mesoscopic variable and the manifold $\mathcal{M}$ for different complex materials

of the molecules in a volume element

$$f(\boldsymbol{m},\boldsymbol{x},t) \equiv f(\cdot), \quad (\cdot) \equiv (\boldsymbol{m},\boldsymbol{x},t) \in \mathcal{M} \times \mathbb{R}^3 \times \mathbb{R}^1 . \tag{1}$$

The MDF describes the distribution of $\boldsymbol{m}$ in a volume element around $\boldsymbol{x}$ at time t, and therefore it is normalized

$$\int f(\boldsymbol{m},\boldsymbol{x},t)\, d\mathcal{M} \;=\; 1. \tag{2}$$

Now the fields as mass density, momentum density, etc. are defined on the mesoscopic space. For distinguishing these fields from the macroscopic ones we add the word "mesoscopic". Consequently the *mesoscopic mass density* is defined by

$$\varrho(\cdot) := \varrho(\boldsymbol{x},t) f(\cdot) \;. \tag{3}$$

Here $\varrho(\boldsymbol{x},t)$ is the macroscopic mass density. By use of (2) we obtain

$$\varrho(\boldsymbol{x},t) = \int \varrho(\boldsymbol{m},\boldsymbol{x},t)\, d\mathcal{M}. \tag{4}$$

This equation shows, that the system can be formally treated as a mixture of components having the partial density $\varrho(\cdot)$ [14]. Here the "component index" $\boldsymbol{m}$ is a continuous one. Because mixture theory is well developed [15], [16] *mesoscopic balance equations* can be written down very easily [17].

Other mesoscopic fields defined on the mesoscopic space are the *mesoscopic material velocity* $\boldsymbol{v}(\cdot)$ of the particles belonging to the mesoscopic variable $\boldsymbol{m}$ at time t in a volume element around $\boldsymbol{x}$, the *external mesoscopic acceleration* $\boldsymbol{k}(\cdot)$, the *mesoscopic stress tensor* $\boldsymbol{T}(\cdot)$, and the *mesoscopic heat flux density* $\boldsymbol{q}(\cdot)$, etc.

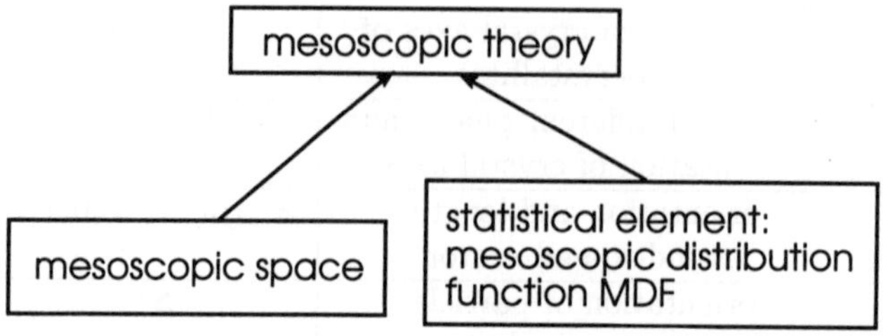

Figure 5. The two main features of a mesoscopic theory are: the domain of the field quantities is enlarged to the mesoscopic space and a distribution function for the values of the mesoscopic variable is introduced.

Macroscopic quantities are obtained from mesoscopic ones as averages with the MDF as probability density:

$$A(\boldsymbol{x},t) = \int_{\mathcal{M}} A(\cdot)f(\cdot)d\boldsymbol{m} \tag{5}$$

This again shows that the complex material can be seen as a mixture of components with different values of the mesoscopic variable.

3 Mesoscopic balance equations

Let G denote a region in $\mathbb{R}^3 \times \mathcal{M}$ and $\boldsymbol{X}$ the density of a balanced quantity. Then the global quantity in the region G changes due to a flux over the boundary of G and due to production and supply within G:

$$\frac{d}{dt}\int_G \boldsymbol{X} d^3x d\boldsymbol{m} = \int_{\partial G} \boldsymbol{\phi}_{\boldsymbol{X}}(\cdot)da + \int_G \Sigma_X(\cdot)d^3x d\boldsymbol{m} \quad . \tag{6}$$

A generalized Reynolds transport theorem in the nematic space [18] is used to transform the time derivative, and Gauss theorem is applied to the boundary integral. This is the point where it is necessary to be able to integrate and to differentiate on the manifold $\mathcal{M}$. Then it results in regular points of the continuum the *local mesoscopic balance* [19]

$$\frac{\partial}{\partial t}\boldsymbol{X}(\cdot) + \nabla_x \cdot [\boldsymbol{v}(\cdot)\boldsymbol{X}(\cdot) - \boldsymbol{S}(\cdot)] + \nabla_m \cdot [\boldsymbol{u}(\cdot)\boldsymbol{X}(\cdot) - \boldsymbol{R}(\cdot)] = \boldsymbol{\Sigma}(\cdot). \tag{7}$$

Here the independent field $\boldsymbol{u}(\cdot)$, defined on the mesoscopic space, describes the change in time of the set of mesoscopic variables: With respect to $\boldsymbol{m}$ the *mesoscopic change velocity* $\boldsymbol{u}(\cdot)$ is the analogue to the mesoscopic material velocity $\boldsymbol{v}(\cdot)$ referring to $\boldsymbol{x}$: If a molecule is characterized by $(\boldsymbol{m}, \boldsymbol{x}, t)$, then for $\Delta t \to +0$ it is characterized by $(\boldsymbol{m}+\boldsymbol{u}(\cdot)\Delta t, \boldsymbol{x}+\boldsymbol{v}(\cdot)\Delta t, t+\Delta t)$. Besides the usual gradient also the gradient with respect to the set of mesoscopic variables appears.

According to equation (3) we obtain from the mesoscopic mass balance (11) a balance of the ODF $f(\cdot)$ by inserting its definition:

$$\frac{\partial}{\partial t}f(\cdot)+\nabla_x\cdot[\boldsymbol{v}(\cdot)f(\cdot)]+\nabla_m\cdot[\boldsymbol{u}(\cdot)f(\cdot)]+ \\ + f(\cdot)\left[\frac{\partial}{\partial t}+\boldsymbol{v}(\cdot)\cdot\nabla_x\right]\ln\varrho(\boldsymbol{x},t)=0. \tag{8}$$

Because this balance equation of the mesoscopic distribution function includes the macroscopic field of the mass density (4), it is not independent of the macroscopic mass balance defined on $\mathbb{R}^3$.

Entropy balance

Besides these mesoscopic balances the entropy balance is neccessary for introducing the second law. Because the production of mesoscopic entropy is not necessaryly positive for each component, the entropy balance is only interesting in its macroscopic form.

$$\frac{\partial}{\partial t}[\varrho(\boldsymbol{x},t)\eta(\boldsymbol{x},t)]+\nabla_x\cdot[\varrho(\boldsymbol{x},t)\eta(\boldsymbol{x},t)\boldsymbol{v}(\boldsymbol{x},t)+\boldsymbol{\phi}(\boldsymbol{x},t)]= \\ =\varrho(\boldsymbol{x},t)\sigma(\boldsymbol{x},t) \tag{9}$$

($\eta(\boldsymbol{x},t)$ = specific entropy, $\boldsymbol{\phi}(\boldsymbol{x},t)$ = entropy flux density, $\sigma(\boldsymbol{x},t)$ = entropy production density). The second law is expressed by the *dissipation inequality*

$$\sigma(\boldsymbol{x},t)\geq 0 \quad . \tag{10}$$

The set of balance equations is not a closed system of equations, but constitutive equations for mesoscopic quantities are needed. The domain of the constitutive mappings is the state space; here a mesoscopic one. There are the possibilities that the mesoscopic state space includes *only* mesoscopic quantities, or that it includes mesoscopic *and* macroscopic quantities. An example of the second kind is discussed for liquid crystals in [20]. Consitutive equations have to be such that the second law of thermodynamics is fulfilled by any solution of the balance equations with the constitutive equations inserted.

The balance equations to be considered in any case are the balances of mass, momentum, and energy. Depending on the internal degrees of freedom additional balance equations have to be taken into account. In the example of liquid crystals this will be the balance of internal angular momentum, connected to rotations of the molecules.

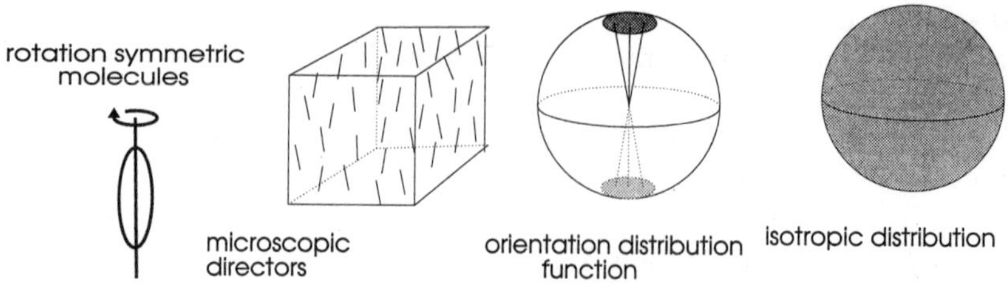

Figure 6. In the isotropic phase all orientations are equally probable, whereas in the liquid crystalline phases the ODF is anisotropic.

4 Example: Mesoscopic theory of uniaxial liquid crystals

An example for $\boldsymbol{m}$ is the *microscopic director* $\boldsymbol{n}$ in mesoscopic liquid crystal theory [17]. This microscopic director is defined as a unit vector pointing into the temporary direction of a needle-shaped rigid particle, or, if the particle is of a plane shape, the microscopic director is perpendicular to the particle. Because the microscopic director is defined on a molecular level it is not a macroscopic field describing the "mean orientation", but it represents a *mesoscopic variable* which spans the 2-dimensional unit sphere $\mathbb{S}^2$. According to table 1 the unit sphere is the manifold $\mathcal{M}$ in this example. The MDF in (1) describes the orientation of the molecules in a volume element. In the case of liquid crystals the MDF is called orientation distribution function (ODF). It is a distribution function on the unit sphere, which is homogenuous in the isotropic phase and more or less concentrated around one direction in the liquid crystalline phases due to the orientational order (see figure 6). It is observed always a so called head-tail-symmetry, i.e. one orientation and the reversed one are equally probable. Consequently the ODF has inversion symmetry.

4.1 Mesoscopic balance equations

The derivative with respect to the mesoscopic variable microscopic director is denoted by $\backslash\nabla_n$. It is the covariant derivative on the unit sphere, and the orientation change velocity $\boldsymbol{u}$ is tangential to the unit sphere.

We obtain the special balances by special identifications of the abstract quantities in $\phi(\cdot)$, $\boldsymbol{\Phi}(\cdot)$, $\boldsymbol{\Sigma}(\cdot)$, $\boldsymbol{S}(\cdot)$, $\boldsymbol{R}(\cdot)$ (7). These balances write [17]:

Mass

$$\frac{\partial}{\partial t}\varrho(\cdot) \;+\; \nabla_x \cdot \{\varrho(\cdot)\boldsymbol{v}(\cdot)\} + \nabla_m \cdot \{\varrho(\cdot)\boldsymbol{u}(\cdot)\} \;=\; 0. \tag{11}$$

Momentum

$$\begin{aligned}\frac{\partial}{\partial t}[\varrho(\cdot)\boldsymbol{v}(\cdot)] + \nabla_x \cdot \left[\boldsymbol{v}(\cdot)\varrho(\cdot)\boldsymbol{v}(\cdot) - \boldsymbol{T}^\top(\cdot)\right] \;+&\\ + \nabla_m \cdot \left[\boldsymbol{u}(\cdot)\varrho(\cdot)\boldsymbol{v}(\cdot) - \boldsymbol{\mathcal{T}}^\top(\cdot)\right] \;=&\; \varrho(\cdot)\boldsymbol{k}(\cdot).\end{aligned} \tag{12}$$

Here $\boldsymbol{k}(\cdot)$ is the external acceleration, $\boldsymbol{T}^\top(\cdot)$ the transposed stress tensor, and $\boldsymbol{\mathcal{T}}^\top(\cdot)$ the transposed stress tensor on $\mathcal{M}$.

Angular Momentum
In addition we have a balance of angular momentum due to rotations of the particles, which are not point like:

$$\boldsymbol{S}(\cdot) := \boldsymbol{x} \times \boldsymbol{v}(\cdot) + \boldsymbol{s}(\cdot), \tag{13}$$

$$\begin{aligned}\frac{\partial}{\partial t}[\varrho(\cdot)\boldsymbol{S}(\cdot)] + \nabla_x \cdot \left[\boldsymbol{v}(\cdot)\varrho(\cdot)\boldsymbol{S}(\cdot) - (\boldsymbol{x} \times \boldsymbol{T}(\cdot))^\top - \boldsymbol{W}^\top(\cdot)\right] \;+&\\ + \nabla_m \cdot \left[\boldsymbol{u}(\cdot)\varrho(\cdot)\boldsymbol{S}(\cdot) - (\boldsymbol{x} \times \boldsymbol{\mathcal{T}}(\cdot))^\top - \boldsymbol{\mathcal{W}}^\top(\cdot)\right] \;=&\\ = \varrho(\cdot)\boldsymbol{x} \times \boldsymbol{k}(\cdot) \;+\; \varrho(\cdot)\boldsymbol{g}.&\end{aligned} \tag{14}$$

Here $\boldsymbol{s}$ is the vector of the specific spin, $\boldsymbol{g}$ the vector of angular momentum exerted by the external forces on $\boldsymbol{m}$, the tensor $\boldsymbol{W}$ is the surface torque, and $\boldsymbol{\mathcal{W}}$ is the analogue to $\boldsymbol{\mathcal{T}}$.

Total Energy

$$e(\cdot) := \frac{1}{2}\boldsymbol{v}^2(\cdot) \;+\; \frac{1}{2}\boldsymbol{s}(\cdot) \cdot \Theta^{-1} \cdot \boldsymbol{s}(\cdot) \;+\; \varepsilon(\cdot), \tag{15}$$

$$\begin{aligned}&\frac{\partial}{\partial t}[\varrho(\cdot)e(\cdot)] \;+\\ &+ \nabla_x \cdot \left[\boldsymbol{v}(\cdot)\varrho(\cdot)e(\cdot) - \boldsymbol{T}(\cdot)^\top \cdot \boldsymbol{v}(\cdot) - \boldsymbol{W}(\cdot)^\top \cdot \Theta^{-1} \cdot \boldsymbol{s}(\cdot) + \boldsymbol{q}(\cdot)\right] +\\ &+ \nabla_m \cdot \left[\boldsymbol{u}(\cdot)\varrho(\cdot)e(\cdot) - \mathcal{T}(\cdot)^\top \cdot \boldsymbol{v}(\cdot) - \mathcal{W}(\cdot)^\top \cdot \Theta^{-1} \cdot \boldsymbol{s}(\cdot) + \boldsymbol{Q}(\cdot)\right] \;=\\ &= \varrho(\cdot)\boldsymbol{k} \cdot \boldsymbol{v}(\cdot) \;+\; \varrho(\cdot)\boldsymbol{g}(\cdot) \cdot \Theta^{-1} \cdot \boldsymbol{s}(\cdot) + \varrho(\cdot)r(\cdot).\end{aligned} \tag{16}$$

Here the absorption supply is r, Θ the moment of inertia tensor, $\boldsymbol{q}$ the heat flux density, and $\boldsymbol{Q}$ the heat flux density on $\mathcal{M}$.

Summary:

- Compared to macroscopic balance equations in the mesoscopic equations there appear additional flux terms in orientation space.
- Compared to simple materials the balance of angular momentum has to be considered in addition. This is clear from the internal degree of freedom, the rotations of particles.

4.2 Macroscopic balance equations

For the extensive quantities of mass density, momentum density, and spin density the macroscopic quantities are integrals of the mesoscopic ones over the unit sphere:

$$\rho(\boldsymbol{x},t) = \oint_{S^2} \rho(\cdot) d^2 n \tag{17}$$

$$\rho(\boldsymbol{x},t)\boldsymbol{v}(\boldsymbol{x},t) = \oint_{S^2} \rho(\cdot)\boldsymbol{v}(\boldsymbol{x},t) d^2 n \tag{18}$$

$$\rho(\boldsymbol{x},t)\boldsymbol{s}(\boldsymbol{x},t) = \oint_{S^2} \rho(\cdot)\boldsymbol{s}(\boldsymbol{x},t) d^2 n \quad . \tag{19}$$

Therefore we obtain the macroscopic balance equations by integrating the mesoscopic ones over all orientations. For the balance of mass we have:

$$\int_{S^2} \frac{\partial}{\partial t}\rho(\cdot) d^2 n + \int_{S^2} \nabla_x \cdot (\boldsymbol{v}(\cdot)\rho(\cdot))\, d^2 n + \int_{S^2} \nabla_n \cdot (\boldsymbol{u}(\cdot)\rho(\cdot))\, d^2 n \tag{20}$$

or

$$\frac{\partial}{\partial t}\rho(\boldsymbol{x},t) + \nabla_x \cdot (\boldsymbol{v}(\boldsymbol{x},t)\rho(\boldsymbol{x},t)) = 0 \quad . \tag{21}$$

The mesoscopic balance equation integrated over the unit sphere shows again the analogy to mixture theory: Summing up the equations for the different components of the mixture (here the different orientations) gives the balance equation of the mixture, and the sum of the productions due to chemical reactions over the components is zero. In the case of liquid crystals chemical reactions are replaced by rotations of particles (the flux in orientation space). It has been used that the flux in orientation space integrated over the whole unit sphere is zero.

Analogously to the macroscopic balance of mass we obtain the other macroscopic balance equations:
Balance of momentum:

$$\frac{\partial}{\partial t}(\rho(\boldsymbol{x},t)\boldsymbol{v}(\boldsymbol{x},t)) + \nabla_x \cdot \left(\boldsymbol{v}(\boldsymbol{x},t)\rho(\boldsymbol{x},t)\boldsymbol{v}(\boldsymbol{x},t) - \boldsymbol{T}^T(\boldsymbol{x},t)\right) = \rho(\boldsymbol{x},t)\boldsymbol{k}(\boldsymbol{x},t) \tag{22}$$

Balance of angular momentum:

$$\frac{\partial}{\partial t}\left(\rho(\boldsymbol{x},t)\left(\boldsymbol{x}\times\boldsymbol{v}(\boldsymbol{x},t)+\boldsymbol{s}(\boldsymbol{x},t)\right)\right)+$$
$$\nabla_x\cdot\Big(\boldsymbol{v}(\boldsymbol{x},t)\rho(\boldsymbol{x},t)\Big(\boldsymbol{x}\times\boldsymbol{v}(\boldsymbol{x},t)+\boldsymbol{s}(\boldsymbol{x},t)-\boldsymbol{T}^T(\boldsymbol{x},t)\Big)\times\boldsymbol{x}-\boldsymbol{W}^T(\boldsymbol{x},t)\Big)$$
$$=\rho(\boldsymbol{x},t)\left(\boldsymbol{x}\times k(\boldsymbol{x},t)+\boldsymbol{g}(\boldsymbol{x},t)\right) \quad (23)$$

and also the balance of energy. These are the balance equations of a micropolar continuum. They have been derived here from the mesoscopic balance equations. The advantage of this derivation is that on the mesoscopic level it is very clear from the internal structure, which balance equations have to be taken into account, and also the form of the balance equations is clear.

4.3 Macroscopic constitutive quantities

From the integrated mesoscopic equations we can identify the macroscopic constitutive quantities in terms of mesoscopic ones. For example for the stress tensor and the internal energy we find

$$\boldsymbol{T}(\boldsymbol{x},t)=\int_{S^2}f(\cdot)\left(\boldsymbol{T}(\cdot)+\rho(\boldsymbol{x},t)\left(\boldsymbol{v}(\boldsymbol{x},t)\boldsymbol{v}(\boldsymbol{x},t)-\boldsymbol{v}(\cdot)\boldsymbol{v}(\cdot)\right)\right)d^2n=$$
$$\langle\boldsymbol{T}(\cdot)\rangle+\rho(\boldsymbol{x},t)\langle\boldsymbol{v}(\boldsymbol{x},t)\boldsymbol{v}(\boldsymbol{x},t)-\boldsymbol{v}(\cdot)\boldsymbol{v}(\cdot)\rangle \quad (24)$$
$$e(\boldsymbol{x},t)=\langle e(\cdot)\rangle+\frac{1}{2}\left(\langle\boldsymbol{v}(\cdot)^2\rangle-\boldsymbol{v}(\boldsymbol{x},t)^2+\langle\theta\boldsymbol{u}(\cdot)^2\rangle-\frac{1}{\theta}\boldsymbol{s}(\boldsymbol{x},t)^2\right) \quad (25)$$

Similiar relations can be derived for the heat flux and for the energy supply.

Summary: For constitutive quantities the macroscopic quantities are in general not simply avarages of the corresponding mesoscopic ones, but there are additional contributions from derivations of the mesoscopic velocity from the macroscopic one and from derivations of the mesoscopic angular velocity from the macroscopic one.

4.4 Order parameters

Using the microscopic director $\boldsymbol{n}$ as a mesoscopic variable we can introduce the family of the macroscopic *fields of order parameters* which is defined by different moments of the ODF.

$$\boldsymbol{a}(\boldsymbol{x},t):=\int_{S^2}f(\cdot)\ \overline{\boldsymbol{n}\boldsymbol{n}}\ d^2n, \quad (26)$$

$$\underline{a}_4(\boldsymbol{x},t) := \int_{S^2} f(\cdot)\ \overline{\boldsymbol{nnnn}}\ d^2n, \tag{27}$$

$$\underline{a}_6(\boldsymbol{x},t) := \int_{S^2} f(\cdot)\ \overline{\boldsymbol{n}\text{...6 times...}\boldsymbol{n}}\ d^2n, \qquad \text{etc.} \tag{28}$$

Here the symbol $\overline{}$ denotes the traceless symmetric part of the tensor in its argument [21]. If the ODF is rotation symmetric with axis of rotation symmetry $\boldsymbol{d}$, the alignment tensors can be written as

$$\boldsymbol{a}_k = S^{(k)}\ \overline{\boldsymbol{d}\ldots\ \text{k times}\ldots\boldsymbol{d}} \tag{29}$$

with scalar order parameters $S^{(k)}$.

These fields of order parameters describe macroscopically the mesoscopic state of the system introduced by $\boldsymbol{n}$. Consequently these fields are the link between the mesoscopic background description of the liquid crystal and its extended description by additional macroscopic fields. In liquid crystal theory the fields of order parameters are called the *alignment tensors* of different orders. Due to the head-tail symetry only even order alignment tensors are nonzero. In the isotropic phase all alignment tensors are zero, whereas in the liquid crystalline phases at least some alignment tensors are nonzero. The alignment tensors are internal variables. In equilibrium they are determined by the equilibrium variables mass density and temperature. The differential equation (8) for the ODF allows to derive a system of differential equations for the alignment tensors. In these equations the alignment tensors of all orders are coupled. These equations are the relaxation equations for the internal variables, which have to be postulated in a purely macroscopic theory.
Summary: The mesoscopic background gives a definition of order parameters, which are the internal variables in a macroscopic description. It also gives equations of motion for these order parameters.

4.5 A constitutive equation from the mesoscopic background: An example

We will discuss very briefly as an example the derivation of the macroscopic constitutive equation for the viscous stress tensor from a mesoscopic one. For the mesoscopic state space we use one including macroscopic and mesoscopic quantities:

$$z = \{\rho(\boldsymbol{x},t), e(\boldsymbol{x},t),\ \overline{\nabla\boldsymbol{v}(\boldsymbol{x},t)}\ , \boldsymbol{n}, \boldsymbol{N} := \dot{\boldsymbol{n}} - (\nabla\times\boldsymbol{v})\times\boldsymbol{n}\} \tag{30}$$

Then a representation theorem linear in the velocity gradient for the stress tensor reads:

$$\boldsymbol{T}(\boldsymbol{n},\boldsymbol{x},t) = \boldsymbol{T}(\rho,\epsilon) + \alpha_1\boldsymbol{nnnn} : \gamma + \alpha_2\boldsymbol{nN} + \alpha_3\boldsymbol{Nn} +$$

$$\alpha_4\gamma + \alpha_5 \boldsymbol{nn}\cdot\gamma + alpha_6\boldsymbol{n}\cdot\gamma\boldsymbol{n} +$$
$$\xi_1\boldsymbol{nn} : \gamma\boldsymbol{\delta} + \xi_2\boldsymbol{nn}\nabla\cdot\boldsymbol{v} + \xi_3\nabla\cdot\boldsymbol{v}\boldsymbol{\delta} \quad (31)$$

where the coefficients α_i may depend on the scalars density and internal energy. If in equation (31) the deviation of the mesoscopic velocity from the macroscopic velocity can be neglected, then the macroscopic stress tensor is the avarage of the mesoscopic one:

$$\boldsymbol{T}(\boldsymbol{x},t) = \{\boldsymbol{T}(\boldsymbol{n},\boldsymbol{x},t)\} = \boldsymbol{T}(\rho,\epsilon) + \alpha_1\{\boldsymbol{nnnn}\} : \gamma + \alpha_2\{\boldsymbol{nN}\} +$$
$$\alpha_3\{\boldsymbol{Nn}\} + \alpha_4\gamma + \alpha_5\{\boldsymbol{nn}\}\cdot\gamma + alpha_6\{\boldsymbol{n}\cdot\gamma\boldsymbol{n}\} +$$
$$\xi_1\{\boldsymbol{nn}\} : \gamma\boldsymbol{\delta} + \xi_2\{\boldsymbol{nn}\}\nabla\cdot\boldsymbol{v} + \xi_3\nabla\cdot\boldsymbol{v}\boldsymbol{\delta} \quad (32)$$

The avarages of products of the microscopic director can be written in terms of the second and the fourth order alignment tensor, or if the distribution function is rotation symmetric, in terms of the macroscopic director and the scalar order parameters $S^{(2)}$ and $S^{(4)}$. The result is a macroscopic constitutive equation giving explicitly the order parameter dependence of the viscosity coefficients. This information is not available from a purely macroscopic theory, and the result of the mesoscopic background theory has been compared to experimental results [22]. The agreement with the measurements is quit good.

5 Summary

The mesoscopic concept introduces field quantities defined on an enlarged domain, the mesoscopic space, which takes into account the internal structure of the complex material. In addition a mesoscopic distribution function is introduced as a statistical element. The advantages of such a mesoscopic background theory over a purely macroscopic description are:

- From the internal structure it is clear, which balance equations have to be taken into account. For liquid crystals for example the balance of angular momentum is also relevant.
- Based on the MDF order parameters are defined, which are the internal variables in a purely macroscopic description. From the mesoscopic balance equations there result equations of motion for the internal variables.
- The mesoscopic constitutive theory gives the order parameter dependence of macroscopic constitutive quantities.
- Different physical systems can be considered within the same or analogous mesoscopic theories. So for instance the orientation of a biaxial molecule

and the orientation of the crystal axes of a microcrystallite are described by the same mesoscopic variable. In addition this mesoscopic variable can be an element of S^{18}, which makes the theory for biaxial liquid crystals analogous to the theory of uniaxial liquid crystals.

Acknowledgements

We thank the Draloric Company, D-95100 Selb, and the Isolde-Dietrich-Stiftung for financial support.

References

1. I. Mueller. *Grundzüge der Thermodynamik.* Springer-Verlag, Berlin, Heidelberg, 1994.
2. M. Doi and S. F. Edwards. *The theory of polymer dynamics.* Clarendon, Oxford, 1986.
3. Martin Kröger. *Rheologie und Struktur von Polymerschmelzen.* W&T Verlag, Berlin, 1995. PhD Thesis, Technische Universität Berlin.
4. P.G. de GENNES and J. Prost. *The Physics of Liquid Crystal, Sect. 2.1.* Clarendon Press, 1993.
5. H. Kelker and R. Hatz. *Handbook of Liquid Crystals.* Verlag Chemie, Weinheim, Deerfield, 1980.
6. W. MUSCHIK. Internal variables in non-equilibrium thermodynamics. *J. Non-Equilib. Thermodyn.*, 15, 1990.
7. G.A. MAUGIN and W. MUSCHIK. Thermodynamics with internal variables. *J. Non-Equilib. Thermodyn.*, 19, 1994.
8. G.A. MAUGIN. *The Thermomechanics of Plasticity and Fracture, Chap. 7.* Cambridge University Press, Cambridge, 1992.
9. C. TRUESDELL and W. Noll. *Non-Linear Field Theories of Mechanics, Encyclopedia of Physics, Vol. III/3, Sect. 98.* Springer Verlag.
10. J.L. ERICKSEN. Anisotropic fluids. *Arch. Rat. Mech. Anal.*, 4, 1960.
11. F.J. LESLIE. Some constitutive equations for liquid crystals. *Arch. Rat. Mech. Anal.*, 28, 1965.
12. S. Hess. Irreversible thermodynamics of nonequilibrium alignment phenomena in molecular liquids and in liquid crystals. *Z. Naturforsch.*, 30a:728–733, 1975.
13. G.A. MAUGIN and R. DROUOT. Thermodynamic modelling of polymers in solution. In Axelrad and W. Muschik, editors, *Constitutive Laws and Microstructure.* Springer Verlag, 1988.

14. S. Blenk, H. Ehrentraut, and W. Muschik. A continuum theory for liquid crystals describing different degrees of orientational order. *Liquid Crystals*, 14(4):1221–1226, 1993.
15. W.H. Müller and W. Muschik. Bilanzgleichungen offener mehrkomponentiger Systeme
I. Massen- und Impulsbilanzen. *J. Non-Equilib. Thermodyn.*, 8:29–46, 1983.
16. W. Muschik and W.H. Müller. Bilanzgleichungen offener mehrkomponentiger Systeme
II. Energie- und Entropiebilanz. *J. Non-Equilib. Thermodyn.*, 8:47–66, 1983.
17. S. Blenk and W. Muschik. Orientational balances for nematic liquid crystals. *J. Non-Equilib. Thermodyn.*, 16:67–87, 1991.
18. H. Ehrentraut. *A Unified Mesoscopic Continuum Theory of Uniaxial and Biaxial Liquid Crystals.* Wissenschaft und Technik Verlag, Berlin, 1996.
19. S. Blenk, H. Ehrentraut, and W. Muschik. Orientation balances for liquid crystals and their representation by alignment tensors. *Mol. Cryst. Liqu. Cryst.*, 204:133–141, 1991.
20. H. Ehrentraut, W. Muschik, and C. Papenfuss. Mesoscopically derived orientation dynamics of liquid crystals. *J. Non-Equilib. Thermodyn.*, 1997.
21. H. EHRENTRAUT BLENK, S. and W. MUSCHIK. Statistical foundation of macroscopic balances for liquid crystals in alignment tensor formulation. *Physica A*, 174, 1991.
22. Harald Ehrentraut and Siegfried Hess. Viscosity coefficients of partially aligned nematic and nematic discotic liquid crystals. *Phys. Rev. E*, 51(3):2203 - 2212, March 1995.

AN INTERACTION OF SIMPLE SOLITARY WAVES WITH PROFILES IN THE FORM OF CHEBYSHOV–HERMITE, MATHIEU, WHITTAKER FUNCTIONS PROPAGATING IN MEDIA WITH THE MICROSTRUCTURE

J.RUSHCHITSKY

Institute of Mechanics, National Academy of Sciences of Ukraine, Nesterov Street 3, Kyiv,252680, Ukraine

E–mail: rushch@imech.freenet.kiev.ua

S.RUSHCHITSKA

National Technical University of Ukraine "KPI", Kyiv

The lecture consists from two parts. The first deals with simple solitary waves, its existing and describing. The second is devoted to the interaction of such waves, which is described by evolution equations, and to the condition such an interaction.

1 Simple solitary waves

First of all, it is need to say some words about microstructural theories of materials. These theories have achieved greatest development in the theory of composite materials. Their allmechanical and allphysical nature has displayed more tardy and today such theories went from the scope of mechanics of composites. However, the first rather complete describing of microstructural theories have been realized in fundamental series of monographs on composite materials. An attempt to qualify microstructural theories has also been made in [1,2].

Composite materials are usually defined as materials consisting of several components with differing physical properties, as a rule, alternate many times in space. The method of alternation, the geometrical shape, and the physical properties of the components determine the microstructure of the composite.

Complexities in the analytical description of composites for exact problem statements have led to the need to construct continual, and by their own nature, approximate models, which retain the existing physical properties of the system on the one hand and at the same time, are rather simple on the other to permit analytical solution of boundary problems for constrained bodies. The increased attention to composite materials, which has evolved for more than thirty years, has led to very rapidly to diverse realizations of above-mentioned need for approximate approaches. A whole series of approximate theories capable of accounting for the microstructure of materials, defining required macroscopic parameters, and solving problems of practical import, have been proposed to date.

One from most developed microstructural theories of materials is the theory, based on mixture model. Let us now briefly formulate the bases of the microstructural theory of solid mixtures. Its basic advantage consists in the fact that it can be constructed similar to the classical theory of elasticity. This provides good possibilities for analytical solutions.

The hypothesis concerning the fact that the microstructure of a multi-component composite (we will later consider two–component composites as the most widely used composites) can be described using continua, the material particles of which simultaneously at each geometric point of a domain and interact one with the other is the basic hypothesis of mixture theory. These are so–called interpenetrating and interacting continua. Each continuum is characterized by its own set of field characterictics of such functions as partial density $\rho_{\alpha\alpha}$, partial vector of the displacements $\vec{u}^{(\alpha)}$, and the partial tensors of the stresses $\sigma_{ik}^{(\alpha)}$, strains $\epsilon_{ik}^{(\alpha)}$, and rotations $\omega_{ik}^{(\alpha)}$. Here and below, the Greek superscripts and Latin subscripts are equal to 1 and 2, and 1,2, and 3, respectively. The parameter is named partial, if it characterizes one phase only.

The basic equations of the theory of mixtures are derived, assuming the laws of mass, momentum, angular momentum, and energy conservation are valid for the mixture.

The second basic feature of this study is its direct connection with the nonlinear acoustics and optics. The classic linear theory of elastic waves was practically formed in the last century. Unfortunately, a development of nonlinear theory was late and some its directions create just now. Nonlinearity in physics is a typical phenomenon. Such a division of physics as celestial mechanics, hydrodynamics, theory of gravitation and a number of others are nonlinear at the bottom. Therefore physics studied nonlinearity for a long time and in its multiform manifestations. That applies to nonlinear oscillation first of all. As long ago as 40s, some principle was formulated in the Muscowite academician Mandelstamm school which was named as *"Principle of isomorphism of generalities of nonlinear physics"*. The principle consists in that generalities of the oscillational motion do not depend on matter of physical phenomenon. Since the understanding of an oscillation problem in each division of physics was different, Mandelstamm named the phenomenon of a transport of achieved understanding of oscillations from one division physics to other as *" Oscillation mutual aid"* [3,4].

That fact from the history turned out characteristic. Very similar situation arised in last decades in the nonlinear wave theory. For example, achieved understanding of wave motion in nonlinear optics was transported to nonlinear physics of plasma or from nonlinear radiophysics to nonlinear optics. Principle of mutual aid works in the nonlinear theory of waves. Strictly speaking, we would like to declarate the fact that achieved in nonlinear optics and nonlinear acoustics understanding of problems of wave interactions must be transported

to nonlinear mechanics of materials.
Nonlinear waves in physics was studying regular. But such a regular investigations of nonlinear waves in solids began in 60ies only. That was sound waves and basic publications belong to representatives of acoustics and not of mechanics. An encyclopedic book of Whitham [5] in the nonlinear wave field, and, yes, a number of highly professional written books on nonlinear waves have not accentuated the attention to waves in solids.
Such a situation is caused, perhaps, by the fact that these waves have some difficult specificity. Especially, on the initial and end stages, i.e. in the statement of a problem and in the commentary on results. Just these stages have circumvented attention in above cited books. Classic books on waves in solids gives any or very small information about nonlinear waves.
Practically the first wave which is studied in the framework of any theory is the plane wave. Therefore we will study plane waves,too. In the mixture a motion is described already by two partial vectors of displacements. In the plane waves these vectors are sure not collinear, but the plane front of a wave must be the same for both vectors. Therefore, the representation of a periodic plane wave is written as

$$\vec{u}^{(\alpha)}(x,t) = \vec{u}^{0(\alpha)} e^{i(\xi - \omega t)},$$

$\vec{u}^{0(\alpha)}$ are arbitrary constant vectors, $\xi = \vec{k} \cdot \vec{r}$; $\vec{r}$ is the radius-vector of the point $x \equiv (x_1, x_2, x_3)$. Let the plane wave propagate in the direction of coordinate axis Ox_1. Then it has the form

$$\vec{u}^{(\alpha)} = \{\vec{u}_k^{(\alpha)}(x_1, t)\}. \tag{1}$$

Let us choose the characteristic for such a theory system of equations [1,2] which describes the propagation of longitudinal plane waves of the type of (5)

$$\rho_{\alpha\alpha} u_{tt}^{(\alpha)} - a_\alpha u_{xx}^{(\alpha)} - a_3 u_{xx}^{(\alpha)} - \beta(u^{(\alpha)} - u^{(\delta)}) = 0. \tag{2}$$

System (6) is some generalization of the classic Klein–Gordon equation from a nonlinear wave theory and has [1] the periodical solution in the form of harmonic dispersive waves

$$u_{long}^{(\alpha)}(x,t) = A_{0long}^{(\alpha)} e^{-\left(k_\alpha^{(long)} x - \omega t\right)} + l\left(k_{(\delta)}^{(long)}\right) A_{0long}^{(\delta)} e^{-\left(k_\delta^{(long)} x - \omega t\right)}. \tag{3}$$

Features of the solution (8) – the features of the plane longitudinal harmonic wave in two-phase elastic mixtures are the following:
1. two modes exist simultaneously, they are distinguished by wave numbers (the subscript α fixes the mode number); 2. both modes are essentially dispersive waves; 3. one from modes is filtered by the mixture, it is blocked for

low frequencies, leadind off with frequency

$$\omega_{cut} = \sqrt{\frac{\beta(\rho_{11} + \rho_{22})}{\rho_{11}\rho_{22}}},$$

which is called the frequency of blocking or cut off frequency; 4. two modes with their own amplitudes propagate in each phase of the mixture, matrixes of amplitude distribution depend on frequency, that display in that with the change of frequency the energy of waves is pumped from one mode to other.
All described above effects has the microstructural nature and waves are linear.
Let us return to basic system (6). For this system the universal D'Alembert solution now may not be written, since the system has nonclassical terms $\pm\beta(u^{(\alpha)} - u^{(\delta)})$. When used in D'Alembert solution arbitrary function F of the phase $z = x - vt$ is not arbitrary and it satisfies an equation

$$F^{''}(z) = f(z)F(z) \qquad (f(z)\text{is known function})$$

then for the system (6) an approximate solution of the D'Alembert type may be obtained according the classical procedure

$$u^{(\alpha)}(x,t) = A^{\alpha)}F^{(\alpha)}(z^{(\alpha)}) + m(z^{(\alpha)})F^{(\delta)}(z^{(\delta)})$$

This solution has a few new properties and it may be treated as the simple Riemann waves. First of all, phase velocity v is already not constant, but it nonlinear depends on the phase z. Secondly, the solution may be considered as following: initial pulse in the form of a function $F(x)$ propagates some time in the form of the simple wave $F[x - v(z)t]$ and accumulate the distortion. Later it distorted in such a way that the approximate solution is not valid.
We considered three cases of the initial pulse presentation :
1. Chebyshov–Hermite functions $\psi_n(z)$ which satisfies [6–8] the equation

$$w^{''} - [x^2/4 - (n + 1/2)]w = 0;$$

2. Mathieu functions $ce_n(x,q), se_n(x,q)$ which satisfies [6–8] the equation

$$w^{''} - (p - 2q\cos 2x)w = 0;$$

3. Whittaker functions $W_{\kappa,\mu}(x)$ which satisfies [6–8] the equation

$$w^{''} - (-1/4 + \kappa/x + (1 - 4\mu^2)/2x^2)w = 0.$$

For these cases the simple wave will be exist with the some restrictions in time. The restrictions have the form

$$1 + tv^{'}(z) \simeq 1; \quad ln[1 + tv^{'}(z)] \simeq 0. \tag{4}$$

We will show the general procedure on the one example. Let us select the initial pulse in the form of bell–shaped function, i.e. the Chebyshov–Hermite function of zero index

$$u^{(\alpha)}(x,0) = A^{(\alpha)}\psi_0(x). \tag{5}$$

With the exception of the small area near cut-off frequency of a second mode, the dispersivity of a mixture is weak. Therefore, the pulse will distort slightly and the solution of basic equation (6) may be seek in the form of bell–shaped function. It can be represented in the form

$$u^{(\alpha)}(x,0) = A^{(\alpha)}\psi_0\big(z^{(\alpha)}\big) + p\big(z^{(\delta)}\big)A^{(\delta)}\psi_0\big(z^{(\delta)}\big) =$$

$$= A^{(\alpha)}e^{-\frac{1}{2}(z^{(\alpha)})^2} + p\big(z^{(\delta)}\big)A^{(\delta)}e^{-\frac{1}{2}(z^{(\delta)})^2}, \tag{6}$$

$$p\big(z^{(\delta)}\big) = \left\{\frac{\lambda_\alpha + 2\mu_\alpha - v\big(z^{(\delta)}\big)\rho_{\alpha\alpha} + \left(\frac{\beta}{1-\left(z^{(\alpha)}\right)^2}\right)}{\lambda_3 + 2\mu_3 - \left(\frac{\beta}{1-\left(z^{(\alpha)}\right)^2}\right)}\right\}^{(-1)^\alpha},$$

$$v\big(z^{(\alpha)}\big) = \left\{K_1 - (-1)^\alpha\sqrt{K_1^2 - K_2}\right\}^{1/2}, \tag{7}$$

$$K_1 = \frac{\lambda_1 + 2\mu_1}{\rho_{11}} + \frac{\lambda_2 + 2\mu_2}{\rho_{22}} - \frac{\rho_{11}+\rho_{22}}{\rho_{11}\rho_{22}}\frac{\beta}{1-\left(z^{(\alpha)}\right)^2},$$

$$K_2 = \frac{\lambda_1 + 2\mu_1}{\rho_{11}}\frac{\lambda_2 + 2\mu_2}{\rho_{22}} - \frac{(\lambda_3 + 2\mu_3)^2}{\rho_{11}\rho_{22}} +$$

$$+\frac{(\lambda_1 + 2\mu_1) + (\lambda_2 + 2\mu_2) + 2(\lambda_3 + 2\mu_3)}{\rho_{11}\rho_{22}}\frac{\beta}{1-\left(z^{(\alpha)}\right)^2}.$$

In this solution a phase acquires the number, since two phase velocities exists in the mixture.

The simplicity in the writting of solution (11) is achieved by two essentially complications - nonlinear dependence of the phase velocity on a phase and approximation character of fulfilling of basic system. The first complication transforms the wave into a simple one. The second is associated with the procedure of calculating the first and second derivatives. If we neglect the denominators in the expressions for these derivatives, i.e., we set the above written restrictions (4), the solution is invalid in the region of variation in x and t, where the latter equations are not fulfilled. The answer follows from analysis of plots of the phase velocities.

Some analogy exists between a condition (12) and a van der Pol condition from the widely used in nonlinear physics method of slowly varying amplitudes. In the van der Pol method is supposed that in the harmonic solution a phase and an amplitude vary with time ,but so slowly, that this varying may be neglected on the one period. In reality, the "accelerations" of amplitudes are neglected, and shorten equations are obtained. Some similar procedure is used here. A condition (12) supposes some time, during of which the "accelerations" of a phase may be neglected, and may be taken into account the "velocities" of phase, only.

The obtained representation of a solution in the form of a simple wave (11) generalizes easily to the case of an arbitrary Chebyshov-Hermite, Mathieu and Whittaker functions. For example, in the first case a solution assumes the form

$$u^{(\alpha)}(x,0) = A^{(\alpha)}\psi_n\big(z^{(\alpha)}\big) + p\big(z^{(\delta)}\big)A^{(\delta)}\psi_n\big(z^{(\delta)}\big), \tag{8}$$

$$p\big(z^{(\delta)}\big) = \left\{\frac{\lambda_\alpha + 2\mu_\alpha - v\big(z^{(\delta)}\big)\rho_{\alpha\alpha} + \left(\frac{\beta}{1+2n-\left(z^{(\alpha)}\right)^2}\right)}{\lambda_3 + 2\mu_3 - \left(\frac{\beta}{1+2n-\left(z^{(\alpha)}\right)^2}\right)}\right\}^{(-1)^\alpha}.$$

Let us now compare solutions (13) and (8). Classical waves (8) are harmonic; this is manifested in the frequency on the phase velocity, and the coefficient β is in conformity for the level of dispersity. Solution (8) also describes the simultaneous existence of two modes distinguished by phase velocities. These basic properties noted before are retained even for both the bell–shaped simple waves (11) and simple waves with more complicated profiles (13). The dispersivity in(8) is transformed into a nonlinear dependence between the phase velocity and the phase in (13). Waves (13) have all the indications of simple waves. Particularly, the characteristics for these waves are always the straight lines along which the wave amplitude $\mid u^{(\alpha)}(x,t) \mid$ is constant. Two modes are also retained here, and the difference between the phase velocities is significant. The very procedure of describing the amplitude distribution between phases is also retained.

The disagreements between solutions (8) and (13) are, however, rather important. Above all,(8) are classical harmonic dispersive waves, whereas(13) are aperiodic solitary simple waves.

That means that, in particular, obtained approximate solution in the form of known bell–shaped signal describes by convenient functional writting with the aim of a Chebyshov–Hermite function the solitary wave with a one hump. Such a writing is valid during some nonsmall time. The reasized computer modelling showed that these properties has the simple wave, expressed by Chebyshov–Hermite function of any subscript.

We would like to repeat that it is easy to see that obtained approximate solution will be valid for small initial period of wave propagation. That will be the moment comparing with the eternity, which is reserved for the harmonic running waves (the last come from minus infinity and go away to infinity).

Computer modeling shown that time of propagating of simple waves studied is not small for some composite materials and distances on which new waves propagates without essential distortions is not small, too. Are small comparing with which distances and time intervals ? Such comparison for studied class of composites and for solitary simple waves gived the positive results. The distance for a comparison for bell–shaped waves, for example, choosed as the distance of the bottom of the "bell". Then the critical distance was 20–30 choosed distances.

2 Solitary waves interaction

In the second part of our lecture we will discuss the nonlinear interaction of above introduced simple solitary waves. The used method – method of slowly varying amplitudes – is taken from nonlinear optics [9,10]. First of all, nonlinear analogy of linear equations (2) has the form

$$\rho_{\alpha\alpha} u_{tt}^{(\alpha)} - a_\alpha u_{xx}^{(\alpha)} - a_3 u_{xx}^{(\alpha)} - \beta(u^{(\alpha)} - u^{(\delta)}) = N_1^\alpha u_{xx}^{(\alpha)} u_x^{(\alpha)}. \tag{9}$$

By usual procedure two groups for each phase of a mixture of evolution equations are obtained. Here, we will write two equations for third waves, only

$$\frac{dA_\alpha^{(3)}}{dt} = \frac{N_1^{(\alpha)}}{2\rho_{\alpha\alpha}} \frac{p(\sigma^{(2)}}{v^{(31)}} \frac{(\sigma^{(1)^2} - 1)\sigma^{(2)} + (\sigma^{(2)^2} - 1)\sigma^{(1)}}{\sigma^{(3)}} A_1^{(1)} A_1^{(2)} e^{\frac{1}{2}(\sigma^{(1)^2} + \sigma^{(2)^2} - \sigma^{(3)^2}} \tag{10}$$

Let us focus attention on the fact that the amplitude evolution of a new third wave in each phase will be its own. Two last equations are not identical. Eqs(11) make it possible to examine the problem of the synchronism of the three waves. The terms "frequency and spatial synchronism" themselves could also be defined more precisely as applies to simple waves. Thus, do not let a third wave be excited initially, i.e., the initial amplitude of the third wave is zero $A_2^{(3)}$ (the first phase can be selected for definiteness). Also let the time interval be selected such that the amplitude $A_2^{(1)}$ in (11) be constant as a result of the slowness of the variation. In the right side of (11), therefore, the expression $\frac{2\sigma^{(1)}(\sigma^{(1)^2}-1)}{v^{(31)}\sigma^{(3)}} = f(\sigma)$ and the exponential expression will vary with time. The solution of (11) will then take on the form

$$A_2^{(3)}(t) = \frac{N^((2)2}{2\rho_{22}} (A_2^{(1)})^2 \int_0^t f(\sigma) e^{-\frac{1}{2}\Delta\sigma} d\tau,$$

$$\Delta\sigma(\tau) = (\sigma^{(3)})^2 - 2(\sigma^{(1)})^2. \tag{11}$$

The integration in last equation will yield an expression like the one under the integral – the product of the fractional–rational function and $e^{-\frac{1}{2}\Delta\sigma}$, which will be a decreasing function, beginning with a certain time t. In effect, consequently, no third wave is excited. If, however, the condition

$$\Delta\sigma(\tau) = 0, \tag{12}$$

is satisfied, expression (12) will assume a somewhat different form - the integral can be calculated; this will be a polynomial with an explicit dependence on time. The amplitude will therefore increase with time as at resonance, i.e., under condition (12), a simple bell–shaped wave will interact with itself and generate a wave with an increasing amplitude. This wave will extract energy from the initial wave.

A condition of a type of (12) for harmonic waves is broken down into two conditions; they provide the basis for us to call them synchronism conditions. The term resonance condition, which is less often encountered in nonlinear physics, could possibly be used for the simple bell–shaped wave.

If, consequently, resonance condition (12) is fulfilled, the equation

$$A_2^{(3)}(t) = \frac{N^((2)2}{2\rho_{22}}(A_2^{(1)})^2 \int_0^t f(\sigma)d\tau, \tag{13}$$

$$\sigma^{(3)} = \pm\sqrt{2}\sigma^{(1)}, \quad v^{(3)} = \sqrt{2}v^{(1)}\sigma^{(1)} - (\sqrt{2}-1)\frac{x}{t},$$

can be derived for calculation of the evolution of amplitude $A_1^{(3)}(t)$.

On passing through a nonlinearly elastic two–phase mixture, therefore, a simple bell–shaped wave will interact with itself and generate a new simple wave of the same form. The amplitude of this new wave will increase to values commensurate with initial amplitude of the primary wave, and in this manner, significantly distort the latter's profile. But, probably, at these moments of time the solution (it is approximate) will be not valid.

Considered problem is generalized on the case, when initial profiles of primary waves is described by Chebyshov–Hermite, Mathieu, Whittaker functions of an arbitrary subscript. In this case the basic participant in the procedure is a simple aperiodic solitary wave with essentially more complex profile than "bell". The second essentially distinction consists in definitionwith what the slowness of a variation of a profile should be compared. It is proposed to compare with "bottom" of profiles, since each Chebyshov–Hermite function is practically nonzero only on some finite symmetric interval around the coordinates origin.

Further the usual procedure is used. The shorten equations are obtained in a similar to above way.In the procedure of obtaining of evolution equations the exponential factor $e^{-\sigma^2/2}$ is extracted from each Chebyshov–Hermite. Let us write here only the evolution equation for a third wave in a first phase of a mixture

$$\frac{dA^{(31)}}{dt} = n_1 \frac{p(\sigma^{(2)})}{v^{(3)}(\sigma^{(3)})} \frac{\tilde{H}_{(11)}}{H^*_{(3)}} A^{(1)} A^{(2)} e^{-\frac{1}{2}(\sigma^{(1)2}+\sigma^{(2)2}-\sigma^{(3)2})}, \qquad (14)$$

$$n_\alpha = \frac{N_1^{(\alpha)}}{2\rho_{\alpha\alpha}}, \quad \tilde{H}_{(km)} = H^*_{(k)} H^{**}_{(m)} + H^*_{(m)} H^{**}_{(k)},$$

$$H^*_{(k)} = nH_{n-1}(\sigma^{(k)}) - H_{n+1}(\sigma^{(k)}), \quad H^{**}_{(k)} = \{(\sigma^{(k)})^2 - 2n - 1\} H_n(\sigma^{(k)}),$$

$H_n(\sigma^{(k)})$ are Chebyshov–Hermite polynomials.
The selection of exponents from Chebyshov–Hermite functions gives a possibility to write the evolution equ-ation for arbitrary profiles formally similar to classical equations. All distinctions are collected in coefficients in amplitudes and in arguments of an exponents. All observation made for a bell–shaped function rest valid in this case, too. First of all, the evolution in different phases will be differing. Equations for amplitudes for different phases are no identical. The distinction of amplitudes in different phases for initial, first and second, waves displays here essentially.

References

1. J.J.Rushchitsky, Nonlinear waves in solid mixtures (review),*Intern. Applied Mechanics*, **33**, No.1, 1-35 (1997).
2. J.J.Rushchitsky, S.I.Tsurpal, *Waves in materials with a microstructure* (In Ukrainian, with English contents),(Institute of Mechanics, Kiev,1998).
3. L.I.Mandelstamm,*Lectures on optics, theory of relativity and quantum mechanics* (In Russian), (Nauka, Moscow, 1972).
4. *Nonlinear waves. Propagation and interaction* (In Russian), (Nauka, Moscow, 1981).
5. G.B.Whitham,*Linear and nonlinear waves*, (John Wiley & Sons,New York–London–Sydney,1974.
6. J.Kampe de Feriet, R.Campbell, G.Petiau, T.Vogel,*Founctions de la physique mathematique*,(Centre national de la recherche scientifique, Paris,1957).
7. F.W.J.Olver,*Asymptotics and special functions*, New York–London, Academic Press, New York–London,1974).
8. M.A.Lavrentiev, B.V.Shabat,*Methodes of the theory of functions of complex variable* (In Russian), Moscow, Nauka, 1963.
9. Y.R.Shen,*The principle of nonlinear optics*,(John Wiley and Sons,New York,1984).
10. M.Schubert, B.Wilgelmi,*Einfuhrung in die nichtlineare Optik, Teil I, Klassische Beschreibung*,(BSB BG Teubner Verlaggesellschaft,Leipzig,1977).

DRYING INDUCED STRESSES AS A FUNCTION OF THE TEMPERATURE OF DRYING MEDIUM

A. RYBICKI
Poznan University of Technology
Institute of Technology and Chemical Engineering
pl.M.Sklodowskiej-Curie 2
60-965 Poznan, Poland
E-mail: ryban@rose.man.poznzn.pl

The influence of the drying medium temperature and in particular the influence of the temperature dependent moisture transport coefficient on the drying induced stresses is analyzed in the paper. The thermomechanical model of drying process proposed by Kowalski[1] was used for the solution of the problem. for the convective drying of an infinite bar with rectangular cross-section is solved for various temperatures and humidities of the drying medium. The problem is described by the coupled system of second-order differential equations, which were solved numerically with the use of the finite element method. The results are presented in the form of plots.

1 Model Presentation

The termomechanical model of drying of capillary-porous materials proposed by Kowalski[1] was used for solution of the initial-boundary value problems. The drying induced stresses σ_{ij} are calculated from the relation (see Kowalski[1]):

$$\sigma_{ij} = M(u_{i,j} + u_{j,i}) + [(A - \frac{\gamma_\Theta^2}{c_\Theta})u_{l,l} - (\gamma_\vartheta - \gamma_\Theta \frac{c_\vartheta}{c_\Theta})\vartheta - \frac{\gamma_\Theta}{c_\Theta}\varrho_o\mu]\delta_{ij} \tag{1}$$

where u_i denotes the displacement, ϑ is the temperature, μ is the moisture potential in dried material, M and A the shear and bulk modulus of the dried material, ϱ_o is the mass density of dry material, $\gamma_\vartheta = (2M + 3A)\alpha_\vartheta$, $\gamma_\Theta = (2M + 3A)\alpha_\Theta$, α_ϑ and α_Θ are the coefficients of linear thermal and humid expansion of the dried material, c_ϑ and c_Θ are the temperature and the moisture content coefficients in the moisture potential function.

The moisture potential μ is considered here as linear function of moisture content X , temperature ϑ , and volumetric strain ε:

$$\mu = [c_\vartheta(\vartheta - \vartheta_r) - \gamma_\Theta\varepsilon - c_X(X - X_r)]/\varrho_s \tag{2}$$

In this paper, we solve a two-dimensional boundary value problem and the displacements $u_x(x, y, t)$, $u_y(x, y, t)$, temperature $\vartheta(x, y, t)$, moisture potential

$\mu(x, y, t)$ are calculated from the following system of differential equations (see Kowalski *et al.*[2]):

$$\begin{aligned} Lu_{x,xx} + Mu_{x,yy} + Nu_{y,xy} &= P\vartheta_{,x} + R\mu_{,x} \\ Lu_{y,yy} + Mu_{y,xx} + Nu_{x,xy} &= P\vartheta_{,y} + R\mu_{,y} \\ K_m(\mu_{,xx} + \mu_{,yy}) &= \mu_{,t} + [\gamma_\Theta(u_{x,xt} + u_{y,yt}) - c_\vartheta \vartheta_{,t}]/\varrho_o \\ K_T(\vartheta_{,xx} + \vartheta_{,yy}) &= \vartheta_{,t} + K_E(u_{x,xt} + u_{y,yt}) - K_\Theta \mu_{,t} \end{aligned} \tag{3}$$

where $K_m = \Lambda_m c_\Theta/\varrho_o^2$, $K_T = \Lambda_T/c_v^*$, $K_E = T_r(\gamma_\vartheta - c_\vartheta\gamma_\Theta/c_\Theta)c_v^*$, $K_\Theta = T_r c_\vartheta \varrho/c_\Theta c_v^*$, $L = M + N$, $N = M + A - \gamma_\Theta^2/c_\Theta$, $P = \gamma_\vartheta - c_\vartheta\gamma_\vartheta/c_\Theta$, $R = \varrho_o\gamma_\Theta/c_\Theta$, $c_v^* = c_v + T_r c_\vartheta^2/c_\Theta$, Λ_T and Λ_m are the thermal and moisture conductivity coefficients, T_r is the reference temperature, c_v is the specific heat of dried material.

The following boundary condition are assumed to be satisfied (see Kowalski *et al.*[2], Kowalski *et al.*[3]):
for the stresses

$$\sigma_{ij} n_j \mid_{\partial B^-} = 0 \tag{4}$$

for the moisture exchange

$$\Lambda_m \ \operatorname{grad} \mu \mid_{\partial B^+} \cdot \mathbf{n} = \pm\alpha_m(\mu \mid_{\partial B^-} - \mu_a) \tag{5}$$

for the heat exchange

$$\Lambda_T \ \operatorname{grad} \vartheta \mid_{\partial B^+} \cdot \mathbf{n} = \alpha_T(\vartheta_a - \vartheta \mid_{\partial B^+}) \mp l\alpha_m(\mu \mid_{\partial B^-} - \mu_a) \tag{6}$$

with l being the latent heat of evaporation.

One assumes, the initial values to be:

$$\sigma_{ij}(x, y, 0) = 0, \quad \mu(x, y, 0) = \mu_o, \quad \vartheta(x, y, 0) = \vartheta_o. \tag{7}$$

In the above condition μ_a and ϑ_a denote the chemical potential and the temperature of the drying medium (air).

The quantities on the left hand side of boundary conditions (4) to (6) are related to the inner side of boundary surface while those on the right hand side to the outer side of this surface. The numerical method for solution of the considered problems is based on the Galerkin discretization (finite elements) in space and the difference approximation for the time derivatives in the final system of ordinary differential equations (see Wait *et al.*[4]).

2 The temperature of the drying medium and the moisture transport coefficient

The motion of the moisture inside the dried material proceed according to the rate equation of the form Kowalski[1]:

$$\mathbf{W} = -\Lambda_m \operatorname{grad} \mu \tag{8}$$

where $\mathbf{W}$ is the moisture flux and Λ_m is the moisture transport coefficient. This equation is non-linear as the moisture potential μ and the moisture transport coefficient Λ_m are generally function of the parameters of state (strain $\mathbf{E}$, temperature T and moisture content X). In order to obtain a simpler (linearized) form of the differential equation, one develops the above quantities into Taylor's series in a chosen point of the thermodynamical state space $(\mathbf{E}, T, X)$ and next confines the developed form to the first terms of the series.

The moisture transport coefficient is not only a quantity characterizing the material structure but also the physical properties of the moisture, as for example, its surface tension and viscosity. Its dependence on the surface tension and the viscosity is as follows, Kneule[5]:

$$\Lambda_m = \frac{\sigma}{\eta} f(X) \tag{9}$$

where σ denotes the surface tension, η is the dynamic viscosity and $f(X)$ is a function of the moisture content. In the above relation the dependency of Λ_m on the temperature is hidden because the viscosity and the surface tension depend evidently on the temperature.

Comparing the moisture transport coefficient for the same material, having the same moisture content but different temperatures ϑ_1 and ϑ_2, one can write:

$$\frac{(\Lambda_m)_2}{(\Lambda_m)_1} = \frac{(\sigma)_2}{(\eta)_2} \frac{(\eta)_1}{(\sigma)_1}$$

where subscripts 1 and 2 refer to two various temperatures.

For distilled water the values of surface tension and the viscosities are: $\sigma_1 = 72.75 \times 10^{-3} N/m$, $\eta_1 = 8.07 \times 10^{-4} Ns/m^2$ for $\vartheta_1 = 30^o C$ and $\sigma_2 = 66.18 \times 10^{-3} N/m$, $\eta_2 = 4.688 \times 10^{-4} Ns/m^2$ for $\vartheta_2 = 60^o C$. Thus the moisture transport coefficient for temperature $\vartheta_1 = 30^o C$ and differ by $\frac{(\Lambda_m)_2}{(\Lambda_m)_1} \cong 1.57$

We see then, that the value of the moisture transport coefficient can have a relevant influence on the moisture distribution and the drying stresses. We shall illustrate this on the example of two drying processes. In the first process the drying parameters are: the temperature $\vartheta_a = 70^o C$, the humidity content $Y_a = 0.075$ and in the second one: the temperature $\vartheta_a = 34^o C$, the humidity content $Y_a = 0.011$. The wet-bulb temperature for the first process reaches the value$\vartheta_{wb} = 59.7^o C$ and in the second one $\vartheta_{wb} = 29.6^o C$. We shall call first process a slow one and the second a fast one.

Figure 1 illustrates the distribution of the moisture content X for these two processes. Note, that for the slow drying process one obtains not so much

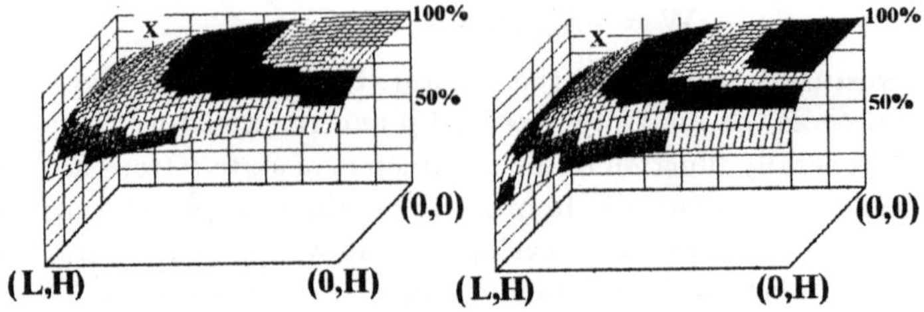

Figure 1: The distribution of the moisture content in the quarter of the bar cross-section at time $t = 3h$ for various parameters of drying medium: a) $\vartheta_{wb} = 59.7°C$, b) $\vartheta_{wb} = 29.6°C$.

differentiated distribution of the moisture content as for the fast one. We can conclude then, that the fast drying processes involve more non-uniform distribution of the moisture content and thus generate more dangerous state of stresses.

3 Stresses in dried materials

The conditions of arising of the shrinkage stresses have been formulated in Rybicki[5]. These conditions point out that main reason for the drying induced stresses are the non-uniform distributions of the moisture content and/or temperature.

During the initial period of the drying process, the temperature ϑ inside of the bar increases very quickly from the initial value to the wet bulb temperature. The change of the moisture potential μ is of little importance during the preheating period, see Figure 2a. The non-uniform distributions of the temperature and moisture content are the reason for the stress generation. In Figure 2b, the normal stresses σ_{yy} are shown. At the beginning of the preheating period the stresses are negative and we identify them as the thermal stresses.

After the preheating period the temperature becomes constant in the whole cross-section, and the non-uniformity in distributions of the moisture content becomes even greater. Figure 3 illustrates the distribution of stresses σ_{yy} for the respective moisture content distribution illustrated in Figure 1.

One can notice, that the stresses reach the maximal value in the point $x = L$, $y = 0$. Figure 4 illustrates the rate of drying W and the evolution of stresses σ_{yy} versus time for two different drying conditions at which the drying rates are almost the same but the temperatures of the drying medium

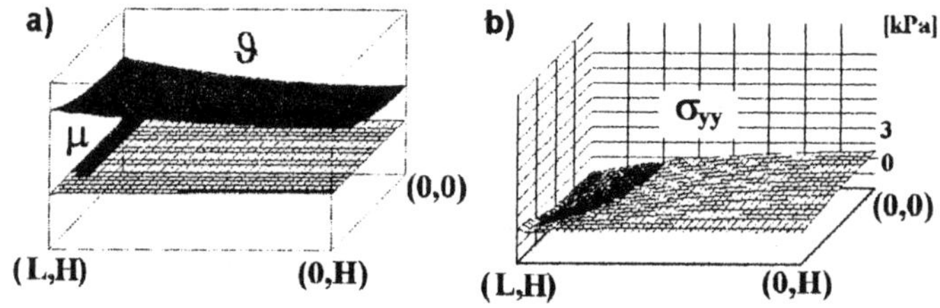

Figure 2: The distribution of the a) temperature ϑ and moisture potential μ ; b) normal stresses σ_{yy} after half of an hour of drying.

are different that is:1) $\vartheta_a = 27^oC$, $Y_a = 0.011$ for which $\vartheta_{wb} = 20.73^oC$ and $\sigma_{\max} = 387kPa$, and 2) $\vartheta_a. = 66^oC$, $Y_a = 0.074$ for which $\vartheta_{wb} = 59.5^oC$ and $\sigma_{\max} = 240kPa$.

4 Final conclusions

1. The considerations indicate a strong dependence of the magnitude of the drying induced stresses on the moisture transport coefficient in dried material.

2. The decrease of the moisture viscosity is one of the methods for increasing of the moisture diffusion rate in dried materials. The magnitude of the drying induced stresses can be reduced while leading the drying process in conditions of the high wet-bulb thermometer temperature.

3. The paper indicates clearly that the alteration of the temperature of the drying process can be also included taken into account by considering a linear model of drying .

Acknowledgments

This work was carried out as a part of research project No3T09C00512 sponsored by the Polish State Committee for Scientific Research in the year 1998.

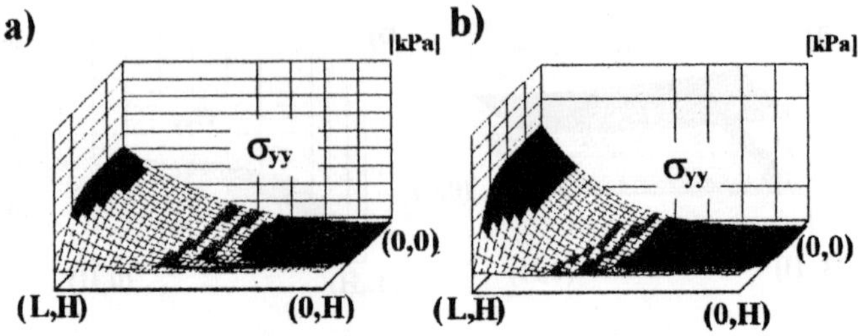

Figure 3: The distribution of the stresses in the quarter of the bar cross-section at time $t = 3h$ for various parameters of drying medium: a) $\vartheta_{wb} = 59.7^oC$, b) $\vartheta_{wb} = 29.6^oC$.

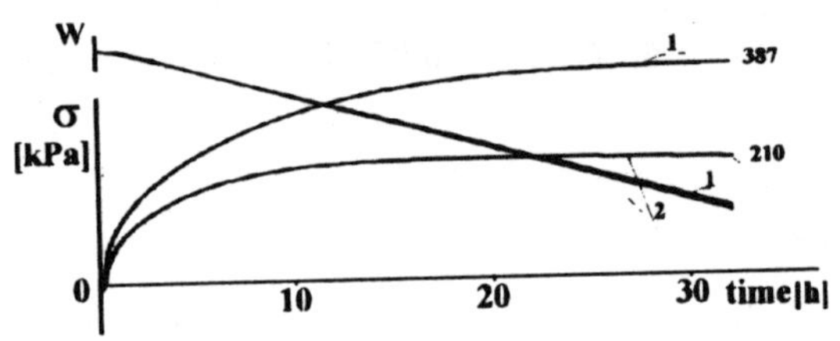

Figure 4: The evolution of the moisture content in the unit length of the bar and the maximal stresses σ_{yy} arisen in point $x = L$, $y = 0$ of the cross-section for the drying process described in the text.

References

1. S.J. Kowalski, Thermomechanics of the Drying Process of Fluid-Saturated Porous Media, *Drying Technology* **Vol.12.** 3, pp .453–482, 1994.
2. S.J.Kowalski and A. Rybicki, Drying Stress Formation Induced by Inhomogeneous Moisture and Temperature Distribution, *Transport in Porous Media*, **Vol. 24**, pp. 139–156, 1996.
3. S.J.Kowalski, A. Rybicki, Computer Simulation of Drying. Optimal Control, *Transport in Porous Media*, **Vol. 26**, 1998.(in print)
4. J.R. Wait and A.R. Mitchell, *Finite Element Analysis and Applications.*, John Wiley & Sons, Chichester, 1986.
5. F. Kneule, Drying, Arkady, 1970, (in Polish).
6. A. Rybicki, Determination of Drying Induced Stresses in a Prismatic Bar, *Engng. Trans.*, **vol. 41**, 2, pp. 139–156, 1993.

TEMPERATURE DEPENDENCE OF THE ELECTRICAL CONDUCTIVITY IN GLASSES WITH TWO TYPES OF HOPPING CENTRES

J. RYBICKI, K. TRZEBIATOWSKI, A. WITKOWSKA

Faculty of Technical Physics and Applied Mathematics, Technical University of Gdansk, Narutowicza 11/12, 80-952 Gdansk, Poland
E-mail: ryba@pg.gda.pl

S. FELIZIANI

Istituto di Matematica e Fisica, Universita' di Camerino, Camerino (MC), Italia

We report in brief on the results of the Monte Carlo calculations of the temperature dependence of hopping electrical conductivity in amorphous systems. In particular, we considered the energetic localised centre distribution $f(\varepsilon)$ consisting of two fractions: a Gaussian distribution g of standard deviation σ, mean energy ε_o, and total concentration c_o, and a single energetic level of concentration c_d and energy $\varepsilon_o + \Delta\varepsilon$ (deeper in the forbidden band), $f(\varepsilon) = c_o\, g\, (\varepsilon - \varepsilon_o; \sigma) + c_d\, \delta\, (\varepsilon - \varepsilon_o - \Delta\varepsilon)$, δ - the Dirac delta function. The probability of the downwards-in-energy jumps is governed by tunnelling factor only, whereas upwards-in-energy jumps are governed also by the Boltzmann factor (thermally activated jumps). The simulations have been performed for the systems in the temperature range of $100\text{K} \leq T \leq 800\text{K}$, with various values of the parameters: $c_o = 0.6 \cdot 10^{27} \div 3.0 \cdot 10^{27}\ \text{m}^{-3}$, $c_d = 0.05 \div 0.3\ c_o$, the Gaussian distribution width $\sigma = 0.0$ and 0.12 eV, and $\Delta\varepsilon = 0.166$ eV, and 0.333 eV. Typical values (common to all simulations) were substituted for parameters like the Bohr radius, and the external field. The conductivity is calculated from the effective time-of-flight of the carriers injected at $t = 0$ from one of the contacts of a thin layer. The simulation results are discussed, and compared to the conductivity measurements in the hydrogen reduced 80% SiO_2 15% Bi_2O_3 5% K_2O glass. The proposed model fully accounts for all qualitative features of the conductivity characteristics in the considered glass.

1 Introduction

Silicate glasses with proper amounts of heavy metal oxides are of great interest because of the possibility of forming surface layers with electron surface conductivities of $10^{-12} \div 10^{-6}\ \text{S}^2$ and a large electron secondary emission coefficient ($3 \div 4$). Properties of these layers remain unchanged even after exposure to the ambient atmosphere. These features make this type of silicate glass very suitable for the production of channeltrons and microchannel plates [1].

Silicate glasses doped with the oxygen co-ordinated Bi, Pb or Sb ions, on annealing in the hydrogen atmosphere at temperatures above 450K undergo a surface (~5μm) modification. At such temperatures there is a certain number of hydrogen molecules (from the high-energy tail of the Maxwell-Boltzmann distribution), which can penetrate into the glass with energies high enough to enter into the reaction with oxygen atoms of SiO_4 tetrahedra, bonded simultaneously with reducible ions (e.g. Bi). This chemical reaction results in the reduction of Bi^{+3} ions to neutral Bi^0 atoms, and in the appearance of H_2O molecules, as well as oxygen vacancies (reduction defects) in the SiO_4 terahedra chains. H_2O molecules evaporate from the glass, leading to a decrease of the sample mass. During the reduction process a certain part of the oxygen vacancies relaxes in such a way, that the nearest non-bonding oxygen from the neighbouring tetrahedra chain achieves the position of the bonding oxygen in the place of the vacancy. As a result in such a material at least two fractions of localised centres exist. The first one is related to the deep structural defects in silica that appear during the glass production, and their concentration does not depend on reduction process. The second fraction of localised centres is related to the described reduction process. The concentration of these centres depends on the reduction conditions, and can change due to various relaxation processes.

The transport mechanism in the considered glasses is not known in detail, and various experiments allow to draw partially contradicting conclusions [2, 3]. In order to gain some insight into the transport mechanism, a simple model distribution of localised centres is proposed. The Monte Carlo simulation technique is applied to investigate the transport properties of the model glass, and the simulation results are compared to the experimental results.

The paper is organised as follows. In Section 2 we describe in short the experiments that have been performed, and in Section 3 we propose a probable model of the energetic distribution of localised centres. The Monte Carlo algorithm is presented in Section 4, whereas the simulation results are described in Section 5. Finally, in Section 6 we compare the numerical predictions with the experimental data on the temperature dependence of the glass conductivity.

2 Experimental

65 SiO_2 15 B_2O_3 15 Bi_2O_3 $5K_2O$ (% mol) glasses [4] were prepared from carefully powdered oxides and nitrates, decomposed at the temperature of ~ 920K. A ceramal prepared in this way was milled again, and placed in a ceramic crucible in a pressed form. The mixture was melted in an electric furnace at the temperature of 1520K for 2 hours. When the molten glass was cooled down to about 1220K, long rods of diameter 1.8 mm were drawn out. A number of experimental methods, as X-ray diffraction, XPS, IR and VIS measurements, AFM microscopy and electrical conductivity measurements (also during the hydrogen reduction process) revealed

conductivity measurements (also during the hydrogen reduction process) revealed a high reproducibility of the samples prepared in the above manner. Such samples were submitted to the hydrogen reduction at temperatures in the range of 470 ÷ 770K for the times from few minutes to several weeks. After the reduction some samples were annealed in vacuum or in the hydrogen atmosphere at temperatures above 670K for the time periods up to several weeks. Then, the DC conductivity measurements were performed on the surface-modified samples. Figure 1 presents two typical conductivity-field characteristics. Curve b corresponds to the sample reduced at 570K for 50 hours, and annealed in vacuum at 720K for 24 hours. Curve a corresponds to the same sample annealed additionally at 770K for another 24 hours. As it is seen, annealing process results in a significant decrease of the conductivity for temperatures lower then 650K. This conductivity decrease must be related to the relaxation of shallow trap states, that occur during the annealing. Since, at sufficiently high temperatures the conductivity is not influenced by the relaxation processes, the concentration of deep structural defects does not change significantly.

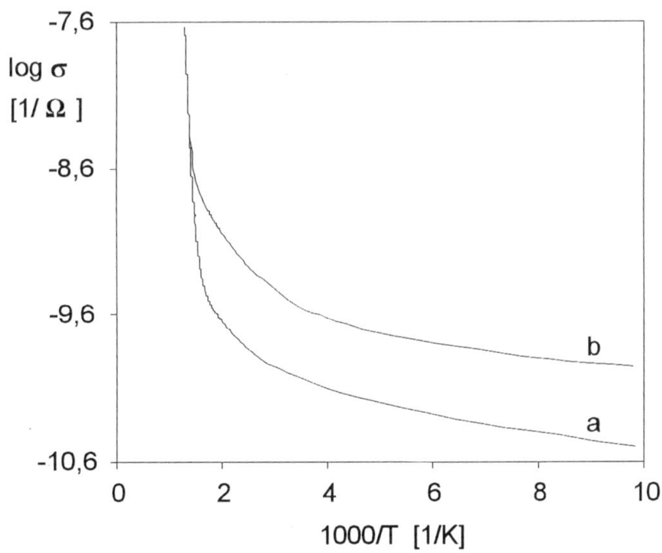

Figure 1: Conductivity - temperature characteristics for 65 SiO_2 15 B_2O_3 15 Bi_2O_3 5K_2O (% mol) glass at two different stages of the annealing process (see text).

3 The glass model

In our numerical modelling we considered the energetic centre distribution $f(\varepsilon)$ consisting of two fractions: a Gaussian distribution g of standard deviation σ, mean energy ε_o, and total concentration c_o, and a single energetic level of concentration c_d and energy $\varepsilon_o + \Delta\varepsilon$ (deeper in the forbidden band), $f(\varepsilon) = c_o\, g\,(\varepsilon - \varepsilon_o; \sigma) + c_d\, \delta\,(\varepsilon - \varepsilon_o - \Delta\varepsilon)$, δ - the Dirac delta function. The two fractions of localised centres represent relatively shallow reduction defects, and deep structural defects, respectively. Only the hopping transport is assumed to be operative. The probability of the downwards-in-energy jumps is governed by tunnelling factor only, whereas upwards-in-energy jumps are governed also by the Boltzmann factor (thermally activated jumps). In particular, the rate ν_{oi} of the carrier jump from the currently occupied site at $\mathbf{r}_0$ to an unoccupied site at $\mathbf{r}_i$ is given by

$$\nu_{oi} = \begin{cases} \nu_0 \exp(-2\alpha|\mathbf{r}_o - \mathbf{r}_i|)\cdot\exp(-\Delta U_{oi}/kT), & \Delta U_{oi} > 0, \\ \nu_0 \exp(-2\alpha|\mathbf{r}_o - \mathbf{r}_i|), & \Delta U_{oi} \le 0, \end{cases} \tag{1}$$

$$\Delta U_{oi} = \varepsilon_i - \varepsilon_o - qE(x_i - x_o)\,. \tag{2}$$

Here α is the reciprocal Bohr radius, ε_o, ε_i, and x_o, x_i are the energies, and x-coordinates of the actually occupied centre, and i-th neighbouring unoccupied centre, respectively, ν_o is the frequency factor, q is the elementary charge, T is the temperature, k is the Boltzmann constant, E is the external field.

4 The simulation algorithm

The conductivity is calculated from the effective time-of-flight of the carriers injected initially from one of the contacts of a planer slab. Thus, the conductivity calculation bases in fact directly on the calculation of transient currents profiles in our model system. Current profiles were calculated using the Monte Carlo method. The applied Monte Carlo algorithm we have applied is similar to that described in [5, 6], and can be summarised as follows. A thin slab of thickness L is placed between the contacts at $x = 0$ and $x = L$. At $t = 0$, an infinitesimally thin sheet of carriers is generated on the left contact ($x = 0$). The applied external electric field E enforces the carrier motion towards the $x = L$ contact. The field E is held constant and uniform, so that no space charge effects are included. The carriers are allowed to jump between localised sites, distributed in space at random, and in energy ε, according to $f(\varepsilon)$, with the average total (integrated over the energy) concentration c. The simulation system is a parallelepiped of dimensions

$M_x \times M_y \times M_z$ sites with periodic boundary conditions imposed in directions perpendicular to the applied field.

The transient currents were calculated from the time and spatial evolution of the injected carrier packet $n(x,t)$, during its motion towards $x = L$, according to the expression:

$$j(t) = \frac{1}{n_0}\frac{d}{dt}\left(\int_0^L n(x,t)dx\right) + \frac{1}{n_0 L}\frac{d}{dt}\left(\int_0^L xn(x,t)dx\right) \tag{3}$$

(e.g. [7]), where $j(t)$ is the particle current per one carrier, n_o is the total number of injected carriers. The applied increment of log t was equal to 0.1 or 0.05. The carrier packets $n(x,t)$ were obtained by averaging the random walks of 3000 individual carriers (20 carriers for each of 150 site generations). The random walk of each individual carrier was started at $x = 0$ and $t = 0$, and finished on arriving to the collecting electrode at $x = L$.
An individual hop from a given occupied centre, say at $\mathbf{r}_0$, to one of the neighbouring empty centres, located at $\mathbf{r}_i$, $i = 1,\dots,$M (M - number of centres within a suitably chosen cut-off radius), has been accomplished as follows. The probability p_{oi} of the hop from the centre at $\mathbf{r}_0$ to the i-th neighbour at $\mathbf{r}_i$ is [6]:

$$p_{oi} = \frac{\nu_{oi}}{\sum_{i=1}^{n} \nu_{oi}},$$

where the jump rate ν_{oi} is given by (1) and (2). According to the probabilities p_{oi}, a corresponding length in a random-number space is given to each site in the neighbourhood of the site at $\mathbf{r}_o$.

5 Simulation results

The Monte Carlo calculations have been performed for the systems in the temperature range of $100\text{K} \le T \le 800\text{K}$, with various values of the parameters: $c_o = 0.6 \cdot 10^{27}$ m^{-3}, $1.5 \cdot 10^{27}$ m^{-3}, and $c_o = 3.0 \cdot 10^{27}$ m^{-3}, $c_d = 0.05 \div 0.3\ c_o$, the width $\sigma = 0.0$ and 0.12 eV, $\Delta\varepsilon = 0.166$ eV, and 0.333 eV.

Figure 2 shows several exemplary families of the MC-calculated transient current profiles. Extensive discussion of the transients shapes, the carrier energy relaxation, and jump statistics, and detailed comparison with the experimental results will be described elsewhere [8]. Here we concentrate ourselves on the temperature dependence of the effective time-of-flight, determined as the intersection of straight lines tangent to the final current decay, and to the current in the region proceeding the final decay. The effective time-of-flight, determined in such a way, is used to calculate the conductivity of the structure. The results are shown in Figure 3. As it is seen, our Monte Carlo calculations show that the conductivity vs $1/T$ curves in high- and low-temperature regions reveal two different activation energies, with their values and transition temperatures strongly dependent on c_d, c_o / c_d ratio, and $\Delta\varepsilon$. The dependencies of σ on c_d, c_o / c_d and $\Delta\varepsilon$ are qualitatively the same as the dependencies on the reduction parameters, which are directly responsible for variations of the hopping centre concentration in hydrogen modified 80% SiO_2 15% Bi_2O_3 5% K_2O glass.

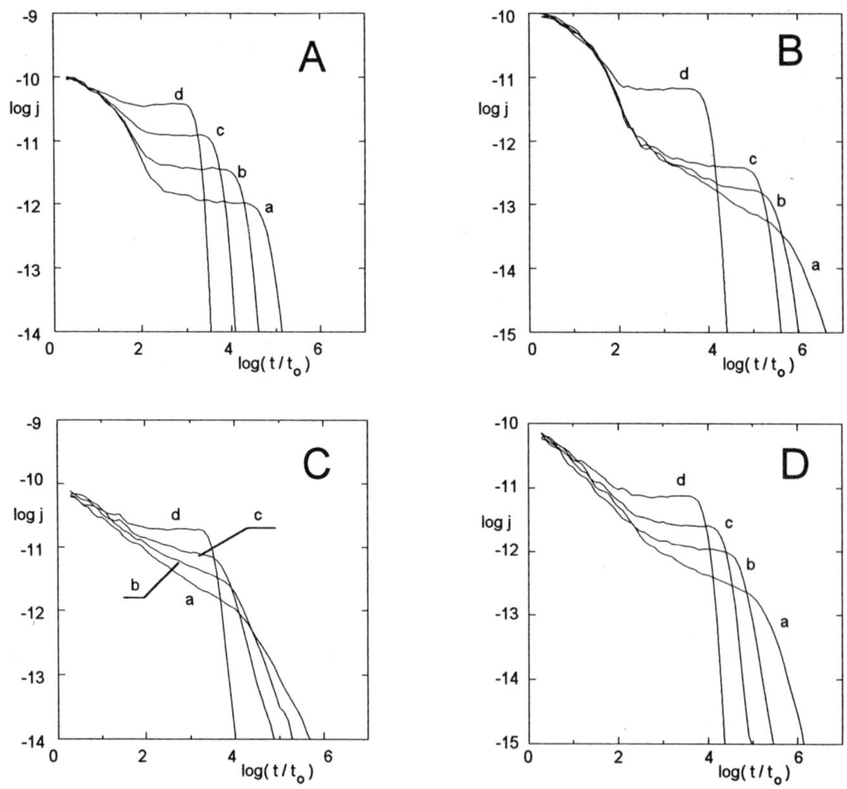

Figure 2: Transient currents in layers with $c_o = 1.5 \cdot 10^{27}$ m^{-1} and $c_d = 0.2\, c_o$. Figure A: $\sigma_o = 0$ eV, $\Delta\varepsilon = 1/6$ eV; a: 100K, b: 200K, c: 333K, d: 800K; Figure B: $\sigma_o = 0$ eV, $\Delta\varepsilon = 1/3$ eV; a: 100K, b: 333K, c: 400K, d: 800K; Figure C: $\sigma_o = 0.12$ eV, $\Delta\varepsilon = 1/6$ eV; a: 100K, b: 250K, c: 400K, d: 800K; Figure D: $\sigma_o = 0.12$ eV, $\Delta\varepsilon = 1/3$ eV; a: 100K, b: 333K, c: 500K, d: 800K. Electric field $E = 10^8$ V/m.

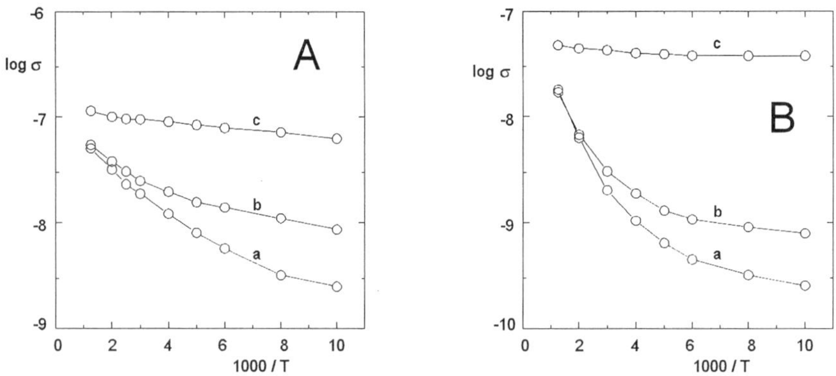

Figure 3: Temperature dependence of the conductivity for $c_o = 1.5 \cdot 10^{27}$ m^{-1}, $\sigma_o = 0.12$ eV. Figure A: $\Delta\varepsilon = 1/6$ eV; Figure B: $\Delta\varepsilon = 1/3$ eV. Curves a: $c_d = 0.1\ c_o$, curves b: $c_d = 0.2\ c_o$, curves c: $c_d = 0.6\ c_o$.

6 Conclusion

The MC simulated conductivity - temperature characteristics reveal a shape which is in general compatible with the corresponding experimental results. The dependence of the simulated characteristics on the parameters such as c_o and c_d is also qualitatively identical as the dependencies of the experimental data on reduction and annealing temperature and time, which directly influence the concentration and energetic distribution of hopping sites. In conclusion we state that all the observed features of the conductivity in hydrogen reduced 80% SiO_2 15% Bi_2O_3 5% K_2O glass can be explained at least qualitatively in terms of the hopping conductivity in solids with two distinct fractions of localised centres.

Acknowledgement

The work has been sponsored by KBN, grant 2 P03 B 167 10, and by CNR, grant 96.00844.ST76. The opportunity to perform our numerical calculations at the TASK Computer Centre (Gdańsk, Poland) is kindly acknowledged.

References

[1] Czajka R, Trzebiatowski K, Polewska W, Kościelska B, Kaszczyszyn S and Susła B 1997 *Vacuum* **48** 213

[2] Then A M and Pantano C G 1990 *J. Non-Cryst. Solids* **120** 178

[3] Ingram M D 1991 *Electrical properties of glasses*, in: *Glasses and amorphous materials*, Zarzycki J (ed), VCH, 715

[4] Trzebiatowski K, Murawski L, Kościelska B, Chybicki M, Gzowski O and Davoli I 1997 Proc. Int. Conf. on Fundamentals of Glass Science and Technology, 9-12 June 1997, Vaxjo, Sweden

[5] Ries B and Bassler H 1987 *Phys. Rev.* **B 35** 2295

[6] Rybicki J, Rybicka A, Feliziani S, Mancini G and Chybicki M 1995 *J.Phys.* **CM 7** 2987; Erratum: *J.Phys.* **CM 7** 5419

[7] Leal Ferreira G F 1977 *Phys. Rev.* **B 16** 4719

[8] Rybicki J, Trzebiatowski K, Witkowska A and Feliziani S 1998 to be submitted to *J. Phys. CM*

REASON OF AMBIGUITY IN ELECTROMAGNETO-MECHANICAL INTERACTIONS

C. RYMARZ
Military University of Technology
S. Kaliski str. 2 01-489 Warsaw POLAND

In this paper some results of studies on the theory of the electromagneto-mechanical coupled (combined) fields formulation are presented. The main attantion has been paid to finding the causes of non-uniqueness of a ponderomotive force, which models the electromagnetic fields action on the condensed matter. The influence of a choice of the system of dimensional units on the uniqueness of the ponderomotive force (PF) has been considered. The reason of the dimensional "degeneration"in CGS system of units has been explained. It is eliminated in the SI system of units.

1. Introduction

Mutual electromagnetic and mechanical interactions in condensed matter play a fundamental role in modern technologies, control systems, and robotics. Therefore they should be properly and uniquely modeled. Modeling of phenomena of interactions is a subject of the theory of Combined of Coupled fields. As we know, mechanical fields act on the electromagnetic (EM) ones by the motion, i.e. by the velocity field, and electromagnetic fields act on the mechanical by the so-called ponderomotive force(PF), which exists only inside the body. Therefore it can be treated as a body force, which acts on its every point. Moreover, it is of a phenomenonlogical nature and differs from the principal electromagnetic forces such as the Coulomb force, describing the interaction of electric charges and the Ampere force, induced by interacting currents. They are nonlocal and singular (1/r). The ponderomotive force is macroscopic and has been introduced to the theory of Electrodynamics of Continuous Media, which is the phenomenological theory of electromagnetism in condensed matter. Nevertheless, it has a good physical background, as the covariant derivative of the four-dimensional tensor of energy-momentum (Minkowski's tensor) and is invariant in respect to the ten-dimensional Galilean group.

In literature one can find different relations for the PF ([1]-[4]). Such a diversity is unacceptable in any theory, since it may lead to some misunderstandings and ambiguites. For example, two different relations for PF are contained in the R.P. Feynman (et.al.) book "*Lectures on Physics*" [1], and in L.I.Siedov's monograph "*Mechanics of Continuous Media*". Another form of relation for PF has been presented by G.A. Maugin and C.A. Eringen in [3].This nonuniqueness of the relatiton for PF is interpreted both by Feynman and Siedov, maintaining that it follows from the some arbitrariness in the definition of the Minkowski's tensor. This problem has been also considered by D.Rogula, S.Kaliski and Z. Płochocki in [5]. This tensor has a bilinear structure, what means that its components are products of the electromagnetic and inductions fields. The Electrodynamics of Continuous media (ECM) allows for some arbitrariness in the choice of these products. Therefore the tensor of energy-momentum is non-uniquely defined. This property is inherent to the theory of EOC and cannot be eliminated, since the Minkowski's tensor is of phenomenological nature and was introduced formally for description of the mechanical phenomena by the EM fields. Thus the " electrodynamic" definition of mechanical values are based on distant physical analogies and dimensional equalities.

For explanation and elimination of the nonuniqueness of the ponderomotive force, the following approaches are proposed:

- experiment,
- construction of the mostly general, phenomenological theory of Combined Fields (CF),
- dimensional analysis based on the principal systems of dimensional units.

Very interesting proposal has been formulated by R.P. Feynman in [1]. He proposed to obtain the unique relation for PF experimentally by joint application of electromagnetic and gravitational fields. It should be treated, however, as a very promising and difficult proposal.

¿From the analysis of the group of invariance of the equations of Combined Fields it follows the existence of the non-trivial subgroups of scaling nature. The properly constructed theory should be invariant with respect to the dimensional transformations belonging to those subgroup This way is promising for construction of a unique theory of Combined Fields, but it should be treated until now as an open problem.

This paper contains the explanation of nonuniqueness of the relation for the ponderomotive force and the proposal how to eliminate it. The considerations are based on dimensional analysis. Following this way, it is possible to establish the cause of the dimensional "degeneration" in the symmetric Gauss system of dimensional units, which generates the investigated nonuniqueness of PF, and to eliminate it the SI system, which has a lower "dimensional" symmetry.

2. Outline of the theory of Combined Fields

Phenomenological theory of Combined Fields consists of:

- equations of Electrodynamics of Continuous Media (ECM),
- equations of Mechanics of Continuous Media (MCM).

1. Set of equations of ECM is the following:

$$rot\mathbf{H} = \frac{1}{c}\frac{\partial \mathbf{D}}{\partial t} + \frac{1}{c}\mathbf{j}, \tag{1}$$

$$rot\mathbf{E} = -\frac{1}{c}\frac{\partial \mathbf{B}}{\partial t}, \tag{2}$$

$$div\mathbf{D} = \varrho_e, \tag{3}$$

$$div\mathbf{B} = \mathbf{0}. \tag{4}$$

The EM fields **E**, **H**, induction fields **D**, **B**, current **j** and electric charge densities satisfy the following additional relations:

$$\mathbf{D} = \mathbf{E} + \mathbf{P}, \tag{5}$$

$$\mathbf{B} = \mathbf{H} + \mathbf{M}, \tag{6}$$

$$\frac{\partial \varrho_e}{\partial t} + div\mathbf{j} = \mathbf{0}, \tag{7}$$

$$div\mathbf{P} + \left(\mathbf{H}\cdot\frac{\partial \mathbf{B}}{\partial t} + \mathbf{E}\cdot\frac{\partial \mathbf{D}}{\partial t}\right) + \mathbf{j}\cdot\mathbf{E} = 0, \tag{8}$$

where the equations (2.7),(2.8) are the charge and energy conservation laws.

2. Set of equations of MCM is the following:

$$\frac{\partial \varrho}{\partial t} + div\,(\varrho\mathbf{v}) = 0, \tag{9}$$

$$\varrho\left(\frac{\partial v_i}{\partial t} + v_j v_{i,j}\right) = F_i + F_i^M + t_{ij,j}\,, \tag{10}$$

$$t_{ij} = t_{ji}, \tag{11}$$

$$\varrho \left(\frac{\partial e}{\partial t} + v_j e_{,j} \right) = t^{ij} v_{j,i} - q_{j,j} + \varrho r. \tag{12}$$

where: ϱ- mass density, v_i- velocity vector, t_{ij}-Cauchy stress tensor, F_i-external body force, F_i^M-ponderomotive force, e- internal energy density, $\mathbf{q}$-heat flux, r-heat sources. The eq. (2.9)-(2.11) are the conservation laws of: mass, momentum, angular momentum and energy in classical formulation.

The above two sets of equations differ significantly.

1. The set of equations of MCM is invariant with respect to the ten-dimensional Galilean group of transformations i.e. under any change of the inertial frames of reference.

2. The set of ECM equations is invariant with respect to the ten-dimensional Lorentz group of transformations. This non-consistency can be eliminated by the construction of the relativistic theory of MCM, or by the asymptotic simplification of the relativistic theory of ECM. Under assumption, that the velocity $\mathbf{V}$ of a medium, or observer satisfies the relation $V << c$, where c is the light velocity, the set of equations of ECM is also invariant with respect to the Galilean group of transformations. Hence the asymptotic theory of Combined Fields can be developed.

In the four-dimensional representation the ponderomotive force is a covariant derivative of the energy-momentum (Minkwski's) tensor S_{ki}.In the three-dimensional representation

$$F_i^M = T_{ji,j} - \frac{1}{c} \frac{\partial p_i^{EM}}{\partial t}, \tag{13}$$

where T_{ji}-the tensor of Maxwell stresses and p_i^{EM}- the density of momentum of EM fields. They are expressed by the EM as follows:

$$\mathbf{p}^{EM} = \frac{1}{c} \mathbf{D} \times \mathbf{B}, \tag{14}$$

$$T_{ij} = E_i D_j + H_i B_j - \frac{1}{2} (\mathbf{DE} + \mathbf{BH}) \, \delta_{ij}. \tag{15}$$

Introducing these two relations into (2.13), we obtain the following formula for PF:

$$F_i^M = \varrho_e E + \frac{1}{c} \epsilon_{ikm} j_k B_m - \frac{1}{2} \left[D_k E_{k,i} - E_k D_{k,i} + B_k H_{k,i} - H_k B_{k,i} \right], \tag{16}$$

This relation contains two parts. The first one, containing two first terms, is the well known Lorentz force, which is exerted on material body by electric charge and current. The second part, containing the derivatives the EM fields, depends on the electric and magnetic anisotropy and homogeneity.

3. Nonuniqueness of relation for ponderomotive force

Let us now investigate the influence of the system of dimensional units on the ambiguity of the PF. We take into account only two of them, i.e. the CGS Gauss symmetric system and the international SI ones. The dimensions of EM fields in both system are given in the following table:

	E	**D**	**H**	**B**
CGS	$M^{1/2}L^{-1/2}T^{-1}$	the same	the same	the same
SI	$MLT^{-3}A^{-1}$	$L^{-2}TA$	$L^{-1}A$	$MT^{-2}A^{-1}$

In the CGS system all EM fields have the same fractal dimensions. Therefore every pair of them is a good "candidate"for the relation of PF. This property follows from a high symmetry of the considered system of dimensional units CGS. This symmetry is caused by the relatively small number of basic units, forming the dimensional space, and by the high symmetries of the used empirical laws. For determination of the dimensions of EM fields in the CGS system, the dimensions of the following physical laws are involved:

¿From the dimension of Coulomb law $[F] = \left[q^2\right] L^2$ and the dimension of the force $[F] = MLT^{-2}$, it follows the dimension of the electric field

$$[E] = M^{\frac{1}{2}} L^{-\frac{1}{2}} T^{-1}. \tag{17}$$

From the relation for the current $[j] = [q]\, T^{-1}$ and from the Ampere law it follows that

$$[B] = [E]\,. \tag{18}$$

The same dimensions can be obtained for **D** and **H.** The considered property of the CGS system is a kind of dimensional "degeneration". It means that different physical values have the same dimension. This degeneration can be eliminated by lowering the existing, dimensional symmetries.

Let us consider now the relation for the PF, applying the SI system of dimensional units. It follows from the presented above table, that each of the EM fields has different dimension. It means that the "degeneration" has been eliminated. It was possible due to the introducing the new, fourth basic dimensional unit-ampere A, and the lowering of the symmetry in the set of physical laws, used for determination of dimensions of the EM in the SI system.

Since the components of the tensor of energy-momentum should be expressed in mechanical units, therefore it can contain only such pairs of fields, for which the ampere unit A is reduced. It follows from the table, that this condition is satisfied by the following pairs: **E D, H B, D B, E H,** and not satisfied by the pairs: **E B, H D.** The dimensions of the allowed pairs are proper for Minkowski's tenor. Therefore it has an invariant structure with respect to the change of system of the dimensional units CGS→SI. For the electromagnetic momentum $\mathbf{p}^{EM}$ only the pair of fields: **D, B** satisfies the dimensional condition. Therefore we get the **unique** form for the ponderomotive force presented in (2.16). This way, in the SI system of dimensional units we eliminated the dimensional degeneration and nonuniqueness of the PF force.

Nevertheless, this result still remains unsatisfactory, since the structure of any properly constructed physical theory should not depend on such a convention as the system dimensional units. Therefore an absolute system of dimensional units should exist, being the "generator" of any system of units, with respect to which, all physical laws should remain invariant. Thus the general problem of construction of the "objective" system of dimensional units remains still an open problem.

4. Conclusions

In this paper:

1. An attempt has been made to explain the reason of nonuniqueness in determination of the ponderomotive force. The unique form of relation for the PF has been obtained only in the SI system of dimensional units.

2. The nonunique relation for the PF in Gauss-symmetric system is caused by the dimensional "degeneration", following from the high symmetry of the principal physic lows used for the calculation of dimensions of the physical values.

3. The obtained results are still physically unsatisfactory, since any well constructed theory should not depend on such a convention as the system of dimensional units.

References

1. R.P. Feynman, R.B. Leighton, M. Sands, *The Feynman Lectures on Physics*, Addison Wesley Publ. Company, London 1964.

2. L.I. Siedov, *Mechanics of Continuous Media (in Russian) tom 1,* Nauka, Moskwa, 1984.

3. G.A. Maugin, A.C. Eringen, J. de Mécanique, **16**,101(1977).

4. C. Rymarz, *Mechanics of Continuous Media (in Polish)*, PWN, Warszawa, 1993.

5. S. Kaliski, Z. Płochocki, D. Rogula, Biul. WAT, **XI**,11(1962).

PROPAGATION OF DISTURBANCES ALONG AXIAL LINES OF TECTONIC FLOWS UNDER CONDITIONS OF CONTINUUM DESTRUCTION

I.SELEZOV
Department of Wave Processes, Inst. of Hydromechanics, NAS of Ukraine, 8/4 Sheliabov Str., 252057 Kiev, Ukraine
E-mail: selezov@wave.pp.kiev.ua

V.SPIRTUS
Department of Seismometry, Inst. of Geophysics, NAS of Ukraine, 32 Palladin Av., 252142 Kiev, Ukraine

A 2D-model is proposed for hydrodynamic flow of geomaterial along the ray tube of tectonic stream. Evolution of the medium damage is governed by a kinetic equation for the damage ratio $\lambda \in [0,1]$. It is shown that the proposed model allows a wave propagation of soliton-like disturbances in a slow background field. This conclusion has been obtained from the analysis of dispersion equation showing that the effect of such a kind can be expected for the narrow wave beams in the vicinity of the critical Morse points of dispersion equation. In a case, when the damage ratio $\lambda \to 1$, the singular degeneration of the system of equations takes place, and this leads to the jumps in structural parameters which can cause earthquakes.

1 Introduction

Dynamical processes connected with tectonic flows in the Earth's lithosphere draws especial attention, particularly in mesozones where along with the flow of geomaterial, the emergence of ruptures can take place. There are several geophysical models but it is not a subject to be considred here except some models.

The first is the diffusion theory of Elsasser [1]. But this model is not able to explain the migration of seismicity to large distances. This is a reason to consider models predicting soliton-like disturbances. Another model developed by Nikolaevsky [2] predicts propagation of tectonic stress disturbances which are like triggers to initiate earthquakes. This model considers the bending and compression of lithosphere plates contacted with the tectonic streams. Inertial forces are neglected in these streams.

Also, it can be noted that earthquake migration effects [3] or tectonic waves propagating with the velocities of 10 - 100 km/h order were earlier understood as connected to the fluid filtration [4] or to interactions between astenosphere and lithosphere [5].

In this paper a new model is developed to predict the possibility of wave propagation. This model is based on a geological conception of tectonic streams as the hydrodynamic structure. The tectonic streams appear at the boundaries of plates as a result of their interactions. These streams are characterized by small velocities (1 cm/year), nonlinear effects, laminar flows, small Reynolds' numbers ($10^{-10} - 10^{-20}$). Typical behaviour of such tectonic streams is observed in seismic regions similar to the Carpathian region [6-8].

The plane problem of 2D-flow of the geomaterial undergoing a lateral pressure and gravitational slipping is considered. In conditions of the continuous destruction [9] the medium damage is described by the structural parameter λ (damage ratio). Following [10] we assume that there is a steadily maintained quasi-balanced thermal regime deteremined by a given temperature $T = T(z)$ depending on the depth. The rheology of tectonic flow is described approximately by the model of viscous Newtonian fluid [11].

Finally, we consider the possibility for soliton-like waves to propagate along the wave tube of tectonic stream in a slow background field. The positive answer gives ground for a new alternative hypothesis about the genesis of some tectonic waves and connected with them earthquakes.

2 Quasistationary flow

In the Cartesian coordinate system $(x_1, x_2, x_3) \leftrightarrow (x, y, z)$ we consider 2D-flow of the medium the state of which is characterized by the vector

$$\vec{Q}^t = \{u_i, \rho, p, \lambda\} \quad (i = 1,3), \tag{1}$$

where $\vec{u} = \left(u_1 = u_x, u_3 = u_z\right)$ is the velocity vector, ρ is the density, p is the pressure, λ is the damage ratio of medium.

Vector $\vec{Q}$ is presented as a superposition of undisturbed and disturbed states

$$\vec{Q}^t = \vec{Q}^0(\vec{x}, t) + \alpha \vec{Q}(\vec{x}, t') \tag{2}$$

corresponding to slow (tectonic) time t and rapid time t', so that $t' << t$. In expression (2) $\vec{Q}^0$ is independent of t' vector which can be considered as a "frozen" background field, α is a small parameter. The field $\vec{Q}$ is considered in the thin layer of a unit width (thin ray tube) (Fig. 1).

Let us introduce the non-dimensional axial and transverse coordinates $\eta = l/l_0$, $\chi = r/h$, where $\eta \in [0,1]$, $\chi \in [-1,1]$. Hereinafter it is assumed that the dynamic viscosity has a local minimum on the axial line, so that $\mu_r = 0$ at $\chi = 0$.

The governing equations are written as follows

$$\frac{D\rho}{Dt} + \rho\ u_{k,k} = 0, \quad k = 1,3, \tag{3}$$

$$\rho \frac{Du_i}{Dt} = -p_{,i} + \tau_{ij,j} + \rho g \delta_{i3}, \quad i,j = 1,3, \tag{4}$$

$$\frac{D\lambda}{Dt} = R(p, \tau_*)(1-\lambda)^{-\beta} \ (\beta > 0), \tag{5}$$

where $\frac{D}{Dt} = \frac{\partial}{\partial t} + (\vec{u} \cdot \vec{\nabla})$, τ_{ij} is the tensor of viscous stresses, $\tau_* = \left(\frac{1}{2}\tau_{ij}\tau_{ij}\right)^{1/2}$ is the intensity of shear stresses. The system (3)-(5) includes the mass conservation law (3), Navier-Stokes equations (4) and kinetic equation (5) [9]. In further, in the stress tensor $\tau_{i,j}$ volume compressibility is not taken into account [12].

From the first law of thermodynamics for a unit mass flow the following equation can be obtained

$$\frac{Ds}{Dt} + \vec{u}\frac{D\vec{u}}{Dt} - gu_3 + \chi_{ext} + \frac{1}{\rho}\ tr(\vec{\tau} \cdot \nabla \vec{u}) = 0 \tag{6}$$

The physical sense of relation (6) lies in the fact that the change of specific enthalpy $s = E + p/\rho$ takes place due to the change of kinetic and potential energy of the flow due to the heat-exchange with the surrounding medium (χ_{ext}),

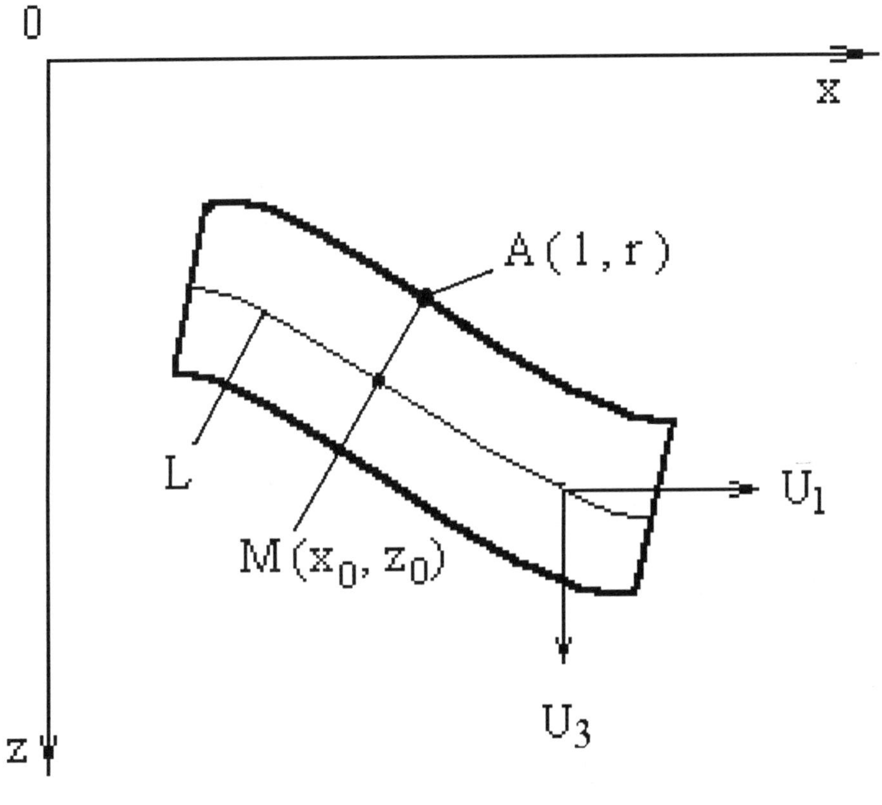

Figure 1

as well as due to the work of shear stresses. The last term, connected with the energy dissipation may be presented in the form

$$\Phi = \frac{1}{\rho} tr(\vec{\tau} \cdot \nabla \vec{u}) = \nu \left[\left(u_{1,x} - u_{3,z} \right)^2 + (u_{3,x} + u_{1,z})^2 \right] , \quad (7)$$

where $\nu = \mu / \rho$ is the kinematic viscosity. Dissipation function Φ characterizes the work of deviator stresses transfering into the internal energy. Using the energy law $E = E(p, T, \lambda)$, it is easy to obtain from (6) the equation

$$q_1 p_t - \frac{p}{\rho^2} \rho_t + \left(\frac{1}{2} u^2 \right)_t + \left(\frac{\partial E}{\partial \lambda} \right)_{p,T} \lambda_t +$$
$$+ q_2 u_3 + q_3 u_1 + \Phi + \chi_{ext} = 0, \quad (8)$$

where

$$q_1 = \left(\frac{\partial E}{\partial p} \right)_{T,\lambda} + \rho^{-1},$$

$$q_2 = -g + c_p T_z + \left(\frac{\partial E}{\partial p} \right)_{T,\lambda} p_x + \left(\frac{\partial E}{\partial \lambda} \right)_{p,T} \lambda_z ,$$

$$q_3 = \left(\frac{\partial E}{\partial p} \right)_{T,\lambda} p_x + \left(\frac{\partial E}{\partial \lambda} \right)_{p,T} \lambda_x .$$

Now the closed system of equations can be written for the undisturbed problem corresponding to steady laminar stream with a "frozen" values $\lambda = \lambda_*$ and negligibly small value $\frac{D\vec{Q}^0(\vec{x},t)}{Dt} << 1$.

Let us introduce instead of u_2 a new variable φ

$$\varphi = \frac{dz}{dx} = \frac{u_3}{u_1} . \quad (9)$$

and new variables

$$\theta = \varphi_x, \quad \psi = u_{1,x}, \quad \Psi = u_{1,z}, \quad \Omega = \varphi_z. \tag{10}$$

For stationary flow at small Reynolds' numbers we may eliminate the inertial term in equation (4), taking $\frac{D\vec{u}}{Dt} = 0$. Then equations (3)-(5) and (8) are reduced to the following equations

$$\rho_x^0 + \rho_z^0 \varphi^0 + \rho^0 \Omega + \frac{\rho^0}{u_1^0}\left(\psi^0 + \varphi^0 \Psi^0\right) = 0 , \tag{11}$$

$$p_x^0 - \mu \ (\psi_x^0 + \Psi_z^0) - \mu_x n_1^0 - \mu_z n_2^0 = 0 , \tag{12}$$

$$p_z^0 - \mu\left[\varphi^0(\psi_x^0 + \Psi_z^0) + 2(\psi^0\theta^0 + \Psi^0\Omega^0) + u_1^0(\theta_x^0 + \Omega_z^0)\right] +$$

$$+\mu_z n_1^0 - \mu_x n_2^0 - \rho g = 0 , \tag{13}$$

$$R(p^0, \tau_*) = 0, \qquad \chi_{ext}^0 + \Phi^0 = 0 , \tag{14}$$

where $n_1^0 = \psi^0 - \Psi^0 \varphi^0 - u_1^0 \Omega^0$, $n_2^0 = \Psi^0 + \psi^0 \varphi^0 + u_1^0 \theta^0$.

Thus, to determine the unknowns u_1, φ, ρ, p and their derivatives $\theta, \psi, \Psi, \Omega$ we obtain eight first order equations. It is easy to show that $\tau_* = \sqrt{\mu \rho^0 \Phi^0}$ and $\Phi^0 = \nu \left[(n_1^0)^2 + (n_2^0)^2\right]$. It is assumed that the value $q_2^0 u_3^0 + q_3^0 u_1^0$ is much lesser than the value $\chi_{ext}^0 + \Phi^0$.

We assume the existence of a stationary solution of undisturbed problem with the function φ independent of z which is in consistency with the second thermodynamic law. This imposes some restrictions on the kinetic equation of damage (5) and equation of state as well.

3 Dispersion equation

Let point A lies on an edge of layer (Fig. 1), so that $A = (l, h(l))$, and point $M = (x_0, z_0) = (l, 0) \in L$. Then the following relationships take place

$$z - z_0 = -\left(x - x_0\right) / \varphi, \tag{15}$$

$$r^2 = (x - x_0)^2\left[1 + (u_1 / u_3)^2\right], \quad r = \pm(x - x_0)l_z , \tag{16}$$

where $l_z = (1+\varphi^{-2})^{1/2} = u / u_3$, $l_x = (1+\varphi^2)^{1/2} = u / u_1$.

For a thin layer $|r / l| << 1$ and $r_x \approx l_z$ approximately , so that

$$\frac{\partial}{\partial x} = \frac{\partial}{\partial l} l_x + \frac{\partial}{\partial r} l_z, \quad \frac{\partial}{\partial z} \cong \frac{\partial}{\partial l} l_z . \tag{17}$$

Taking into account (17) we obtain the following system of equations for five unknown components of $\vec{Q}$

$$b_0 u_1 + \rho_l l_z u_3 + \rho\, l_x u_{1,\,l} + \rho\, l_z u_{3,\,l} + \rho_{t'} = 0,$$

$$\rho u_{1,\,t'} + b_l u_{1,\,l} - \mu b_2 u_{1,\,ll} + l_z(\mu_x - \varphi\mu_z) u_{3,\,l} + l_x p_l = 0,$$

$$\rho u_{3,\,t'} + b_l u_{3,\,l} - \mu b_2 u_{3,\,ll} - l_z(\mu_x - \varphi\mu_z) u_{1,\,l} - g\,\rho + l_z p_l = 0,$$

$$-R_p p + (1-\lambda)^\beta \lambda_{t'} = 0, \quad -\frac{p}{\rho^2} p_{t'} + q_1 p_{t'} + \left[\left(\frac{\partial E}{\partial \lambda}\right)_{p,T}\right]_{t'} \lambda_{t'} = 0 , \tag{18}$$

where

$$b_0 = \rho_l l_x + \rho_r l_z , \quad b_l = -\mu\, l_{xx} - l_z(\mu_x - \varphi\mu_z), \quad b_2 = l_x^2 + l_z^2 .$$

Due to smallness of u_1, u_3 and their gradients for slow tectonic motions (characteristic velocities of 1 cm/year order) the terms with such coefficients in the equations (18) are dropped.

Undisturbed state functions $Q_r^0 (r = \overline{1,5})$ depend on $\vec{x}$ and slow time $t' << t$, so that analysing disturbed state for fast time t' it is possible to consider "frozen" state depending only on space coordinates x, z (or l_τ, l_n). However, in this case the solution for travelling waves should be presented in the form $f(l)e^{i(kl-\omega t)}$. This problem is of great complexity. In further, we consider the uniform state with constant values Q_r^0.

Travelling waves along the streamline are characterized by the wave number vector $\vec{k}$ and the tangent vector $\vec{l} = l_\tau \vec{e}_\tau$ (in the case of unit orthonormal basis vector $|\vec{e}_\tau| = 1$). We consider travelling waves along the straight streamline coinciding with a tangent at the point under consideration.

Now, the solutions can be presented in the form of one-dimensional travelling waves along the tangent direction

$$(u_1,\ u_3,\ \rho,\ p,\ \lambda) = (\hat{u}_1, \hat{u}_3, \hat{\rho}, \hat{p}, \hat{\lambda}) e^{i(\kappa l - \omega t')} . \tag{19}$$

Substituting (19) into the system (18) leads to the dispersion equation

$$b_3 \omega^2 \kappa^2 - q_1 \rho i\omega^3 + b_1 q_1 i\omega^2 \kappa + (b_2 \frac{p}{\rho} + \xi_* \mu b_2) i\omega \kappa^2 +$$

$$+\rho \xi_* \omega^2 + (b_5 - b_1 \xi_*) \omega \kappa + q_1 b_4 i\omega - b_6 \xi_* i\kappa - b_4 \xi_* = 0 , \tag{20}$$

where

$$b_3 = \mu b_2 q_1 \ , \ b_4 = g\rho_l l_z \ , \ b_5 = \frac{p}{\rho^2}(b_0 l_x + \rho_l l_z^2) - b_6 q_1 \ ,$$

$$b_6 = g l_z \rho \ , \quad \xi_* = \left(\frac{\partial E}{\partial \lambda}\right)_{p,T} R_p (1-\lambda)^{-\beta} .$$

It should be noted that all coefficients in (20) and in further are constant in accordance with two above assumptions about uniform background field and straight streamline.

4 Analysis in the vicinity of the critical Morse points (CMP)

Following [13] we introduce the parameter ξ governing the local evolution processes of the mapping $F(\kappa, \omega, \xi) : C \times C \times C \to C$. It is convenient to choose the parameter ξ as "dispersion" of the parameter ξ_*^{-1}, according to

$$\xi_*^{-1} = \xi_0 + \xi^2 . \tag{21}$$

Then we have

$$F(\kappa, \omega, \xi) = F_1(\kappa, \omega) + \xi^2 F_2(\kappa, \omega), \tag{22}$$

where

$$F_1(\kappa,\omega)=b_3\xi_0\omega^2\kappa^2-q_1\rho\xi_0 i\omega^3+b_7 i\omega\kappa^2+b_9 i\omega^2\kappa+$$
$$+\rho\omega^2+b_8\omega\kappa+q_1 b_4\xi_0 i\omega-b_6 i\kappa-b_4,$$
$$F_2(\kappa,\omega)=b_3\omega^2\kappa^2-q_1\rho i\omega^3+$$
$$+b_2\frac{p}{\rho}i\omega\kappa^2+b_1 q_1 i\omega^2\kappa+b_5\omega\kappa+q_1 b_4 i\omega,$$

$$b_7=b_2\frac{p}{\rho}\xi_0+\mu b_2,\ b_8=b_5\xi_0-b_1\ ,b_9=b_1 q_1\xi_0.$$

In an isolated nondegenerate CMP $W_m\in C^3$

$$F_\kappa\Big|_{W_m}=F_\omega\Big|_{W_m}=F_\xi\Big|_{W_m}=0,\quad \det J(\kappa,\omega,\xi)\Big|_{W_m}\neq 0, \tag{23}$$

where J is the Hessian of the mapping F [14].

Providing that $F_2\neq 0$, $\xi=0$ in points W_m. Then, if $W_m=(\kappa_0,\omega_0,0)$, from (23) we have for κ_0,ω_0 the following system of equations

$$F_{1,\kappa}=C_3\kappa_0+iC_4=0,\qquad F_{1,\omega}=C_7\kappa_0^2+C_5\kappa_0+C_6=0, \tag{24}$$

where

$$C_3=2(b_3\xi_0\omega_0^2+b_7 i\omega_0),$$
$$C_4=b_9\omega_0^2-b_8 i\omega_0-b_6,$$
$$C_5=b_8+2ib_9\omega_0,$$
$$C_6=-3q_1\rho i\omega_0^2\xi_0+2\rho\omega_0+ib_4 q_1\xi_0,$$
$$C_7=2b_3\xi_0\omega_0+b_7 i,$$
$$\xi_0=\xi_*^{-1}=\left(\frac{\partial E}{\partial\lambda}\right)^{-1}_{p,T}R_P^{-1}(1-\lambda)^\beta,$$

and we obtain the relationships

$$\kappa_0=-i\frac{C_4}{C_3},\quad C_6C_3^2-iC_3C_4C_5-C_7C_4^2=0. \tag{25}$$

The values $k_0^{(j)}$ corresponding to six complex roots $\omega_0^{(j)}$ of algebraic equation (25) $(j=\overline{(1,6)}$ are found from the first formula of (25).

It can be easily proved that the critical points W of the mapping F are degenerate points only in some special cases. The complex hypersurface in the vicinity of MCP on the basis of the Morse lemma can be presented in the form [14]

$$\lambda_1\hat{\kappa}^2+\lambda_2\hat{\omega}^2+\lambda_3\hat{\xi}^2+2\delta=0\,. \tag{26}$$

Passing from the dispersion equation (26) onto the corresponding linear partial differential equations is realized according to the formulae $\hat{\kappa}=-i\partial/\partial s$, $\hat{\omega}=i\partial/\partial\tau$ where s and τ are already not the spatial and time coordinates but their linear combinations [14]. Thus, in the vicinity of CMP we obtain the linear Klein-Gordon equation

$$\lambda_1\frac{\partial^2 u}{\partial s^2}-\lambda_2\frac{\partial^2 u}{\partial\tau^2}+\gamma(\xi)u=0,\quad \gamma(\xi)=\lambda_3\hat{\xi}^2+2\delta. \tag{27}$$

If the signs of λ_1, λ_2, λ_3 and δ are the same, then equation (27) is of a hyperbolic type and describes the wave propagation. In the case of weak nonlinearities, the equation (27) can be transformed onto nonlinear evolution equations which allow the formation of solitons and kinks, as well as their transmutations [15, 16].

Taking into account above mentioned remarks and having made certain simplifications, we may assume the possibility of propagation of soliton-like disturbances in the form of narrow wave beams with $\kappa\approx\kappa_0^{(j)}$ that corresponds to the roots of equation (25), that is in the vicinity of CMP.

It shoud be noted that in the nonlinear optics for narrow beams such effects as self-compression and self-focusing of wave packets are investigated theoretically. The instability of one-dimensional wave packet as a result of self-compression leads to its decomposition onto separate wave clusters.

Consider the most important case

$$(1-\lambda)\to 0\,, \tag{28}$$

corresponding to pre-ruptured state of the medium. It is transparent that when $\xi_0\to 0$ the equations (25) take the form

$$\kappa_0=\frac{b_6\omega_0^{-1}-ib_1}{2\mu b_2},\qquad 8\rho\mu b_2 i\omega_0^3-b_1^2\omega_0^2-b_6^2=0. \tag{29}$$

Thus, in the limiting cases no more than three CMP exist. Similarly to the behaviour of Van der Pol oscillator, it can be expected that under condition (28), the system of equations (18) containing the small parameter in the derivative of $\tilde{\lambda}_{t}$, will predict the jumps typical for relaxation oscillations. According to [17] some standing waves correspond to pure imaginary frequencies and wave numbers in equations (29). For these spectral parameters, dependence on the equations of state and kinetic equation disapppears.

Conclusions

1. A new model to describe the pre-ruptured state of tectonic flows is proposed. The model is based on hydrodynamic and thermodynamic equations and kinetic equation for damage ratio $\lambda \in [0,1]$, as well as on the conception of streamline structure of tectonic flows.
2. Analysis of the vicinity of spectral parameters corresponding to critical Morse's points (CMP) allows to discover the influence of geometric, kinematic and physical parameters on a tectonic flow evolution.
3. It is shown that near the pre-ruptured state when $(1-\lambda) \to 0$ there exist no more than three CMP. Due to singular degeneration of the system of governing equations, in this case the jumps of parameters appear which are typical for relaxation oscillations to cause earthquakes.

References

1. W.M. Elsasser, Convectional stress propagation in the upper mantle, in *Appl. Modern Physics. Earth,* Planets Interior (New York, Wiley, 1969).
2. V.N. Nikolaevsky and T.K. Ramasanov, *Appl. Mathematics and Mechanics* (in Russian) **49**, 462 (1985).
3. K. Kasahara, *Earthquake mechanics.* (Cambridge University Press, 1981).
4. A.Nur and J.R.Booker , *Science,* **175**,.885 (1975).
5. J.R Rice, The mechanics of earthquake rupture, in *Physics of the Earth's Interior,* 555 (1980).
6. E.I. Patalakha, A.I. Lukienko and V.V. Gonchar, *Tectonic flows as the basis of understanding geological structures* (in Russian) (Kiev, Naukova Dumka, 1995).
7. M.V.Komarova and E.S.Shtengelov, *Geophys. J.* (in Russian), 2(4), 63 (1980).
8. O.Cadek, Yuen D.A. and H.Cizkova, *University of Minnesota Supercomputing Institute Research Report UMSI* 97/118, (1977).
9. V.L, Kondaurov et al, *Mechanics of destruction of mountaing rocks* (in Russian) (Moscow, Nauka, 1987).
10. Z. Reches , G. Schubert and C. Anderson, *J. Geophys. Res.*, 99, 983 (1994).
11. V.E,Kulchitsky and V.B.Spirtus, *Geophys. J.* 16, 701 (1997).

12. G. Astarita and G. Marrucci, *Pronciples of non-Newtonian fluid mechanics.* (McGraw-Hill, 1974).
13. V.P. Shestopalov, *Reports of Russian Acad. Sci.* 335, 440 (1994).
14. J. Milnor, *Singular points of complex hypersurfaces* (Prinston University Press, 1968).
15. V.I.Karpman, *Nonlinear waves in dispersive media* (in Russian) (Nauka, Moscow, 1973)
16. I.T.Selezov and S.V.Korsunsky, *Nonstationary and nonlinear waves in electrically conducting media* (in Russian) (Kiev, NaukovaDumka, 1991).
17. P.L. Bhatnagar, *Nonlinear waves in one-dimensional dispersive systems* (Clarendon Press, Oxford, 1979).

A CELL DYNAMICAL SYSTEM MODEL FOR SIMULATION OF CONTINUUM DYNAMICS OF TURBULENT FLUID FLOWS

A.M.SELVAM AND SUVARNA FADNAVIS

Indian Institute of Tropical Meteorology, Pune 411 008, India

Email: selvam@tropmet.ernet.in

Website: http://www.geocities.com/CapeCanaveral/Lab/5833

Mathematical models for the simulation of turbulent fluid flows based on Newtonian continuum dynamics are nonlinear and finite precision computer realizations give unrealistic solutions because of sensitive dependence on initial conditions identified as *deterministic chaos*. Sensitive dependence on initial conditions implies long-range spatiotemporal correlations. Such non-local connections are also exhibited by dynamical systems in nature, identified as *self-organized criticality* . It has been shown that deterministic chaos is a consequence of finite precision related round-off error growth in iterative computations in numerical solutions. A non-deterministic cell dynamical system model has been developed which predicts the observed self-organized criticality as a direct consequence of quantumlike mechanics governing turbulent flow dynamics of the atmosphere. The model concepts lead to the following predictions. (1) The flow structure consist of an overall logarithmic spiral trajectory with the quasiperiodic Penrose tiling pattern for the internal structure. Conventional power spectral analysis will resolve such a spiral trajectory as a continuum of eddies with progressive increase in phase. (2) The eddy energy spectrum self-organizes to form statistical normal distribution. (3) The universal algorithm for self-organized criticality is expressed in term of the universal Feigenbaum's constant a and d as $2a^2 = \pi d$ where πd represents the relative volume intermittency of occurrence which contributes to total variance $2a^2$ of the fractal structures. (4) The observationally determined quantum mechanical constants, *fine structure constant* equal to about *137* and *ratio of proton mass to electron mass* equal to about *1840* are now derived from the theory for macroscale flow dynamics.

1. INTRODUCTION

Atmospheric flows exhibit long-range spatiotemporal correlations manifested as the *fractal* geometry to the global cloud cover pattern concomitant with inverse power-law form for power spectra of temporal fluctuations of all scales ranging from turbulence (millimeters-seconds) to climate (thousands of kilometers-years) [22]. Long-range spatiotemporal correlations are ubiquitous to dynamical systems in nature and are identified as signatures of *self-organized criticality* [1]. Standard models for turbulent fluid flows in meteorological theory cannot explain satisfactorily the observed *multifractal* (space-time) structures in atmospheric flows. Numerical models for simulation and prediction of atmospheric flows are subject to *deterministic chaos* and give unrealistic solutions. *Deterministic chaos* is a direct consequence of round-off error growth in iterative computations. Round-off error of finite precision computations doubles on an average at each step of iterative computations[13]. Round-off error will propagate to the mainstream computation and give unrealistic solutions in numerical weather prediction (*NWP*) and climate models which incorporate thousands of iterative computations in long-term numerical integration schemes. A recently developed non-deterministic cell dynamical system model for atmospheric flows[12,14] predicts the observed self-organized criticality as intrinsic to quantumlike mechanics governing flow dynamics. Further, the fractal space-time structure to the stringlike atmospheric flow trajectory is resolved into a continuum of eddies. The eddy circulations obey *Kepler*'s third law of planetary motion and therefore eddy inertial masses obey *Newton's inverse square law of gravitation* on all scales from microscopic to macroscale[19,20]. El Naschie[6] has discussed, the fractal structure to space-time and also states that fractalisation of microspace is the origin of gravity.

2. A Non-Deterministic Cell Dynamical System Model For Fluid Flows: A String Theory For Fractal Spacetime

Based on Townsend's [23] concept that large eddies are the envelopes of enclosed turbulent eddy circulations, the relationship between the large and turbulent eddy circulation speeds (W and w_*) and radii (R and r) respectively is given as

$$W^2 = \frac{2}{\pi}\frac{r}{R}w_*^2 \tag{1}$$

Since the large eddy is the integrated mean of enclosed turbulent eddy circulations, the eddy energy (kinetic) spectrum follows statistical normal distribution. Therefore, square of the eddy amplitude or the variance represents the probability. Such a result that the additive amplitudes of eddies, when squared, represent the probability densities is obtained for the subatomic dynamics of quantum systems such as the electron or photon [11]. Atmospheric flows, therefore, follow quantumlike mechanical laws. Incidentally, one of the strangest things about physics is that we seem to need two different kinds of mechanics, quantum mechanics for microscopic dynamics of quantum systems and classical mechanics for macroscale phenomena [18]. The above visualization of the unified network of atmospheric flows as a quantum system is consistent with Grossing's [9] concept of quantum systems *as order out of chaos* phenomena. Order and chaos have been reported in strong fields in quantum systems [3].

The square of the eddy amplitude W^2 represents the kinetic energy E given as (from Equation 1)

$$E=H\nu \tag{2}$$

Where ν (proportional to $1/R$) is the frequency of the large eddy and H is a constant equal to $\frac{2}{\pi}r\,w_*^2$ for growth of large eddies sustained by constant energy input proportional to w_*^2 from fixed primary small scale eddy fluctuations. Energy content of eddies is therefore similar to quantum systems which can possess only discrete quanta or packets of energy content $h\nu$ where h is a universal constant of nature (*Planck's constant*) and ν is the frequency in cycles per second of the electromagnetic radiation. The relative phase angle between large and turbulent eddies is equal to r/R and is directly proportional to W^2 (Equation 1). The phase angle therefore represents variance and also there is progressive increase in phase with increase in wavelength. The above relationship between phase angle, variance and frequency has been identified as *Berry's Phase* [2] in the subatomic dynamics of quantum systems. Berry's phase has been identified in atmospheric flows [14,15,19].

Writing Equation (1) in terms of the periodicities T and t of large and small eddies respectively, where

$$T = \frac{2\pi R}{W}$$

and

$$t = \frac{2\pi r}{w_*}$$

We obtain

$$\frac{R^3}{T^2} = \frac{2}{\pi}\frac{r^3}{t^2} \tag{3}$$

Equation (3) is analogous to *Kepler*'s third law of planetary motion, namely, the square of the planet's year (period) to the cube of the planet's mean distance from the *Sun* is the same for all planets [24]. Newton developed the idea of an inverse square law for gravitation in order to explain Kepler's laws, in particular, the third law. Kepler's laws were formulated on the basis of observational data and therefore are of empirical nature. A basic physical theory for the inverse square law of gravitation applicable to all objects, from macroscale astronomical objects to microscopic scale quantum systems is still lacking. The model concepts are analogous to a string theory[10] where, superposition of different modes of vibration in stringlike energy flow patterns result in material phenomena with intrinsic quantumlike mechanical laws which incorporate inverse square law for inertial forces, the equivalent of gravitational forces, on all scales of eddy fluctuations from macro- to microscopic scales. The cumulative sum of centripetal forces in a hierarchy of vortex circulations may result in the observed inverse square law form for gravitational attraction between inertial masses (of the eddies).

The apparent paradox of wave-particle duality in microscopic scale quantum systems [18] is however physically consistent in the context of macroscale atmospheric flows since the bi-directional energy flow structure of a complete atmospheric eddy results in the formation of clouds in updraft regions and dissipation of clouds in downdraft regions. The commonplace occurrence of clouds in a row is a manifestation of *wave-particle duality* in the macroscale quantum system of atmospheric flows (Fig. 1).

Figure 1. Wave-particle duality in atmospheric flows.

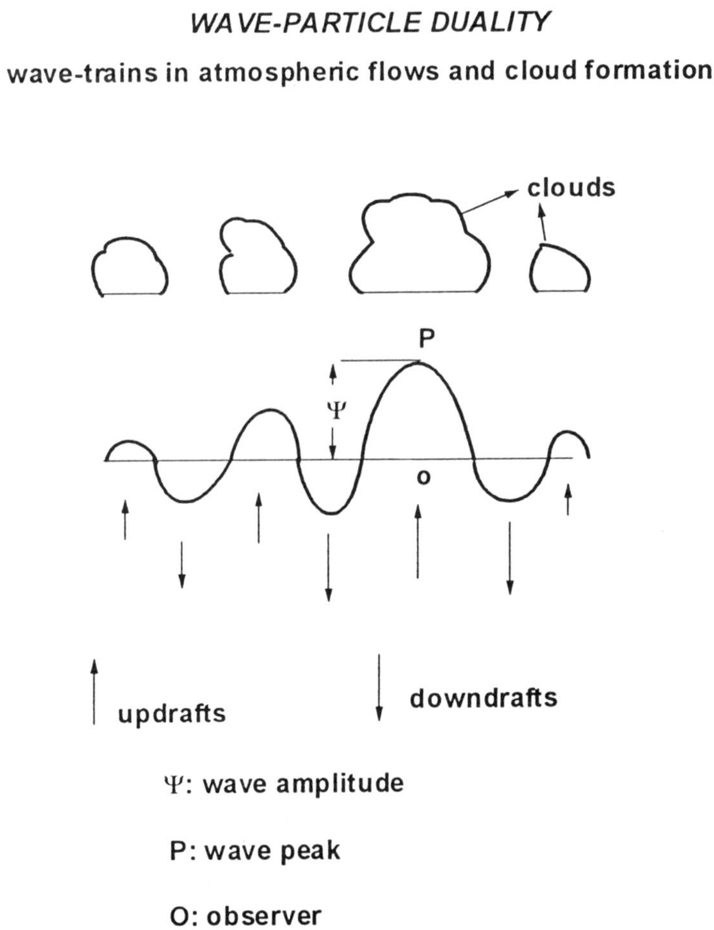

The above-described analogy of quantumlike mechanics for atmospheric flows is similar to the concept of a subquantum level of fluctuations whose space-time organization gives rise to the observed manifestation of subatomic phenomena, i.e. quantum systems as order out of chaos phenomena.

Puthoff [17] has also put forth the concept of "gravity as a zero-point fluctuation force". The vacuum zero-point fluctuation (electromagnetic) energy is manifested in the *Casimir effect* is analogous to the turbulent scale fluctuations whose spatial integration results in coherent large eddy structures. El Naschie has also proposed[6] that *Cantorian-fractal* conception of spacetime may effect reconciliation between quantum mechanics and gravity.

2.1. *Model Predictions*

(a) Atmospheric flows trace an overall logarithmic spiral trajectory $OR_0R_1R_2R_3R_4R_5$ simultaneously in clockwise and anti-clockwise directions with the quasi-periodic *Penrose tiling pattern* [21] for the internal structure Fig. 2.

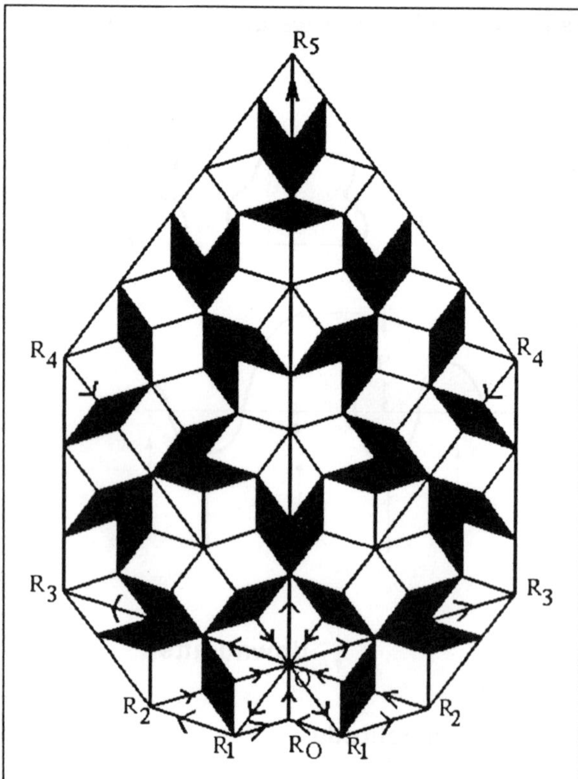

Figure 2. The quasiperiodic *Penrose* tiling pattern.

(b) Conventional continuous periodogram power spectral analyses of such spiral trajectories will reveal a continuum of periodicities with progressive increase in phase.

(c) The broadband power spectrum will have embedded dominant wave-bands the bandwidth increasing with period length. The peak periods E_n in the dominant wavebands will be given by the relation

$$E_n = T_S(2+\tau)\tau^n \tag{4}$$

Where τ is the *golden mean* equal to $(1+\sqrt{5})/2$ [$\cong 1.618$] and T_s is the primary perturbation time period, for example, the solar powered annual cycle (summer to winter) of solar heating in a study of interannual climate variability. Ghil [8] reports that the most striking feature in climate variability on all time scales is the presence of sharp peaks superimposed on a continuous background. The model predicted periodicities are *2.2, 3.6, 5.8, 9.5, 15.3, 24.8, 40.1* and *64.9* years for values of n ranging -1 to 6. Periodicities close to model predicted have been reported [4].

(d) The overall logarithmic spiral flow structure is given by the relation

$$W = \frac{w_*}{k} \ln Z \tag{5}$$

Where the constant k is the steady state fractional volume dilution of large eddy by inherent turbulent eddy fluctuations. The constant k is equal to $1/\tau^2$ ($\cong 0.382$) and is identified as the universal constant for deterministic chaos in fluid flows. The steady state emergence of fractal structures is therefore equal to

$$1/k \cong 2.62 \tag{6}$$

The model predicted logarithmic wind profile relationship such as equation (5) is a long-established (observational) feature of atmospheric flows in the boundary layer, the constant k, called the *Von Karman* 's constant has the value equal to *0.38* as determined from observations. Historically, equation (5), basically an empirical law known as the *universal logarithmic law of the wall* , first proposed in the early 1930s by pioneering aerodynamicists *Theodor von Karman* and *Ludwig Prandtl*, describes shear forces exerted by turbulent flows at boundaries such as wings or fan blades or the interior wall of a pipe. *The law of the wall* has been used for decades by engineers in the design of aircraft, pipelines and other structures (Cipra,1996)[5].

In equation (5), W represents the standard deviation of eddy fluctuations, since W is computed as the instantaneous r.m.s. (root mean square) eddy perturbation amplitude with reference to the earlier step of eddy growth. For two successive stages of eddy growth starting from primary perturbation w_*, the ratio of the standard deviations W_{n+1} and W_n is given from equation (5) as $(n+1)/n$. Denoting by σ the standard deviation of eddy fluctuations at the reference level $(n=1)$ the standard deviations of eddy fluctuations for successive stages of eddy growth are given as integer multiples of σ, i.e., σ, 2σ, 3σ, etc. and correspond respectively to

Statistical normalized standard deviation $t = 0,1,2,3,$ **etc.** (7)

The conventional power spectrum plotted as the variance versus the frequency in log-log scale will now represent the eddy probability density on logarithmic scale versus the standard deviation of the eddy fluctuations on linear scale since the logarithm of the eddy wavelength represents the standard deviation, i.e. the r.m.s. value of eddy fluctuations (5). The r.m.s. value of eddy fluctuations can be represented in terms of statistical normal distribution as follows. A normalized standard deviation $t=0$ corresponds to cumulative percentage probability density equal to *50* for the mean value of the distribution. Since the logarithm of the wavelength represents the r.m.s. value of eddy fluctuations the normalized standard deviation t is defined for the eddy energy as

$$t = \frac{logL}{logT_{50}} - 1 \tag{8}$$

Where L is the period in years and T_{50} is the period up to which the cumulative percentage contribution to total variance is equal to 50 and $t = 0$. The variable $logT_{50}$ also represents the mean value for the r.m.s. eddy fluctuations and is consistent with the concept of the mean level represented by r.m.s. eddy fluctuations. Spectra of time series of meteorological parameters when plotted as cumulative percentage contribution to total variance versus t have been shown to follow the model predicted universal spectrum[19].

(e) Mary Selvam [13] has shown that Equation (1) represents the universal algorithm for deterministic chaos in dynamical systems and is expressed in terms of the universal *Feigenbaum*'s [7] *constants* a and d as follows. The successive length step growths generating the eddy continuum $OR_0R_1R_2R_3R_4R_5$ analogous to the period doubling route to chaos (growth) is initiated and sustained by the turbulent (fine scale) eddy acceleration w_* which then propagates by the inherent property of inertia of the medium of propagation. Therefore, the statistical parameters *mean* , *variance* , *skewness* and *kurtosis* of the perturbation field in the medium of propagation are given by w_* , w_*^2 , w_*^3 and w_*^4 respectively. The associated dynamics of the perturbation field can be described by the following parameters. The perturbation speed w_* (motion) per second (unit time) sustained by its inertia represents the mass, w_*^2 the acceleration or force, w_*^3 the angular momentum or potential energy, and w_*^4 the spin angular momentum, since an eddy motion has an inherent curvature to its trajectory.

It is shown that *Feigenbaum's* constant a is equal to [13]

$$a = \frac{W_2 R_2}{W_1 R_1} \tag{9}$$

Where the subscripts 1 and 2 refer to two successive stages of eddy growth. *Feigenbaum's* constant a as defined above represents the steady state emergence of fractional *Euclidean* structures. Considering dynamical eddy growth processes, *Feigenbaum's* constant a also represents the steady state fractional outward mass dispersion rate and a^2 represents the energy flux into the environment generated by the persistent primary perturbation w_*. Considering both clockwise and counterclockwise rotations, the total energy flux into the environment is equal to $2a^2$. In statistical terminology, $2a^2$ represents the variance of fractal structures for both clockwise and counterclockwise rotation directions.

The *Feigenbaum's* constant d is shown to be equal to [13]

$$d = \frac{W_2^4 R_2^4}{W_1^4 R_1^4} \tag{10}$$

and represents the fractional volume intermittency of occurrence of fractal structures for each length step growth. *Feigenbaum's* constant ***d*** also represents the relative spin angular momentum of the growing large eddy structures as explained earlier.

Equation (1). may now be written as

$$2\frac{W^2 R^2}{w_*^2 \left(dR^2\right)} = \pi \frac{W^4 R^3}{w_*^4 \left(dR^3\right)} \tag{11}$$

Where ***dR*** equal to ***r*** represents the incremental growth in radius for each length step growth, i.e. ***r*** relates to the earlier stage of eddy growth.

Substituting the *Feigenbaum's constants* ***a*** and ***d*** defined above (Equations 9 and 10) Equation (11) can be written as

$$2a^2 = \pi d \tag{12}$$

Where πd, the relative volume intermittency of occurrence contributes to the total variance $2a^2$ of fractal structures.

In terms of eddy dynamics, the above equation states that during each length step growth, the energy flux into the environment equal to $2a^2$ contributes to generate relative spin angular momentum equal to πd of the growing fractal structures.

It was shown at Equation (6) above that the steady state emergence of fractal structures in fluid flows is equal to $1/k$ ($= \tau^2$) and therefore the *Feigenbaum's constant* ***a*** is equal to

$$a = \tau^2 = 1/k = 2.62 \tag{13}$$

(f) The relationship between *Feigenbaum's constant a* and statistical normal distribution for power spectra is derived in the following.

The steady state emergence of fractal structures is equal to the *Feigenbaum's constant* ***a*** (Equation 6). The relative variance of fractal structure for each length step growth is then equal to a^2. The normalized variance $1/a^{2n}$ will now represent the statistical normal probability density for the nth step growth according to model predicted quantumlike mechanics for fluid flows . Model predicted probability density values ***P*** are computed as

$$P = \tau^{-4n} \tag{14}$$

or

$$P = \tau^{-4t} \tag{15}$$

Where t is the normalized standard deviation (Equation 7) and are in agreement with statistical normal distribution as shown in Table 1.

Table 1

Model predicted and statistical

normal probability density distributions

growth step	normalized std dev	probability densities	
n	t	model predicted $P = \tau^{-4t}$	statistical normal distribution
1	1	.1459	.1587
2	2	.0213	.0228
3	3	.0031	.0013

(g) The power spectra of fluctuations in fluid flows can now be quantified in terms of universal *Feigenbaum's constant* a as follows.

The normalized variance and therefore the statistical normal distribution is represented by (from Equation 14)

$$P = a^{-2t} \tag{16}$$

Where P is the probability density corresponding to normalized standard deviation t. The graph of P versus t will represent the power spectrum. The slope S of the power spectrum is equal to

$$S = \frac{dP}{dt} \approx -P \tag{17}$$

The power spectrum therefore follows inverse power law form, the slope decreasing with increase in t. Increase in t corresponds to large eddies (low frequencies) and is consistent with observed decrease in slope at low frequencies in dynamical systems.

(h) The fractal dimension D can be expressed as a function of the universal *Feigenbaum's constant* a as follows.

The steady state emergence of fractal structures is equal to a for each length step growth (Equations 7 & 13) and therefore the fractal structure domain is equal to a^m at m^{th} growth step starting from unit perturbation. Starting from unit perturbation, the fractal object occupies spatial (two dimensional) domain a^m associated with radial extent τ^m since successive radii follow *Fibonacci* number series. The fractal dimension D is defined as

$$D = \frac{d\ln M}{d\ln R}$$

where M is the mass contained within a distance R from a point in the fractal object. Considering growth from n^{th} to $(n+m)^{th}$ step

$$dlnM = \frac{dM}{M} = \frac{a^{n+m} - a^{n}}{a^{n}} = a^{m} - 1 \tag{18}$$

Similarly

$$dlnR = \frac{dR}{R} = \frac{\tau^{n+m} - \tau^{n}}{\tau^{n}} = \tau^{m} - 1 \tag{19}$$

Therefore the fractal dimension D is given as

$$D = \frac{\tau^{2m} - 1}{\tau^{m} - 1} = \tau^{m} + 1 \tag{20}$$

The fractal dimension increases with the number of growth steps. The dominant wavebands increase in length with successive growth steps. The fractal dimension D indicates the number of periodicities which superimpose to give the observed four dimensional space-time structure to the flow pattern. The above concept of dimension for real world spacetime patterns is consistent with El Naschie's [6] interpretation of dimensions for superstring theories in particle physics, namely a string rotates in ordinary space and only uses the extra dimensions for vibrations which simulate particle masses.

(i) The relationship between *fine structure constant*, i.e. the eddy energy ratio between successive dominant eddies and *Feigenbaum*'s constant a is derived as follows.

$2a^2$ = relative variance of fractal structure (both clockwise and anticlockwise rotation) for each growth step.

For one dominant large eddy (Fig. 2) $OR_0R_1R_2R_3R_4R_5$ comprising of five growth steps each for clockwise and counterclockwise rotation, the total variance is equal to

$$2a^2 \times 10 = 137.07 \tag{21}$$

For each complete cycle (comprising of five growth steps each) in simultaneous clockwise and counterclockwise rotations, the relative energy increase is equal to *137.07* and represents the *fine structure constant* for eddy energy structure.

Incidentally, the *fine structure constant* in atomic physics [16], designated as α^{-1} , a dimensionless number equal to *137.03604*, is very close to that derived above for atmospheric eddy energy structure. This fundamental constant has attracted much attention and it is felt that quantum mechanics cannot be interpreted properly until such time as we can derive this physical constant from a more basic theory.

(j) The ratio of proton mass M to electron mass m_e , i.e. M/m_e is another fundamental dimensionless number which also awaits derivation from a physically consistent theory. M/m_e determined by observation is equal to about *2000*. In the following it is shown that ratio of energy content of large to small eddies for specific length scale ratios is equivalent to M/m_e.

From Equation (21),

The energy ratio for two successive dominant eddy growth = $(2a^2 \times 10)^2$

Since each large eddy consists of five growth steps each for clockwise and anticlockwise rotation,
The relative energy content of large eddy with respect to primary circulation structure inside this large eddy

$$= (2a^2 \times 10)^2/10$$
$$\cong 1879$$

The cell dynamical system model concepts therefore enable physically consistent derivation of fundamental constants which define the basic structure of quantum systems. These two fundamental constants could not be derived so far from a basic theory in traditional quantum mechanics for subatomic dynamics [16].

4. CONCLUSION

The cell dynamical system model presented in the paper is basically a string theory applicable to all dynamical systems ranging from macroscale atmospheric flows to subatomic scale quantum systems. The four dimensional real world spacetime continuum fluctuation are manifestation of the superimposition of a hierarchical continuum of eddy circulations, whose centripetal acceleration add cumulatively to represent the inertial mass, which is equivalent to gravitational mass.

Acknowledgements

The authors express their gratitude to Dr. A. S. R. Murty for his keen interest and encouragement during the course of this study.

References

1. Bak P. C, Tang C. and Wiesenfeld K., Self-organized criticality, *Phys. Rev.* **A 38**, (1988) pp. 364-374.
2. Berry M. V., The geometric phase, Sci. Amer. Dec., (1988) pp. 26-32.
3. Brown J., Where two worlds meet, New Scientist,18 May, (1996) pp. 26-30.
4. Burroughs W. J., Weather Cycles: Real or Imaginary?. (Cambridge University Press, Cambridge 1992).
5. Cipra B., A new theory of turbulence causes a stir among experts. *Science* **272**,(1996) pp. 951.
6. El Naschie M. .S., Remarks on superstrings, fractal gravity, Nagasawa's diffusion and Cantorian space-time, Chaos, Solitons and Fractals 8(11), (1997) pp. 1873-1886 .
7. Feigenbaum M. J., Universal behavior in nonlinear systems, Los Alamos Sci., 1, (1980) pp. 4-27.

8. Ghil M., Cryothermodynamics : The chaotic dynamics of paleoclimate, Physica D 77, (1994) pp. 130-159.
9. Grossing G., Quantum systems as order out of chaos phenomena, Il Nuovo Cimento, 103B , (1989) pp. 497-510.
10. Kaku M., Into the eleventh dimension , New Scientist, 18 Jan (1997) pp. 32-36.
11. Maddox J., Licence to slang Copenhagen ? *Nature* **332**, (1988) pp.581.
12. Mary Selvam A., Deterministic chaos, fractals and quantumlike mechanics in atmospheric flows , Can. J. Phys. 68, (1990) pp. 831-841.
13. Mary Selvam A., Universal quantification for deterministic chaos in dynamical systems, Applied Math. Modelling, 17, (1993) pp. 642-649 .
14. Mary Selvam A., Pethkar J. S., Kulkarni M. K. and Vijayakumar R., Signatures of a universal spectrum for atmospheric interannual variability in COADS surface pressure time series, Int J. Climol. 116, (1996) pp.1-11.
15. Mary Selvam A., Quasicrystalline pattern formation in fluid substrates and phyllotaxis, pp. 795-809, In: Symmetry in Plants. D. Barabe and R. V. Jean (eds.), World Scientific Series No 4. in Mathematical Biology and Medicine, (World Scientific, Singapore, 1998).
16. Omnes R., The Interpretation of Quantum Mechanics. (Princeton University. Press, Princeton, NJ , 1994).
17. Puthoff H. E., Gravity as a zero-point fluctuation force, *Phys. Rev A* **39**, (1989) pp. 2333.
18. Rae, A., Quantum-Physics: Illusion or Reality ? (Cambridge University Press, New York ,1988).
19. Selvam A.M. and Suvarna Fadnavis, Signatures of a universal spectrum for atmospheric interannual variability in some disparate climatic regimes, *Meteorology and Atmospheric Physics* **66**, (1998a) pp. 87-112 .
20. Selvam A. M. and Suvarna. Fadnavis, Superstrings, Cantorian-fractal Spacetime and Quantum-like Chaos in Atmospheric Flows , *Chaos, Solitons & Fractals* (in Press)(1998b)
21. Steinhardt P., 1997: Crazy crystals, *New Scientist* **25 Jan**, (1997) pp. 32-35.
22. Tessier Y., Lovejoy S., Hubert P., Schertzer D. and Pecknold S. Multifractal analysis and modeling of rainfall and river flows and scaling, casual transfer functions, *J. Geophys. Res.* 101(**D21**), (1996) pp. 26427-26440 .
23. Townsend A. A., The Structure of Turbulent Shear Flow. (Cambridge University Press, London, U.K. , 1956).
24. Weinberg S.,. Dreams of a Final Theory. (Vintage, 1993).

WORK CRITERIA FOR CREATION AND DESTRUCTION OF NON-EQUILIBRIUM STRUCTURES STEMMING FROM THERMODYNAMICS

S. SIENIUTYCZ
Faculty of Chemical Engineering, Warsaw Polytechnic University,
00-645 Warsaw, 1 Waryńskiego Street, Poland
E-mail: sieniutycz@ichip.pw.edu.pl

A general optimal control framework is outlined for multistage finite-rate processes which extremize work. Equations of dynamics, which follow from energy and matter balances and transfer equations, are difference constraints in optimization. Work functionals are formulated for systems with pure heat transfer or simultaneous heat and mass transfer. Dynamic programming method is suitable either to construct Bellman's recurrence equation or to arrive at a discrete maximum principle of Pontryagin's type in which a Hamiltonian is maximized with respect to controls. Classical static limits are shown to be too far from reality to be useful; enhanced bounds for work released from an engine system or added to a heat-pump system are found. These bounds diverge, i.e. they differ for a process and its inversion. From a general viewpoint, the divergence of bounds proves that processes with creation of a non-equilibrium structure require more work than those with destruction of this structure. In effect, a finite-time process of the structure creation consumes more mechanical energy than the corresponding inverse process can deliver during the structure destruction. On this ground permissible and forbidden evolutions are determined for irreversible thermodynamic cycles.

1 Introduction

Recently it has been recognized that approaches which use discrete formalisms of non-equilibrium thermodynamics are capable of providing quite realistic performance criteria and bounds for real processes occurring in finite time and for practical systems of finite size. Especially finite-time thermodynamics and thermoeconomics, both related in spirit to an older field called Second Law analysis, have been developed to aid in the search for the optimum ways to operate machines and processes, either by finding best or optimum values of key parameters or by determining optimum pathways of operation associated with various thermal engines and unit operations in thermal and chemical plants. One of the reasons for the attractiveness of these newer branches of thermodynamics over older ones is that they admit explicitly that for given process requirements, operations must occur with finite intensities, and that inherent loss mechanisms need to be taken into account.

Thanks to the synthesizing nature of these approaches, as well as to their effectiveness, significant progress has been achieved in the design of new or improved thermal, separation, chemical and radiative systems. With the help of thermodynamic approaches, a variety of practical and industrial systems can now be modeled and optimized, including thermal and solar-driven engines, heat exchangers, diffusional separators, semiconductor devices, etc. These applications, on the one hand, result in a deeper understanding of the theory and, on the other, lead to further improvements in the design of practical devices. The unifying factor for these subjects follows from the way most of them deal with thermodynamic systems which have some control that can be adjusted to achieve the best or extremal performance. Thus, in essence, these approaches focus primarily on the optimization of energy conversion and transmission systems using thermodynamic methods. Having developed formulations for evaluating system performance, design variables and costs, researchers can now address various problems of system control and optimization. Dynamic bounds can be determined which are usually functions of operational constraints consistent with finite rates of operation for the process. These bounds can be contrasted with the static bounds of classical thermodynamics pertinent to infinitely slow processes.

For optimization purposes, system models are being developed which contain, in general, control (decision) variables, state variables and (usually) some uncontrolled or fixed parameters. These models incorporate diverse process characteristics such as finite heat conductances, semiconductor bandgaps, diffusion and beam transfer channels, friction, heat loss, chemical resistances, and other factors which are essential in real energy conversion and transmission processes. Approaches from classical and finite-time thermodynamics and exergy analysis search for the best or optimum values of the principal parameters of various engines (thermal, solar, combustion, etc.) and unit operations or processes (distillation, chemical reactions, etc.) as well as of some combined structures operating in thermal or chemical plants under specified operational constraints. Optimization usually yields a few basic recommendations on how to run a practical system. Optimal paths and optimal controls (for instance, driving heat fluxes maximizing work produced) are determined. The role of mathematical programming and optimal control theory is essential when solving these problems.

2 Outline of Basic Modeling

In a series of our recent papers[1-3] we have developed a general optimal control framework for a difficult class of problems of work maximization in multistage endoreversible processes[4-6] which yield mechanical work with finite rates and may be characterized by multiple (vectorial) efficiencies. Methods originally designed to processes with pure heat transfer[1-3] have then been extended to those with simultaneous heat and mass transfer[6]. The main purpose of the extension was to develop an analysis which would be a suitable starting point to investigate non-isothermal chemical engines. To date the only acknowledged theory is that dealing with isothermal chemical engines[7,8].

Equations of dynamics which follow from energy and matter balances and transfer equations are difference constraints for optimized work. Irreversibilities caused by the energy and mass transport are essential. Variation of efficiencies takes place due to finite heat and mass fluxes as natural control variables. With our modeling enhanced bounds for the work released from an engine system or added to a heat-pump system can be evaluated. Lagrangians and Hamiltonians of work functionals and discrete canonical equations are effective; they reach their continuous counterparts in the limit of an infinite number of stages. Bellman's dynamic programming is an efficient method either to construct his recurrence equation or to arrive at a discrete maximum principle of Pontryagin's type, in which a Hamiltonian is maximized with respect to controls[9]. Both these algorithms are powerful computational tools which serve to maximize power output and evaluate optimal controls. For a finite-time passage of a resource fluid between two given thermodynamic states, an optimal process is shown to be irreversible. Its optimal intensity is characterized well by the Hamiltonian of the optimization problem, H. Characteristic functions which describe extremal work are found numerically in terms of final states, process duration and number of stages[9].

An extension of the classical exergy and its underlying work potential I to thermal and separation systems with a finite number of stages and a finite holdup time of the resource fluid is one of the main results of such analyses[1-3]. This extended exergy simplifies to the classical thermal exergy in the limit of infinite duration and an infinite number of stages. The extended exergy exhibits a hysteretic property as a decrease of maximum work received from a multistage engine system and an increase of minimum work added to a heat-pump system, two properties which are particularly important in high-rate regimes. Our purpose here is, in particular, to apply the extended exergy and its underlying work potential to irreversible finite-period cycles in thermodynamic spaces.

The theory of infinite sequence of infinitesimal Curzon-Ahlborn-Novikov processes (CAN processes) has been applied[1-3] to describe the active (work producing) exchange of heat between two fluids, in particular fluid and bath, thus leading to the dissipative extension of the corresponding reversible problem. Both in the irreversible problem and in its reversible prototype, a maximum of work from a finite resource is sought. In the case of the reversible process, the maximum work is the well known exergy or available energy of the resource. In this work we show that the CAN sequence is the basic theoretical tool to define a duration-dependent work potential and related available energy (exergy) which generalizes the classical exergy for finite time processes with dissipation occurring in associated resistances. Earlier generalizations reviewed to date[10] have missed related functional statements and associated time evolution, and did not make distinction between the processes approaching and leaving the equilibrium. The distinction has been emphasized only recently, and the analysis explained that the distinction becomes unnecessary only for reversible, quasistatic process, when the effect of resistances vanish, and the nonideal exergy simplifies to the classical reversible exergy, the latter being known from many textbooks. The classical thermal exergy can also be obtained in a less standard way, as the as the limiting work received from the sequence of a finite number of Carnot cycles at the limit when the number of these cycles tends to infinity. For the purpose of the standard exergy, the commencement of the theoretical scheme with a finite-stage model is unnecessary though; the traditional model of infinitesimal stages is sufficient. Indeed, the presence of *reversible* cycles fixes automatically the first-law efficiency of each infinitesimal stage at the known (Carnot) level, $\eta = 1 - T^e/T$, where T is the instantaneous temperature of the resource and T^e is the temperature of the environment or an infinite bath. Since the unit mass of resource releases the heat $dQ = - cdT$, where c is the specific heat, the classical thermal exergy E_X follows by integration of the product $- c\eta dT = - c(1 - T^e/T)dT$ between the limits T and T^e. The integration yields the well-known classical expression for the thermal exergy in the quite simple way.

However, the sequential cycles approach to the exergy becomes nontrivial in the irreversible (finite rate) situation, since in this case the instantaneous efficiencies h of the system differ from those of Carnot at each time instant, and their prior evaluation is necessary before the maximizing and integration of a related work integral can be made. Moreover, these efficiencies have to be evaluated in a proper form, as certain functions of the process state T and the corresponding rate dT/dt, to assure the genuine functional property (path dependence) for the work integral. In the case of more state variables (for example T, X) vectorial generalizations of the theory have to be taken into account[6]. The work integration must be proceeded by an optimization procedure which maximizes the work integral and assures the choice of an optimal path.

Following our previous works[1-3] we now outline the derivation of the power flux and a related work functional for a steady process of work production in a sequence of infinitesimal Curzon-Ahlborn-Novikov engines. We introduce the differential of the overall conductance γ associated with the overall transfer coefficient α'

$$d\gamma \equiv \frac{d\gamma_1 d\gamma_2}{d\gamma_1 + d\gamma_2} = \alpha' dA = \alpha' a_v F dx . \tag{1}$$

We also introduce a spatial scale for the overall transfer process. It is identical with the so-called 'height of the heat transfer unit' (H_{TU})

$$\frac{Gc}{\alpha' a_v F} = H_{TU} = \frac{\rho c}{\alpha' a_v} v \equiv \chi v . \tag{2}$$

Since the mass flux G has units of mass per unit time and the volumetric heat transfer coefficient $\alpha' a_v$ has units $J/(m^3 Ks)$, the quantity H_{TU} has units of the length. In the above equation, the H_{TU} is referred to the driving fluid (fluid 1). This equation also contains the quantity $\chi = \rho c/(\alpha' a_v)$, the ratio of the volumetric capacity ρc to $\alpha' a_v$, which has units of time, and plays the role of a time constant for the driving fluid flow. The quantity χ remains a finite constant at the limit of vanishing velocities v since the transfer coefficients approach in this limiting case certain nonvanishing constant values governed by the fluid conductivity in absence of the convection. In the case of finite α', the quasistatic limit can be attained by pushing the resource fluid infinitely slowly through the system.

The process evolution is traced in terms of a nondimensional quantity τ, equation (3), related to t or x, which is defined as the ratio x/H_{TU}. It represents a nondimensional length which is known as the 'number of transfer units'. It is the length x scaled in units of the H_{TU}. Since τ is a measure of both the extent of the system and of the fluid's contact time with the energy exchange area, it plays also the role of a nondimensional time, and this is why it is designated by τ. The variable τ can be defined by several expressions

$$\tau \equiv \frac{x}{H_{TU}} = \frac{\alpha' a_v}{\rho c v} x = \frac{x}{\chi v} = \frac{t}{\chi} , \tag{3}$$

the latter being restricted to the case of the constant velocity v. Then it can be shown [1-3] that the heat balance for the driving fluid yields

$$\frac{dQ_1}{d\gamma} \equiv - u = - \frac{GcdT}{\alpha' dA} = - \chi v \frac{dT}{dx} = - \chi \frac{dT}{dt} = - \frac{dT}{d\tau} \tag{4}$$

and the power differential in terms of the driving heat differential dQ_1 is

$$dW = (1 - \frac{T^e}{T - dQ_1/d\gamma}) dQ_1 = - Gc(1 - \frac{T^e}{T + u}) dT . \tag{5}$$

Consequently the corresponding power integral per unit mass flow or the specific work of the fluid at flow is

$$W \equiv \frac{W}{G} = - \int_{\tau^i}^{\tau^f} L d\tau = - \int_{T^i}^{T^f} c(1 - \frac{T^e}{T + u}) u d\tau . \tag{6}$$

This equation defines the process Lagrangian L or the integrand of the work functional for the case pure heat transfer. A formula of this sort can also be found for an unsteady counterpart of the process with the active heat exchange between a body and the thermal reservoir.

When simultaneous transfer of heat and mass occurs the form of L is much more involved. In this case, L depends not only on the rates of temperature change but also on the rates of change of all independent concentrations. For an active evaporation process the derivatives $u = dT/d\tau$ and $v = dX/d\tau$ can be accepted as controls[6]. These controls are related to fluxes of heat and mole numbers, q and n, by the expressions q = -gcu and n = -gv.

The negative ratio of the one-stage power w to the mass conductance g defines the Lagrangian L of the problem in terms of the controls u and v. In the present case it is more suitable to use L that has units of energy per mole. One may also use the quantity $f_0 = -L$ which is the power output per unit molar conductance $g = G\theta$; then

$$dW/d\tau \;=\; w/g \;=\; w/(\theta G) \;\equiv\; f_0 = -L\,. \tag{7}$$

To obtain the work produced per unit mole of the fluid, we integrate over τ the function $f_0 = -L$ expressed in terms of controls and state. For simultaneous transfer of heat and mass satisfying the Lewis analogy

$$\begin{aligned}-L(u,v) &= \frac{w}{g} = c_1 u - c_1 T_1 v + \frac{g_2}{g}c_2 T_2 - \left(\frac{g_2}{g}c_2 - c_p v\right)T_2\Bigg\{\left(\frac{T_1 + \left(c_1\frac{g_1}{g} + c_p v\right)^{-1} c_1 u}{T_1 + \frac{g_1}{g}\left(c_1\frac{g_1}{g} + c_p v\right)^{-1} c_1 u}\right)\\ &\left(\frac{(1+X_1)^{(1+X_1)}\left(X_1 + \frac{gv}{g_1}\right)^{\left(X_1 + \frac{gv}{g_1}\right)}}{X_1^{X_1}\left(1+X_1+\frac{gv}{g_1}\right)^{\left(1+X_1+\frac{gv}{g_1}\right)}}\right)^{\left(\frac{g_1 R}{g_2 c_2}\right)}\left(\frac{1+X_2)^{(1+X_2)}\left(X_2 - \frac{gv}{g_2}\right)^{\left(X_2 - \frac{gv}{g_2}\right)}}{X_2^{X_2}\left(1+X_2-\frac{gv}{g_2}\right)^{\left(1+X_2-\frac{gv}{g_2}\right)}}\right)^{\left(-\frac{g_2 R}{g_2 c_2}\right)}\Bigg\}\,.\end{aligned} \tag{8}$$

This equation allows us to find a maximum for the cumulative mechanical work W when a finite-resource fluid changes its thermodynamic parameters in a finite time between two assumed states. It may be shown that this power formula reduces *exactly* to that of pure heat transfer when n = 0. The methodology of the optimization approach is preserved even for more complicated Lagrangians.

It may be shown that the basic functional for the process with pure heat transfer, equation (6), can be written in a general form which explicitly contains the entropy production, i.e.

$$W = -\int_{T^i}^{T^f} c\left(1 - \frac{T^e}{T}\right)dT \;-\; T^e S_\sigma \;=\; W^{rev} - T^e S_\sigma\,. \tag{9}$$

An analogous form holds for the mass transfer functional derived from the Lagarangian (8). The first or classical term of the work integral has the potential or path-independent property. Consequently, this term does not have any influence on the family of extremal trajectories obtained as solutions of corresponding Euler-Lagrange equations. It is the entropy production which causes the (second) non-potential component of the work integral. The equivalence of the work and the entropy production extremizations, stemming from equation (9), is shown here for the realm of functionals rather than functions, thus representing the second law formulation of Guy and Stodola[2,9] in infinitedimensional spaces.

Whenever a quasistatic limit is achieved for the process of pure heat transfer, in which case the rates $\dot{T}$ vanish, the work functional leads to the *classical thermal exergy*

$$E_X = -\int_T^{T^e} c(1 - \frac{T^e}{T})dT = c(T - T^e) - cT^e\ln(\frac{T}{T^e}) . \quad (10)$$

For a quasistatic process with simultaneous heat and mass transfer, a generalization of the limiting formula (10) involves the concentration of the active component, X. Again, in the quasistatic case classical exergy follows.

3 Enhanced Thermodynamic Bounds and Difference Between Work of Creation and Work of Destruction

Variet of optimization methods can be used to solve the problem of extremum work. They include maximum principles[11-15] which apply a Hamiltonian and canonical equations or dynamic programming algorithms which make use Bellman's recurrence equation[16,17]. The connection between these methods is now well understood[15,18]. The results of the optimal work should be presented in the form of a generalized potential function which depends on the end states and duration. For the purpose of direct evaluation of the optimal work potential either the so called Hamilton-Jacobi-Bellman equation, known from the optimal control theory, or the related Hamilton-Jacobi equation, known from the variational methods of classical mechanics, has to be found and then solved[9,15]. But, especially in the case of vectorial efficiencies, when the analytical solutions are difficult to find, discrete counterparts of these fundamental equations are necessary to find the optimization solutions by numerical methods. An approach which uses the dynamic programming remains then virtually the sole suitable method. The numerical solution of Belman's equation is achieved with the help of a computer, a very suitable tool, because the iterative nature of operations involved.

In the case of pure heat transfer, the extremal specific work can be found analytically for every process mode; the result is

$$\begin{aligned} I \equiv\ & c(T^i - T^f) - cT^e\ln\frac{T^i}{T^f} + cT^e(\frac{\xi}{1+\xi})\ln\frac{T^i}{T^f} \\ =\ & c(T^i - T^f) - cT^e\ln\frac{T^i}{T^f} - cT^e(H/cT^2)^{1/2}\ln\frac{T^i}{T^f} \equiv - R . \end{aligned} \quad (11)$$

The principal function I, or its negative R ≡ - I, constitute the solution of the Hamilton-Jacobi equation for the work optimization problem. The particular extremal work which describes the (generalized or classical) exergy should contain the environment temperature as one of the boundary states. The finite-time exergy is the maximal work $W_{max} = I(T^i, \tau^i, T^f, \tau^f)$ with $T^i = T$ and $T^f = T^e$ for the engine mode, and the negative minimal work $(-W)_{min} = R(T^i, t^i, T^f, \tau^f)$ with $T^i = T^e$ and $T^f = T$ for the heat-pump mode. For the vanishing intensities H or ξ the change of the classical thermal exergy is recovered.

A simple formula for the dissipative thermal exergy is obtained in terms of the Hamiltonian

$$\begin{aligned} E_x &= c(T - T^e) - cT^e \ln\frac{T}{T^e} \pm cT^e(H/cT^e)^{1/2} \ln\frac{T}{T^e} \\ &= E_x(T, T^e, 0) \pm T^e S_\sigma , \end{aligned} \tag{12}$$

where $E_x(T, T^e, \infty)$ is the classical exergy, and $S_\sigma = \min S_\sigma$ is the minimal entropy production. The upper sign refers to the heat-pump mode and the lower sign -- to the engine mode.

In the case of simultaneous heat and mass transfer in an evaporation proces the following molar exergy was found[6]

$$\begin{aligned} E_x = (c_g + Xc_p)[(T - T^e) - T^e \ln\frac{T}{T^e}] + RT^e \{X \ln[\frac{X(1 + X^e)}{(1 + X)X^e}] \\ + \ln[\frac{1 + X^e}{1 + X}]\} \pm T^e S_\sigma(T, X, T^e, X^e, \tau^f) . \end{aligned} \tag{13}$$

This exergy also corresponds with underlying functions, $I(T, X, T^e, X^e, \tau^f)$ and $R(T, X, T^e, X^e, \tau^f)$, which are extensions of those described by equation (11). As it is very easy to find the functions $I(T, X, T^e, X^e, \tau^f)$ and $R(T, X, T^e, X^e, \tau^f)$ from equation (13) we do not write down here the respective formulae; yet, in analysis that follows, we assume that these formulae are known. Of the two parts of the finite-time exergy, the classical one is known from reversible thermodynamics. Thus one may use tables of E_x to evaluate the associated mimimal entropy production for the process. These results are both in the spirit of the finite-time thermodynamics[1,4,5,10] and the entropy generation minimization[19-21]. Both steady-state processes and their unsteady counterparts can effectively be modeled[22].

It follows that the classical exergy suffices for an exact estimation of the extremal work for small H_{TU}, i.e. for the excellent transfer conditions, or for infinitely long times of the energy exchange. The generalized exergy E_x serves to explain the restrictive applicability of the classical thermodynamic bounds when they are applied to real processes, and to show that these bounds should be replaced by stronger bounds obtained from nonequilibrium thermodynamics.

The classical exergy is the quantity which defines bounds on work delivered from (or supplied to) very slow, reversible processes. These classical bounds are reversible in the sense that the magnitude of the work delivered during the reversible approaching of the system to equilibrium is equal to the magnitude of the work supplied, after the initial and final states are inverted, i.e. when the second process reverses to the initial state of the first. Our results (12) and (13) generalize the classical exergy for finite rate transitions.

Let us now discuss main physical properties of such transitions. During the approach to the equilibrium the so-called *engine mode* of the system takes place in which the work is released, during the departure from the equilibrium the so-called *heat-pump mode* occurs in which the work is supplied. The work W delivered in the engine mode is positive by assumption. In the heat-pump mode the work W is negative, which means that the positive work (-W) must be supplied to the system. The optimization problems, which yield the generalized exergies and their underlying potential functions I and R, involve the maximum of the work delivered [max W] and the minimum of the work supplied [min (-W)]. We have shown that while the reversibility property is lost for such exergy, its thermokinetic bounds are stronger and hence more useful than classical thermostatic bounds. This substantiates role of the dissipative exergy for evaluation of energy limits in practical systems.

The reversibility of bounds in classical thermodynamics means that with the classical exergy, thermostatics simultaneously provides the lower bound on the real work which should be supplied to the system and the upper bound on the work which can be released by the system. The work-producing process is the inversion of the work-consuming process (the final state of the second process is the initial state of the first, and conversely), and the duration of each process is infinitely long. In thermostatics the two bounds on the work coincide. From a more general viewpoint, the two processes (direct and inverse) represent the destruction and creation of a non-equilibrium structure. Thus, in classical thermostatics, the two bounds, the one on the structure destruction and the one on the structure creation, coincide.

However such static limits are often too far from reality to be very useful. Whenever one takes into account necessity of termination of the process in a finite time and inherent role of resistances as dissipative parts of the system (in boundary layers, in particular) the finite-rate exergy provides stronger (lower and upper) bounds on the real work which should be supplied to the system or delivered from it. These bounds do not coincide, but they are more realistic than the quasistatic bounds of classical thermostatics. It is the hysteretic effect, caused by the dissipation, i.e. an increase of the exergy supplied in the pump mode of the system (work consumption) and a decrease of the exergy released in the engine mode of the system (work delivery), which makes thermokinetic bounds stronger than thermostatic bounds. The divergence of the bounds proves that a large number of processes which are permitted by classical thermostatics is excluded by limitations inherent in the process kinetics. Also, from the more general viewpoint formulated above, the divergence of the bounds proves that processes of creation of a non-equilibrium structures require larger magnitudes of mechanical energy than processes of destruction of this structure. In the latter only a part of the available mechanical energy can be released in a finite time.

In effect, the creation processes consume always more mechanical energy than the corresponding destruction processes with inverted thermodynamic states of the system.

Summing up, for a process and its inversion, the two bounds which coincide in thermostatics diverge in thermokinetics, and the divergence grows with the rate indices of the process. This means that, for the sufficiently high rate indices, one can obtain quite high lower bound on the supplied work and even vanishing upper bound on the released work. The finite rate processes always increase the absolute value of the extremal work supplied in processes departing from the equilibrium, and decrease the corresponding work produced in processes approaching the equilibrium. These conclusions, along with the quantitative analysis presented in the series of works[1-3], provide a means for improved evaluation of the mechanical energy limits in practical systems.

In particular, we can apply the above general results along with equations (11)-(13) and principal functions $I(T, X, T^e, X^e, \tau^f)$ and $R(T, X, T^e, X^e, \tau^f)$ to the important problem of irreversible cycles undergoing with finite periods. These cycles can be described in standard thermodynamic spaces or in extended spaces including the physical time or its monotonic measure (such as, for example our τ or R) as an additional variable, Fig. 1.

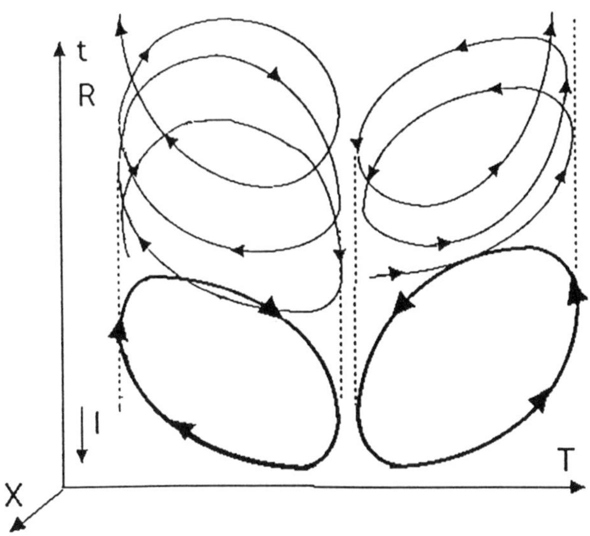

Fig. 1 Admissible screwlines and cycles for simultaneous heat and mass transfer

The irreversible cycles constitute closed loops in thermodynamic spaces. However, in the extended spaces which contain time as an extra coordinate, these cycles are no longer represented by closed loops, but by screwlines. Thermodynamic cycles are projections of these screwlines on the space of thermodynamic parameters. As this work shows, not all of mathematically possible screwlines are admissible from the viewpoint of the second law. Fig. 1

illustrates types of screwlines admissible in the space-time T-X-τ. Their basic property is determined by the requirement that in a real cycle parts corresponding to work consumption must assure larger magnitude of work that parts corresponding to work production.

In conclusion, enhanced bounds are formed by the generalized exergy for the work released (destruction) and consumed (creation) in real processes, in terms of the residual generated entropy as the rate penalty. The hysteretic properties of the generalized exergy as a finite-time work function are important. They are associated with different values of the extremal work obtained when the process which leaves the equilibrium is compared with its inverse, which approaches the equilibrium. While in the classical reversible thermodynamics these two modes are accomplished with exactly the same magnitude of work, in the dissipative theory the work of creation and work of destruction are no longer equal. A significant decrease of the maximal work received from the destruction of a system and an increase of the minimal work that has to be added to create a system is shown in the high-rate regimes and for short durations of thermodynamic processes. This property makes it possible to exclude definite cyclic evolutions as forbidden by the second law, which is the result important for the theory of irreversible processes.

Acknowledgments

The author acknowledges a support from the EC program 'Carnet'.

Notation

a_v - specific area of heat exchange (per unit volume)
c- specific heat capacity
E_x - exergy function
f_0- work generation function, negative Lagrangian
g_1, g_2 - partial conductances for mass transfer
g- overall conductance for mass transfer
G- flux of resource fluid
H- hamiltonian function of the optimization problem
H_{TU} - height of transfer unit
I- optimal performance function of the profit type
n- molar flux of active component
Q-cumulative heat exchange
R- optimal performance function of the cost type
t- physical time
T^e -constant temperature of infinite reservoir
T - temperature of fluid driving the engine process
u, v - rates of temperature and concentration change as control variables
W - cumulative power delivered from the overall process
W- total specific work, total power per unit mass flux
X- concentration of active component in moles per unit mole of inert

X^e constant reference concentration in infinite reservoir
x- length coordinate
α' - heat transfer coefficient
γ- cumulative conductance
η = thermal efficiency
τ -dimensionless time (x/H_{TU})
θ- time interval at stage n, nondimensional conductance

References

1. S. Sieniutycz, *J. of Non-Equilib. Thermodyn.* **22**, 260 (1997).
2. S. Sieniutycz, *Phys. Rev. E* **56**, 5051 (1997).
3. S. Sieniutycz, *Intern. J. Eng. Science* **36**, 577 (1998).
4. A. de Vos, *Endoreversible Thermodynamics of Solar Energy Conversion* (Clarendon Press, Oxford, 1992).
5. F. L. Curzon and B. Ahlborn, *Amer. J. Phys.***43**, 22 (1975).
6. S. Sieniutycz, Optimal control for multistage endoreversible engines with heat and mass transfer, *J. of Non-Equilib. Thermodyn.*, in print (1998).
7. J. M. Gordon and V. Orlov, *J. Appl. Phys.* **74**, 5303 (1993).
8. L. Chen *et al*, *Energy Convers. Mgmt* **38**, 1841 (1997).
9. S. Sieniutycz, *Optimization in Process Engineering* (Wydawnictwa Naukowo Techniczne, Warszawa, 1978).
10. B. Andresen, Finite-Time Thermodynamics (University of Copenhagen, Copenhagen, 1983)
11. S. Sieniutycz and Z. Szwast, *Intern. J. Chem. Engng.* **23**, 155 (1983).
12. L. T. Fan and C. S. Wang, *The Discrete Maximum Principle, A Study of Multistage System Optimization* (Wiley, New York, 1964).
13. V. G. Boltyanski, *Optimal Control of Discrete Systems* (Nauka, Moscow, 1973).
14. L. T. Fan, *The Continuous Maximum Principle, A Study of Complex System Optimization* (Wiley, New York, 1966).
15. W. Findeisen, J. Szymanowski, and A. Wierzbicki, *Theory and Computational Methods of Optimization* (Państwowe Wydawnictwa Naukowe, Warsaw, 1980).
16. R. E. Bellman, *Adaptive Control Processes: a Guided Tour*, (Princeton University Press, Princeton, 1961).
17. R. Aris, *Discrete Dynamic Programming* (Blaisdell, New York, 1964).
18. V. G. Boltyanski, *Maximum Principle and Optimal Control of Continuous Systems* (Nauka, Moscow, 1971)
19. A. Bejan, *Am. J. Phys.* **64** (1996) 1054-1059.
20. A. Bejan, J. *Appl. Phys.* **79** (1996) 1191-1218.
21. A. Bejan, *Entropy Generation Minimization: The Method of Thermodynamic Optimnization of Finite-Size Systems and Finite-time Processes* (CRC Press, Boca Raton, 1995).
22. S. Sieniutycz, *Intern. J. of Applied Thermodyn.* **1** , 45 (1998).

IMPACT OF STRUCTURAL DAMAGE TO SURFACE MORPHOLOGY OF THE THORATIC AORTA WALL AFFECTED BY HYPODYNAMIC STRESS

V. SNITKA, M. RACKAITIS

Research Center "Vibrotechnika", Kaunas University of Technology, Studentu 65, LT-3031 Kaunas, Lithuania
E-mail: vsnitka@VB.KTU.LT

M. KUSLEIKAITE

Institute of Cardiology, Kaunas Medical Academy, Sukileliu 17, LT-3005 Kaunas, Lithuania

Current paper investigates possibility to use fractal dimension for the evaluation of thoratic aorta smooth muscle structural damage affected by hypodynamic stress affect. Experiments show that the fractal dimension of surface cross - section curve decreases from 1.6 to 1.4. This show that the surface structure alter. Comparison of height histogram and FFT analysis is made with fractal analysis. Although height histogram and FFT analysis exhibit certain features it is difficult to interpret them. Fractal dimension variation has similar character as muscle contratility.

1 Introduction

It is known that the development and complication of ishemic heart disease are related with disorder of circulation of the blood in coronary arteries. On the other hand, circulation of the blood in arteries depend on the cell's wall's structural damage and its influence to the contractility of the smooth muscles. Physical inactivity is also important risk factor for the heart disease as hypertension, high serum cholesterol and smoking [1]. Thus it is important to know structural damage in the wall of thoratic aorta in case of hypodynamic stress. It is also very important to track structural alteration of the smooth muscles which are dominant in our system. According previously published results [2,3], in condition of hypodynamic stress of duration of 5 and 28 days the reaction of smooth muscles of thoratic aorta increased under the effect of adrenaline and acetylcholine. However in case of hypodynamic stress of 36 days decrease of smooth muscle reaction to the adrenaline and acetylcholine was detected as compared to the control group.

Usually structural damage of arteries is evaluated by electron, polarized microscopy using phase contrast light, scanning electron microscopy and other methods. However, those methods cannot give quantitative evaluation of the artery's wall structural damage or determine contraction of the smooth muscles which compose the arterial wall.

Among other data processing techniques fractal dimension determination was used for the quantitative evaluation of overall structural damage. Its use as a classification tool holds promise because it is relatively easy to estimate and the value of the dimension gives us indication of what type of system is under investigation. Fractal dimension is an indication of how much space the surface "comes near" [4].

Much work has been done investigating the fractals [4, 5, 6, 7]. Most of the work is dedicated to the behaviour of dynamical systems which forms fractal patterns in a phase space. There is also a lot of work done for the investigation of dendrite growth phenomena [8]. However most of the work are the applications of the Diffusion Limited (DLA) and Dielectric Breakdown Models (DBM) [9]. These models can explain the origin of fractal structures in a variety of processes. They are considered by most workers in the field as the prototype fractal growth models [10].

Fractal technique was used for a number of biological applications such as bacterial growth form evaluation, DNA information coding and heart beat time series analysis [4]. There are also publications which study fractal dimension usage for the quantitative distinguishing of medical images [11].

This paper seeks to find relationship between the surface fractal dimension and structural damage of thoratic aorta on purpose to evaluate structural damage quantitatively.

Correlation between the fractal dimension and the contractility variation character [2,3] promises the fractal dimension usage for evaluation of the muscle surface.

2 Experimental materials and methods

Hypodynamic stress of duration of 5 (quantity n=9), 28 (n=9) and 36 (n=9) days was induced in Chinchilla rabbits by placing them in tight metal hutches (according Fiodorov B.M., 1991). Rabbits (n=9) of the control group that had no intervention were kept under vivarium conditions. After the stress thoratic aorta specimens were cut using microtome and fixed in formalin.

Morphological alteration images were obtained by optical microscope with CCD camera. The stress influence on the muscle surface morphology were analysed using statistical analysis. Height histogram calculations were performed using graphical image processing package (Paintshop Pro). The program calculates quantity of points having the same gray level. That corresponds to the surface height. Fast Fourier Transformations (FFT) of cross - sections' profile were calculated using Matlab signal processing toolbox. Fractal dimensions were calculated using FD3 software [12]. A box counting method was used for the calculations of capacity and correlation fractal dimensions [4, 6].

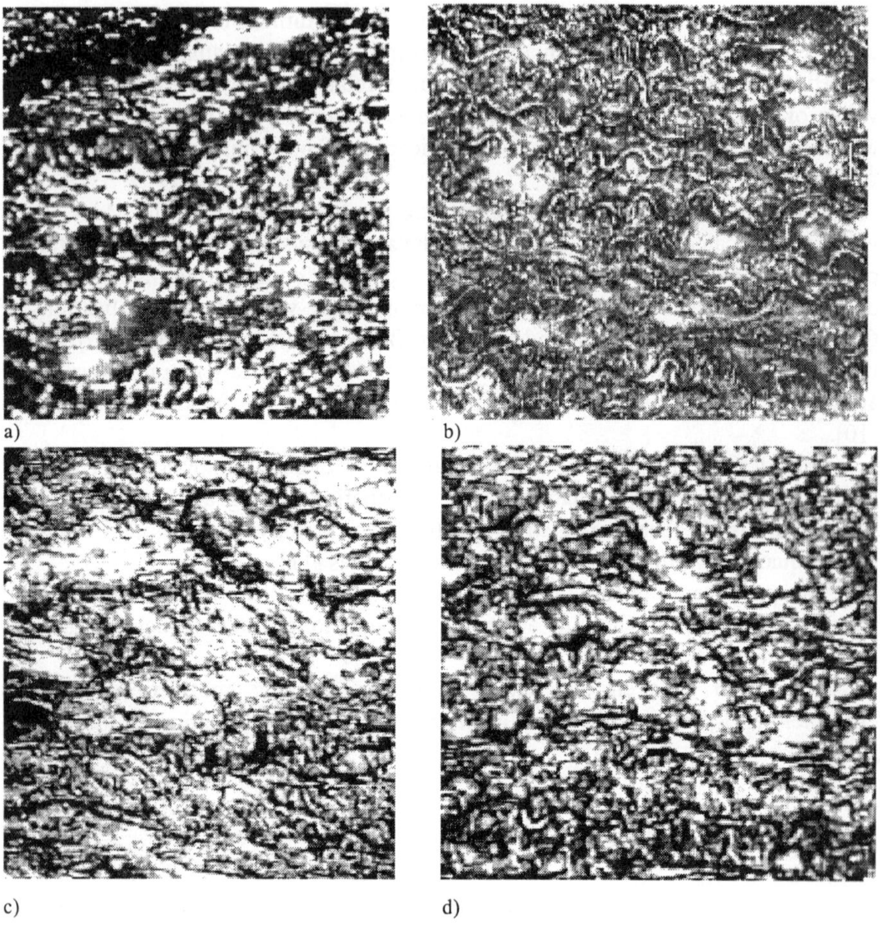

Figure 1. Morphology of the wall cross section of thoratic aorta (in media): a) - control group; b) - after 5 days of hypodynamic stress, c) - 28 days and d) - 36 days.

Capacity dimension indicates non-uniformity of distribution of data points in space. Correlation dimension is closely related to the correlation of two points at a given distance. When the distance r between the points comes close to the zero then the correlation exhibit power law dependence on r. The exponent in the power law dependence is the correlation dimension. Thus correlation dimension indicates correlation between data points. Calculated fractal dimension is a measure of the cross section curve, which is associated with the surface fractal dimension. This dimension reduction technique is widely used and it is based on an extension of theorem by Marstrand [6, 11].

3 Results and discussion

Morphology images are given in a fig. 1. One can see that the surface is losing its structure. If at the beginning (5 days) of the stress some order in a structure, at the end (36 days) the structure is altered. The surface becomes amorphous and image loses its contrast.

Height histograms and FFT spectrum of cross sections are given in a fig. 2 and fig. 3 respectively. One can notice that of histograms at the beginning of stress (5 days) and at the end of it (36 days) has more expressed peaks than in the middle. At the beginning of stress (5 days) the surface has its original structure and the cells form some pattern which has its characteristic height. At the end of the stress (36 days) the structure is altered and surface becomes smooth, so it also has the characteristic height. In the middle we see that the cell structure is in an alteration process so cells are constantly changing and we cannot distinguish any characteristic height.

FFT spectrum analysis does not exhibit any particular peaks which could have an information about surface changes. Although there are some peaks that could be used for evaluation but the spectrum looks as a chaotic structure's spectrum.

Fractals are curves or surfaces which have complex structure no matter what scale is used to examine them. Since we have very complex biological surface which is made of cells, it is convenient to have one measure for the determination of the surface smoothness. Standard surface roughness characteristics such as RMS

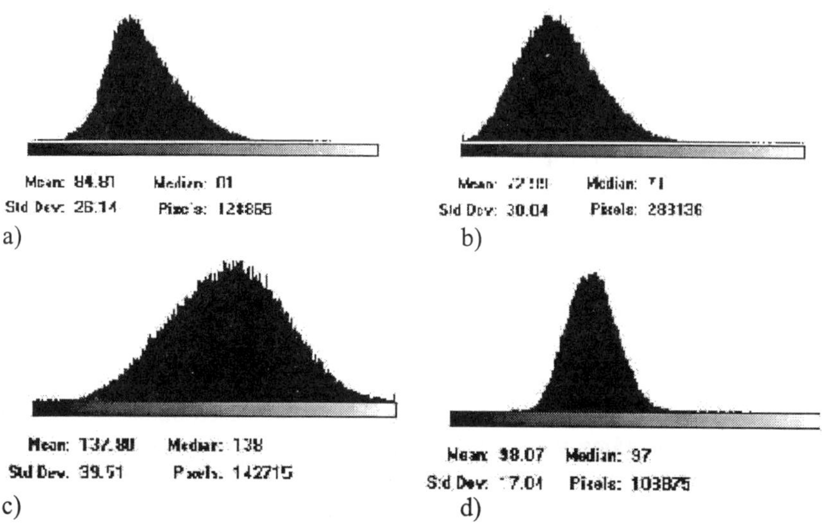

Figure 2. Cross section's of thoratic aorta surface height histograms a) - control group; b) - after 5 days of hypodynamic stress, c) - 28 days and d) - 36 days.

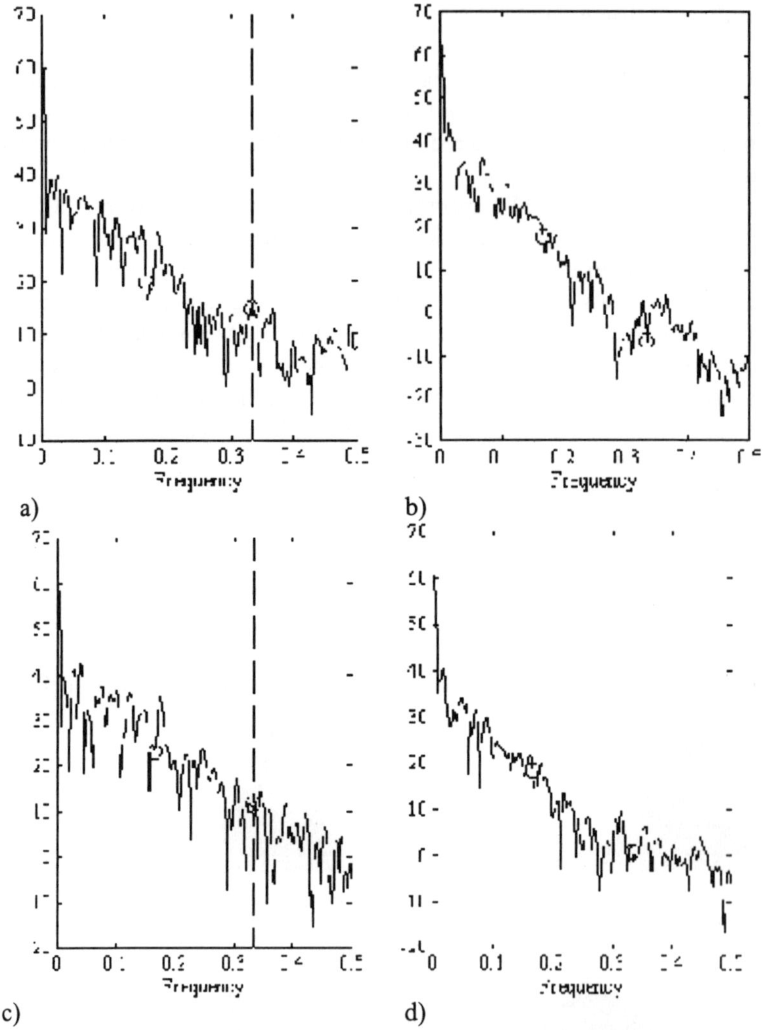

Figure 3. FFT of cross - section's of thoratic aorta a) - control group; b) - after 5 days of hypodynamic stress, c) - 28 days and d) - 36 days.

and others are not so informative because calculations are based on statistical data which do not say anything about surface shape or of cross - section . Fractal dimension depends on the shape of a curve and indicates power law dependence of the curve's mass on linear measure, where fractal dimension is an exponent in the power law. Therefore even small deviations in fractal dimension result in considerable changes in curve shape.

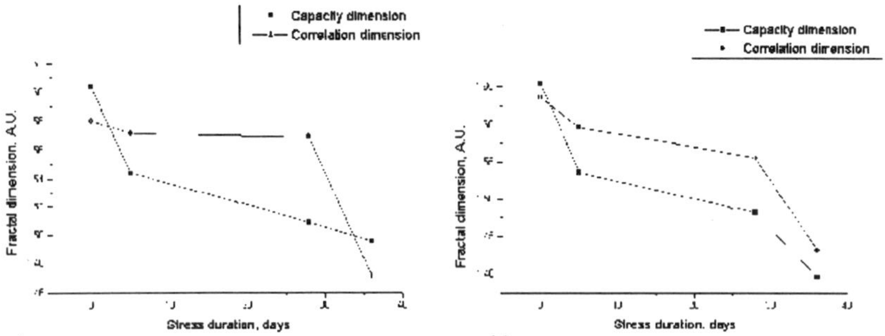

a) b)

Figure 4. Cross section's of thoratic aorta fractal dimension; a) when the cross section is made along the smooth muscle structure; b) when cross section is made across the smooth muscle structure.

Looking to our results given in a fig. 4 one sees that the fractal dimensions of the cross section of the measured profile decrease during the experiment. Cross sections were made along vertical and horizontal directions of the cell structure (it is not the same directions in a world coordinates). Capacity dimension decreased from 1.6038 to 1.49517 in the case of horizontal cross section and from 1.65422 to 1.39585 in the case of vertical cross section. Correlation dimension decreased from 1.57983 to 1.47169 in the case of horizontal cross section and from 1.63571 to 1.43172 in the case of vertical cross section. Biggest drop of the capacity dimension value happens at the very beginning of the experiment. This shows that the structure of the tissue is altered at the very beginning of the stress and later on its alteration increases not so fast. From the figures it is clear that the capacity dimension decreases faster at the beginning. This leads to the idea that at first cell structure changes. The correlation dimension does not exhibit the same fast decrease at the first part of the experiment. This shows that although the cell's physical properties are changing, the interconnections of the cells and their parts tend to stay the same. But of course, when we continue our stress impact the cells lose those interconnections and the correlation dimensions fall very fast. Total change in fractal dimensions show that the structure has some prevailed direction of damage (in our experiments it was vertical direction).

The results have shown that the fractal dimensions of various types can be used for the evaluation not only surface or curve smoothness, but also the interconnections between the structural components. Experimental results show that capacity and correlation dimensions could be used for the indication of different properties of biological system. Student's paired t test was applied when performing statistical analysis and establishing differences in p values ($p<0.05$ was considered as to be significant). Muscle cell structure investigations by Atomic Force Microscope (AFM) were also performed. Results exhibit very similar relationship between the stress duration and surface morphology. Detailed discussion will be published elsewhere.

4 Conclusions

Presented paper proposes estimation of fractal dimension for the quantitative evaluation of the structural damage in the arteries' wall. In case of hypodynamic stress of duration of 5 and 28 days correlation fractal dimension changed little, but for the stress duration of 36 days the fractal dimension statistically significantly decreased as compared to the control group. Besides fractal analysis of the smooth muscle surface, histogram and FFT analysis results are presented in the paper. Height histogram and FFT carry little information about structure damage. On the other hand, fractal dimension exhibit similar dependence on stress duration as compared to contractility measurements [2,3]. Contractility tests show that the smooth muscle contraction at the beginning of the stress (5 days) is higher than that of the control group and it remains higher up to the 28 th day of stress. After that if stress is continued the contraction of the smooth muscles significantly decreases. From the figure 4 it is seen that the correlation dimension little changes until 28 th day of stress and after that decreases.

Interconnections between adjacent structural units (smooth muscle cells in our case) are described by correlation dimension. Its dependence on stress duration exhibit that although the surface becomes smoother (decreases capacity dimension) at the beginning of the stress (5 days) correlation between the structural units little changes. This shows that at first cells tend to change their location or form and keep the interconnections at the same level as long as it is possible (28 days).

Our experiments show that the fractal dimension is suitable for the evaluation not only of the surface curvature but also geometric correlation between the structural units. Therefore, the fractal dimension could be used as a quantitative measure of muscle structural damage.

5 Acknowledgements

This work was supported by the Lithuanian State Science and Studies Foundation (grant No. 98090).

References

1. Physical Activity, health and Well-being: world forum on Physical Activity and Sport / UNESCO, - Quebec, 1995.

2. Kusleikaite M., Civinskiene G., Reigardiene D., Relation of growth hormone and bioelements with contractility of smooth muscles of thoratic aorta, morphology of myocardium and data of electrocardiograms in case of hypodynamic stress; Acta Medica Lithuanica, 1996; 3, p. 50-54.
3. Kulsleikaite M., Civinskiene G., Stonkus S., Biochemical, physiological and structural changes in organism in case of hypodynamic stress; Medicina, 1997, 33(8), p.701-707.
4. Fractals in Science; eds. A. Bunde, S. Havlin, Springer-Verlag, Berlin Heidelberg, 1994.
5. C. Tricot, Curves and Fractal Dimension, Springer-Verlag, New York, 1995.
6. T. Vicsek, Fractal Growth Phenomena, World Scientific, Singapore, 1992.
7. F.C. Moon, Chaotic Vibrations, Wiley, New York, 1987.
8. A. Arneodo, F. Argoul, Y. Couder, M. Rabaud, in : Growth and Form, eds. M. Ben Amar et al., Plenum Press, New York, 1991, p.297-315.
9. B. Mandelbrot, in Fractals in Physics, Proceedings of the sixth Trieste international symposium on fractals in physics, eds. L. Pietronero, E. Tosatti, ICTP, Trieste, Italy, July 9-12, 1985.
10. L. Pietronero, in : Growth and Form, eds. M. Ben Amar et al., Plenum Press, New York, 1991, p.345-359.
11. A. I. Penn, M. H. Loew, Estimating fractal dimension of medical images with fractal interpolation function models, http://www/seas/gwu.edu/seas/institutes/ medimage/.
12. http://spanky.triumf.ca/pub/fractals/PROGRAMS/IBMPC/

THE ACOUSTIC SPECTROSCOPY OF AQUEOUS SOLUTIONS OF ISOPROPANOL

V.S. SPERCACH, L.A. BULAVIN, A.O. SEMENOV, V.P. KOPYLCHUK
Physical Faculty, Kyiv Taras Shevchenko National University,
Kyiv, UA 252022, Ukraine
E-mail: baziv@hq.ups.kiev.ua

Many properties of alcohol-water mixtures have been studied extensively. Since in [1,2] the Relay light scattering experiments on ethanol, isopropanol, *n*-propanol and tertiary butanol have been carried out. The additional maximums of intensity of light scattering was observed on such concentrations: x_2=0.09; 0.06; 0.05; 0.03 (x_2 - moll fractions of alcohols) accordingly. Was accertaiment, that in system isopropyl alcohol-water (heavy water) the concentration maximum is absent. Was shown that erasing of addition maximum is not caused by fluctuations of concentration. The review of different treatments of results have shown that they predict the possibility of processes like to second type phase transition [1,2] or forming of molecules complexes like clusters.

The aim of this work was obtaining of kinetic data of isopropyl alcohol-water binary liquid system in width range of frequencies.

Methods

The present work deals with determination of density, ρ, coefficient of shear viscosity, η_s, sound attenuation, αf^{-2}, and velocity of sound propagation, c, in aqueous solutions of isopropyl alcohol-water at temperature 293 K. For investigation the samples of follow concentration of alcohol was taken: 0.03; 0.058; 0.14; 0.15; 0.18; 0.25; 0.5 (moll fractions of alcohol). The value of density was determined by pycnometry method, with an accuracy of about 0.05%, and the coefficient of shear viscosity was measured using capillary viscometry with a total error amounting to about 0.5%. Sound attenuation was determined using pulse method technique in frequency range 3-110 MHz, with error (2 - 5)%. The velocity of sound was measured using pulse-phase technique with error 0.1%, in temperature range 293-343 K and frequency range pointed above. Besides the sample with concentration of alcohol 0.13 moll. frac. was studied separately in more width frequency range - from 3 to 1500 MHz. The methods of these measurements is fully described in [3].

The isopropyl alcohol was taken from Chemical Faculty of Kyiv Taras Shevchenko National University with purity 99.5 %. Twice distilled water was used for preparing of samples.

Results and discussion

The measurements of density and shear viscosity of solutions at temperature 293 K have shown that the density of solutions decrease with increasing of concentration. Besides in concentration range 0 - 0.25 (moll. frac.) the dependence of density on concentration is linear and in the range 0.25-1 (moll. frac.) the essential deviation from linear law was observed.

The dependence of shear viscosity on concentration have shown presence of width maximum at concentration of alcohol near 0.2 (moll. frac.).

The analysis of experimental data obtained from measurements of velocity of sound have demonstrated that this value depend on concentration and temperature. At the concentrations of alcohol smaller 0.058 (moll. frac.) the temperature dependence of sound propagation velocity go through maximum, like in water. But if concentration of alcohol is more 0.058, the maximum disappear and velocity decrease with increasing of temperature according to linear law. Such behavior of dependence, c(T), caused by structural changes of solution with changing of concentration of alcohol.

The dependence of sound velocity on concentration go through maximum at concentration x_2=0.06 (moll. frac.). The dependence of velocity value is not symmetrical relatively this maximum (Fig. 1).

The determining of value of sound attenuation at constant temperature have shown the dependence attenuation from frequency, thus, the acoustical relaxation was observed. This relaxation can be described by [4]:

$$\frac{\alpha}{f^2} = \frac{2\pi^2 c}{c_0^2} \sum \frac{b_i \tau_i}{1 + (\omega \tau_i)^2} + B, \tag{1}$$

here c - the velocity of sound on frequency $\omega=2\pi f$; c_0 - velocity of sound at $\omega\tau_{PS} \ll 1$; b_i and τ_i - relaxation forces and times of i-th simple relaxation region accordingly; B is high frequency ($\omega\tau_{PS} \gg 1$) limit of value α/f^2.

The analysis of value of sound absorption for frequencies $\omega\tau_{PS} \ll 1$ at temperature 293 K for concentrations pointed above have shown presence nonsymmetrical maximum at the concentration 0.13 (moll. frac.) (Fig. 1). After that the solution of this concentration was investigated in more width frequency diapason from 3 to 1500 MHz in temperature range 293-434 K. The obtained experimental data demonstrated the presence of two region of relaxation. The calculation of acoustic spectra was carried out using the empirical equation:

$$\alpha \cdot f^{-2} = \frac{A_1}{1+\left(\frac{f}{f_1}\right)^2} + \frac{A_2}{1+\left(\frac{f}{f_2}\right)} + B \tag{2}$$

The parameters A_1, A_2 and B was calculated by method of iteration approximations. The results of calculations are given in Table 1. There are given the data upon sound attenuation, based on data of density, shear viscosity and velocity of sound, too. The sound attenuation is caused by coefficient of shear viscosity as:

$$\frac{\alpha_{cl}}{f^2} = 26.3\frac{\eta_s}{\rho c^3} \tag{3}$$

The results of calculation of relaxation forces (b_i), was obtained by expression:

$$b_i = \frac{A_i c_0 f}{\pi}, \quad i=1,2. \tag{4}$$

and are presented in Table 1.

Table 1. The acoustic parameters of isopropyl alcohol-water system with concentration 0.13 moll fractions at temperature T=293 K.

T	η_s	ρ	A_1	A_2	B	α_{cl}/f^2	f_1	f_2	$b_1 \cdot 10^2$	$b_2 \cdot 10^2$	c
K	mPa	kg/m^3	$\times 10^{-15} m^{-1} s$				MHz				m/s
283	5,73	955,2	800		400	39,6	42		17		1585
293	3,41	947,4	470	200	30	25,4	55	650	13	6,5	1550
303	2,26	939,6	250		170	18,2	70		8,4		1515
313	1,61	931,7	120		140	14,0	90		5,1		1480
323	1,20	923,9	60		100	11,3	115		3,2		1445

Thus we can conclude, that value of A, and b, monotony decrease with increasing of temperature. Such behavior of these value is typically for solutions, where acoustical relaxation is caused by processes of reorganisation of structure or conformational transformations of molecules [5].

The discussion relatively ability of two region of acoustic relaxation in aqueous solutions of isopropyl alcohol and coincidence of peaks of sound velocity and attenuation on concentration scale will be present on next works. Now the treatment of Inelastic Neutron Scattering data for this systems is continue too.

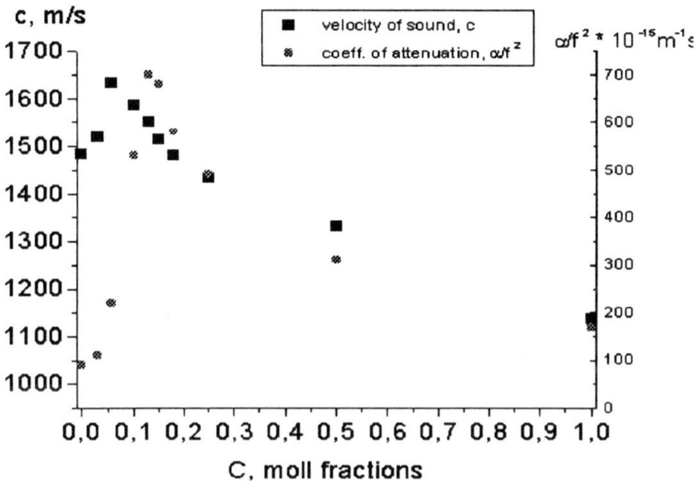

Fig.1. The dependences of sound velocity (left scale) and coefficient of sound absorption (right scale) from concentration.

References

1. Wooks, *The Light Scattering in gases, liquids and solutions* (Leningrad State Univ., 1997).
2. Anisimov, *The Critical Phenomenas in liquids and liquid crystals*, (Nauka, Moscow, 1987).
3. V. S. Spercach, T. O. Gushcha, P. F. Cholpan, *The Experimental Method of determination of sound attenuation in liquids* (UMK VO Minvuz USSR, Kiev, 1989).
4. Mihaylov, V. A. Solov`ev, Yu. P. Syrnikov, *The Principles of molecular acoustic*, (Nauka, Moscow, 1964).
5. Shahparonov, *High School,* Moscow, 350-354, (1980).

EXTENDED IRREVERSIBLE THERMODYNAMICS AND THE MOTION OF CONTINUA

J. VERHÁS

Technical University of Budapest, Budapest, Hungary, H-1521
E-mail: Verhas@phy.bme.hu

Non-equilibrium thermodynamics supplemented with dynamic degrees of freedom gives the frame of the theories of rheology, liquid crystals, fiber reinforced materials, as well as, fluid turbulence.

1 Introduction

A lot of materials not obeying the classical rules of continuum mechanics are applied in modern industry. The demand of a more accurate description of mechanical motion puts constitutive modeling into limelight. A firm and systematic foundation of the applied sciences is obviously needed. Modern non-equilibrium thermodynamics—departing from the principle of local equilibrium—may be the needed theory.[1,2]

Here we try to give a short survey on the thermodynamic methods.

2 Non-equilibrium thermodynamics

Modern thermodynamics[3–9] starts with the conservation laws. The conservation of mass reads

$$\frac{d\rho}{dt} + \rho \operatorname{div} \mathbf{v} = 0. \tag{1}$$

The form of the balance equation of the linear momentum is

$$\rho \frac{d\mathbf{v}}{dt} = \operatorname{Div} \boldsymbol{t} + \rho \mathbf{f}. \tag{2}$$

Here ρ is the density, $\mathbf{v}$ the macroscopic velocity of the material, $\boldsymbol{t}$ is Cauchy's stress, and $\mathbf{f}$ is the body force. The balance equation for the moment of momentum reads

$$\rho \frac{d\mathbf{l}^i}{dt} = 2\vec{w}(\boldsymbol{t}) + \operatorname{Div} \mathbf{\Pi} + \rho \vec{\tau}, \tag{3}$$

where $\vec{w}(\boldsymbol{t})$ is the vector invariant of Cauchy's stress (not symmetric in general), $\mathbf{\Pi}$ is the couple stress, and $\vec{\tau}$ is the torque, $\mathbf{l}^i$ is the internal moment of momentum defined by

$$\mathbf{l} = \mathbf{r} \times \mathbf{v} + \mathbf{l}^i. \tag{4}$$

The most prominent conservation law is that of the energy. Unfortunately, the energy is not an objective quantity, it depends on the frame of reference. For the sake of objectivity, the idea of the internal energy is introduced. The specific energy is split into tree parts, the translational kinetic energy ($\epsilon_t = \mathbf{v}^2/2$), the rotational kinetic energy,—which is in close relation to the internal moment of momentum—and the internal energy;

$$\epsilon = \epsilon_t + \epsilon_r + u. \tag{5}$$

The internal energy is a frame independent property of the material but it is not preserved. The balance equation is

$$\rho \frac{du}{dt} + \operatorname{div} \mathbf{J}_u = \boldsymbol{t} : \mathring{\boldsymbol{d}} + \boldsymbol{\Pi} : \operatorname{Grad} \vec{\omega} + r, \tag{6}$$

where $\mathring{\boldsymbol{d}}$ is the symmetric part of the velocity gradient and $\vec{\omega}$ is its vector invariant ($\operatorname{Grad} \mathbf{v} = \mathring{\boldsymbol{d}} + \vec{\omega} \times \boldsymbol{\delta}$). The rotational part of the kinetic energy has been supposed to obey the $d\epsilon_r/dt = \vec{\omega} d\mathbf{l}^i/dt$ relation, which is analogous to the $d\epsilon_t/dt = \mathbf{v} d\mathbf{v}/dt$ relation but is not firmly established. The structure of the balance equation of the energy shows that no rotational kinetic energy has to be assumed if no internal angular momentum is supposed. In the case of magnetic polarization (Einstein–de Haas effect), the problem of the rotational kinetic energy must be reconsidered.[a]

3 Entropy

The second law of thermodynamics says that entropy s exists and obeys a balance equation of form

$$\rho \frac{ds}{dt} + \operatorname{div} \mathbf{J}_s = r_s + \sigma_s \tag{7}$$

The entropy is a "state function" but the second law does not determine how and what independent variables it depends on. The entropy flow $\mathbf{J}_s$ and the entropy supply r_s give account on entropy decrease somewhere and increase elsewhere.[b]

[a] Here, we mention that $\mathbf{J}_u$ is sometime referred to as heat flow and r as heat supply; this wording may be misleading; they may contain both heat and work.

[b] If the increase and the decrease of the entropy are separated in space $\mathbf{J}_s$ is actual; if the separation is mental (e.g., in the material and the field on the same place) r_s is active.

The entropy production density σ_s is always positive if anything is happening and zero in equilibrium. The entropy production is small and negligible in some considerations; then we speak of reversible processes.

The actual form of the entropy function, the entropy flow, and the entropy supply are established by modeling. Whatever the above quantities are, the entropy production has a form linear in the process rates J_i, which are referred to as fluxes[c]:

$$\sigma_s = \sum_i X_i J_i. \tag{8}$$

The coefficients of the fluxes are called "thermodynamic forces". The second law is satisfied only if

1. the fluxes depend on the thermodynamic forces and
2. the only solution of the equations

$$J_i(X_1, X_2, \ldots, X_n) = 0 \quad \{i = 1, 2, \ldots, n\} \tag{9}$$

is

$$X_i = 0 \qquad \{i = 1, 2, \ldots, n\} \tag{10}$$

i.e., the necessary and sufficient condition for an equilibrium is the vanishing of all the forces.

The material equations determining the process rates have to be of the form

$$J_i = \sum_{k=1}^{n} L_{ik} X_k \qquad \{i = 1, 2, \ldots, n\} \tag{11}$$

where the coefficients L_{ik} are smooth enough functions of the local state variables and the thermodynamic forces.[10–13] We speak of Onsager's linear theory if the coefficients L_{ik} do not depend on the thermodynamic forces. The conductivity coefficients make a positive definite form (entropy inequality) in the sense of algebra. It can be proved that any subset of the thermodynamic forces can be expressed explicitly from the linear equations as a linear function on the conjugated fluxes and the other forces. It means that the linear equations can be read in 2^n equivalent form. For any of them, the set of the independent variables contains one element for each term in the entropy production that may be either the force X_i or the flux J_i. In this sense, we change the notation and X_i will stand for the independent variable and J_i for the dependent

---

[c]We put aside the problem what we call process rate.

one. This way, any of the above mentioned variants of the linear laws reads in the form of equations (11).[14] If any of them is assumed they satisfy the so called Onsager–Casimir reciprocal relations. Sometime they have simple form but a general wording is rather complicated. Perhaps, Meixner's principle of "macroscopic reversibility" is easier.[15]

The irreversibility of the natural processes means that if we try to build up a transformation that may be called time inversion[d] the effort fails. Meixner's principle says that a time inversion shows the form (8) of the entropy production to be β-type—i.e., changing sign under time inversion—and to be α-type—i.e., invariant under time inversion—after having introduced any variant of the linear equations (11) into it. Usually, the variables appearing in a theory are α-type or β-type.[e]

The Onsager–Casimir reciprocal relations can be formulated in the form

$$L_{ik} = \pm L^{*}_{ki}, \tag{12}$$

where the positive sign has to be taken if X_i and X_k are of the same type (i.e., both are α-type or both are β-type) else the negative. The * means that the time inversion has to be applied on the independent variables the L_{ki} coefficient depends on. If the positive sign is valid we speak of Onsager's reciprocal relation and of Casimir's one if the negative.[f]

If we know the forms of the entropy function, the entropy flow, the entropy supply, and the material equations (11) then the theory is complete and can answer any question.[g] Usually, we are not in the position like that; the above correlations have to be find out[h] during which the material symmetries, the main features of material behavior, and other principles are of great merit.

4 Local equilibrium

The earliest—and best known—applications of non-equilibrium thermodynamics assume systems in local equilibrium. It means that the independent state variables and the form of the entropy function are the same in an equilibrium and out of equilibrium. The equilibrium state variables have to pick up

[d]Not having enough time and room, we put aside the requirements a time inversion has to satisfy. Keep in mind that time inversion is a human imagination; time can not be inverted.
[e]In principle, we can not exclude the possibility of a variable neither α nor β-type.[13, 16]
[f]We mention that Onsager's reciprocal relation means symmetry and Casimir's one skew-symmetry if and only if the coefficients L_{ik} are invariant under time inversion.
[g]Supposing that we can solve the—usually partial—differential equations resulted.
[h]The name of guessing is constitutive modeling.

in accordance to the energy balance (6).[i] In rheology or continuum mechanics, the independent variables are the internal energy u and some measure for the deformation[j] $\boldsymbol{d}$, moreover, a vector or a second order tensor giving account on the orientation if the material is anisotropic.[k]

$$s = s^e(u, \boldsymbol{d}). \tag{13}$$

The time derivative of the entropy is arranged into the form

$$\frac{ds}{dt} = \frac{1}{T}\frac{du}{dt} + \boldsymbol{s}_d : \mathring{\boldsymbol{d}}, \tag{14}$$

where $\mathring{\boldsymbol{d}}$ is the symmetric part of the velocity gradient.[l] In the most simple theory, the couple stress, the energy supply, and the entropy supply are dropped and $\mathbf{J}_u$ in (6) is taken the heat flow, moreover, the classical relation

$$\mathbf{J}_s = \frac{1}{T}\mathbf{J}_u \tag{15}$$

for the entropy flow is accepted. The entropy production results in the form

$$\sigma_s = \frac{1}{T}\left\{(\boldsymbol{t} + \rho T \boldsymbol{s}_d) : \mathring{\boldsymbol{d}} - \mathbf{J}_u \frac{1}{T}\operatorname{grad} T\right\}. \tag{16}$$

From here, we get the conditions of an equilibrium: the uniformity of the temperature ($\operatorname{grad} T = 0$) and the correlation

$$\boldsymbol{t} = \boldsymbol{t}^e(T, \boldsymbol{d}) = -\rho T \boldsymbol{s}_d,$$

which also gives the physical meaning of the tensor $\boldsymbol{s}_d$. The Onsager equations result from here;

$$\begin{aligned} \boldsymbol{t} - \boldsymbol{t}^e(T, \boldsymbol{d}) &= \frac{1}{T}L_{11}\mathring{\boldsymbol{d}} - L_{12}\frac{1}{T}\operatorname{grad} T \\ \mathbf{J}_u &= \frac{1}{T}L_{21}\mathring{\boldsymbol{d}} - L_{22}\frac{1}{T}\operatorname{grad} T. \end{aligned} \tag{17}$$

[i]In thermostatics, the independent state variables may be chosen several ways. (The frequent change of the independent variables makes thermodynamics difficult for beginners.) Out of equilibrium, the suitable choice is nearly unique, one of the independent variables is the internal energy and the others are the ones the time derivative of which are present in balance equation (6) for the internal energy. This is the way we can get Gibbs' relation.

[j]A lot of deformation tensors are known in the literature; equivalent from a general point of view but introduced with different mathematics and suggesting different visualization of the motion and, of course, different ways of thinking.

[k]The variable accounting on the orientation will be omitted from the notation except for liquid crystals as dealing with the theory of crystals is not intended.

[l]Nevertheless, several deformation tensors are introduced, the symmetric part of the velocity gradient is widely accepted for a measure of the deformation rate. How it relates to the a deformation tensor is a long story and not understood even by excellent experts on the field.

The coefficients L are second, third, and forth order tensors reflecting the symmetry of the material. As they are material properties, they have to be invariant under any isotropy transformation of the particular material.[m] The equations contain complete information on elasticity, viscosity, as well as, on thermo- and pyro-elasticity.

For isotropic materials and linear approximation, the structure of the equations becomes simpler and more transparent[n]; they read

$$\begin{aligned} \boldsymbol{t} - \boldsymbol{t}^e &= 2\eta \overset{\circ}{\boldsymbol{d}} + \eta_v \operatorname{div} \mathbf{v} \boldsymbol{\delta} \\ \mathbf{J}_u &= -\lambda \operatorname{grad} T. \end{aligned} \tag{18}$$

In an isotropic material, there is no pyro-elasticity in linear order and the spherical and the deviatoric part of the second order tensors are not coupled. The viscoelastic properties of a material in local equilibrium are those of a Kelvin–Voigt medium. The equations reduce to Newton's viscosity law for an incompressible fluid. As we know from experiments and practice that the above models are much simpler than the reality,[18–21] we conclude that the hypothesis of local equilibrium must be dropped.

During irreversible processes the structure of materials changes; the local structure may be essentially different from that in an equilibrium. The structural changes must be reflected by the entropy function.[1,2,8,22] It is not a strong requirement if we suppose that such a structure can be accounted on with sufficient accuracy by some new variables the number of which is finite—say n.

The entropy function turns into

$$s = s(u, \boldsymbol{d}, \xi_1, \xi_2, \ldots, \xi_n). \tag{19}$$

During a no work no heat process—i.e., with constant u and $\boldsymbol{d}$—the material tends to an equilibrium, in which the entropy is a maximum. In an equilibrium, the new variables are determined by the equations

$$\frac{\partial s}{\partial \xi_i} = 0 \qquad \{i = 1, 2, \ldots, n\}. \tag{20}$$

As the entropy function is a maximum in an equilibrium state, the dynamic variables can be chosen so that the entropy function be of the form

$$s = s^e(u, \boldsymbol{d}) - \sum_i \xi_i^2 \tag{21}$$

[m]In a non-linear theory, they are covariant functions.

[n]Use the theorem of Smith and Rivlin[17] on isotropic tensors to obtain the form of the coefficients L.

according to the Morse lemma.[23] A rather general form of the entropy production reads

$$\sigma_s = \frac{1}{T}\left\{(\boldsymbol{t}-\boldsymbol{t}^e) : \mathring{\boldsymbol{d}} + \boldsymbol{\Pi} : \mathrm{Grad}\,\vec{\omega} - \mathbf{J}_u \frac{1}{T}\,\mathrm{grad}\,T\right\} - \sum_i \xi_i \mathring{\xi}_i; \tag{22}$$

which is valid if no energy supply or entropy supply is present and the relation (15) holds.[o] The quantity T is the temperature of the material in an equilibrium state with the same internal energy and deformation.[p] The form of the material equations is complicated and it is not worth putting them down in general. The small circle above the symbol ξ means corotational time derivative; one of the possible objective time derivatives. An objective time derivative[q] is convenient to define process rate as several tensor component may be present among the dynamic variables.[r]

In continuum mechanical applications, the term due to heat flow is usually dropped.

5 Rheology

To obtain the usual models in rheology suppose that no couple stress is present and the dynamic variables are α-type second order tensors.[2] The material equations read

$$\begin{aligned} \boldsymbol{t} - \boldsymbol{t}^e &= \frac{1}{T} L_{00} \mathring{\boldsymbol{d}} + \sum_i L_{0i} \boldsymbol{\xi}_i, \\ -\mathring{\boldsymbol{\xi}}_i &= \frac{1}{T} L_{i0} \mathring{\boldsymbol{d}} + L_i \boldsymbol{\xi}_i. \end{aligned} \tag{23}$$

Here the coefficients L are forth order tensor valued functions reducing to numbers if concerning the linear approximation for isotropic material and separating the trace of the tensors from the deviatoric (i.e., traceless) parts. The material equations for volumetric- and deviatoric effects (quantities) are separate and of the form of equations (23) but with different coefficients. Notice that no cross effect is between the dynamic variables. The reason is

[o]The latter relation can not be of general validity; see, e.g., in reference.[24]

[p]The problems of non-equilibrium temperature is another complicated story[24, 25] not dealt with here.

[q]The use of an objective time derivative is optional and results that the coefficients L are independent of the angular velocity of the material if the Coriolis–force and other "gravitational" effects are negligible.

[r]The symbol $\mathring{\boldsymbol{d}}$ is generally not an objective time derivative but for small deformations or in the case of fluids. For the latter, the deformation tensor $\boldsymbol{d}$ is replaced by the relative deformation gradient, i.e., the deformation with respect to the present configuration.

that an orthogonal transformation of the dynamic variables leaves the entropy function invariant and a suitable one takes the matrix of the conductivity coefficients into diagonal form.[s]

Nevertheless, the deviatoric equations are linear in the Onsagerian sense, they are not linear at all in the unknown quantities we usually look for because of the objective time derivatives.

The theory is the simplest for fluids; the deformation tensor is replaced by the specific volume $\mathcal{V}$ and the deviatoric part of the equilibrium stress is zero (Pascal's law).

The reduction of the equations to particular motions gives the usual rheological models. For small amplitude oscillations they reduce to the equations of linear viscoelasticity and any body-model consisting of dashpots and springs can be founded.[t]

For viscometric flow, the equations give account on the viscometric functions; any monotonically decreasing shear viscosity function is approximated for any prescribed accuracy. The normal stress functions are also put forward but—with corotational derivatives—no normal stress effect perpendicular to the plain of the motion is forecasted, moreover, the shear stress function and the normal stress functions are related each to the other by Hilbert transformation. The shear viscosity is the same function on shear rate as the real part of complex viscosity on the circular frequency; the empirical Cox–Merz rule[26,27] has been founded. The state variables determine material properties; rheo-optics is based on optical parameters depending on the dynamic variables.

Some parameters result a domain of the shear rate in which the shear stress function is decreasing; the viscometric solution of the equations is unstable, (shear-banding).[28] For the same material, also the elongational flow is unstable for high load; another stationary solution replaces the unstable one; the displayed effect models the experimental background of the theory of plasticity without a postulated yield condition including also creep and viscoelastic deformations before and after plastic flow.[29,30]

The inertial element (together with an increasing section of the shear viscosity) —what appears in some theoretical works but rarely if ever reported by experiments —has also room in the theory if β-type dynamic variables are also allowed.

Among the non-linear theories—i.e., when the coefficients L are isotropic tensor functions on the dynamic variables and the deformation rate—one can

[s]The coefficients L_{ik} form a symmetric matrix due to Onsager's reciprocal relations.

[t]For a unified linear theory of fluids and solids, equation (23) is linearized.

find all the theories of plasticity. The Mieses theory is founded without dynamic degrees of freedom.[31] A non-linear theory is able to give account on tixotropy and rheopexy, as well. It has to be mentioned that the heuristic power of non-equilibrium thermodynamics is weaker for a non-linear case than in the linear case.[u]

The models mentioned up to now fit to the idea of "simple materials", for which the stress is determined by the history of the deformation tensor.[v] The model of "simple materials" can not bear non-local interactions even week non-locality, i.e., when the gradient of the deformation tensor or that of the angular velocity are involved into the considerations.

6 Some weekly non-local materials

The materials to be mentioned are the melts of fiber reinforced materials, liquid crystals, and turbulent fluids.

6.1 Pseudo tensors

The considerations of section 5 can not do with the skew-symmetric stress not entering the entropy production. Skew-symmetric stress is generated by electromagnetic polarization (torque and internal moment of momentum as in the Einstein–de Haas effect) or by couple stress. The couple stress does enter into the entropy production. The gradient of the angular velocity is closely related to the gradient of the deformation rate; $(\mathrm{Grad}\,\vec{\omega} = \mathrm{Rot}\,\mathring{\boldsymbol{d}})$. The model needs dynamic variables that are second order symmetric or skew-symmetric pseudo tensors. If they are of α-type the Onsager equations are

$$\begin{aligned} \boldsymbol{\Pi}^d &= L_{00}^d \frac{1}{T} (\mathrm{Grad}\,\vec{\omega})^d + \sum_i L_{01}^d \boldsymbol{\xi}_i^d \\ -\mathring{\boldsymbol{\xi}}_i^d &= -L_{i0}^d \frac{1}{T} (\mathrm{Grad}\,\vec{\omega})^d + L_i^d \boldsymbol{\xi}_i^d \end{aligned} \tag{24}$$

for the deviatoric parts and

$$\begin{aligned} \vec{w}(\boldsymbol{\Pi}) &= L_{00}^a \,\mathrm{rot}\,\mathrm{rot}\,\mathbf{v} + \sum_i L_{0i}^a \vec{\xi}_i^a \\ -\mathring{\vec{\xi}}_i^a &= -L_{0i}^a \frac{1}{2T} \,\mathrm{rot}\,\mathrm{rot}\,\mathbf{v} + L_i^a \vec{\xi}_i^a \end{aligned} \tag{25}$$

---

[u] A result containing arbitrary functions is of less practical value than another containing arbitrary positive numbers.

[v] Usually, the history of the deformation gradient tensor is referred to and the "simple fluid" is defined.

for the vector invariants of the skew-symmetric parts. The dynamic degrees of freedom give account on the deformation of fibers. As the vector invariant of a skew-symmetric pseudo tensor is a polar vector, the possibility of a cross effect with heat conduction has to be examined. Introducing the above equations into Cauchy's equation of motion, the presence of pseudo tensors involve terms containing the forth partial derivatives of the velocity.

6.2 Liquid crystals

Liquid crystals are anisotropic fluids; the entropy depends on the internal energy, the specific volume, and a vector or second order tensor for the orientation.[32] Couple stress is assumed and—in some works—internal angular momentum.

Several classes of liquid crystals are known; the material symmetry influences the form of the polynomial. Dynamic degrees of freedom are not introduced because of the low symmetry results a complicated enough theory without them. Even the local-equilibrium theory is reduced[33,34] to small gradients of the orientation to obtain the Leslie theory.[35]

6.3 Fluid turbulence

The thermodynamic theory of fluid turbulence has not been worked out. Dynamic degrees of freedom are surely needed for the intensity of turbulence. The relation (15) for the entropy flow can not hold as it leads to "simple fluid", which satisfies Reynolds' similarity only for laminar flow. The linear Onsager equations assuming also the transport of dynamic variables combined with Reynolds' similarity lead to laminar flow. The simplest promising equations (one dynamic degree of freedom and non-linear Onsager-equations) give back the essential features of fluid turbulence but the differential equations proved very hard and the theory has not been succeeded in quantitative checking.

Non-equilibrium thermodynamics without the hypothesis of local equilibrium gives a large enough frame to develop a quantitative theory of any macroscopic phenomenon in a moving continuum. The rich arsenal has not been exhausted.

Acknowledgments

This work has been supported by the Hungarian National Scientific Research Funds, OTKA (T-17000) and MKM (FKFP 0287/1997), and the EC (CARNET Contract No: ICOP-DISS-2168-96)

References

1. D. Jou, J. Casas-Vázques, and G. Lebon. *Extended Irreversible Thermodynamics.* Springer, New York, Berlin, Heidelberg, London,Paris, Tokyo, Hong Kong, Barcelona, Budapest, 1993.
2. J. Verhás. *Thermodynamics and Rheology.* Akadémiai Kiadó and Kluwer Academic Publishers, Budapest and Dordrecht, 1997.
3. L. Onsager. Reciprocal relations in irreversible processes. I. *Phys. Rev.*, 37:405–426, 1931.
4. L. Onsager. Reciprocal relations in irreversible processes. II. *Phys. Rev.*, 38:2265–2279, 1931.
5. L. Onsager and S. Maclup. Fluctuations and irreversible processes. *Phys. Rev.*, 91:1505–1512, 1953.
6. S. Machlup and L. Onsager. Fluctuations and irreversible processes. II. systems with kinetic energy. *Phys. Rev.*, 91:1512–1515, 1953.
7. S. R. De Groot and P. Mazur. *Non-Equilibrium Thermodynamics.* North-Holland Publ. Co., Amsterdam, 1962.
8. I. Müller and T. Ruggeri. *Extended Thermodynamics.* Springer, Berlin, New York, 1993.
9. I. Gyarmati. *Non-Equilibrium Thermodynamics.* Springer, Berlin, 1970.
10. J. Hurley and C. Garrod. Generalization of the Onsager reciprocity theorem. *Phys. Rev. Lett.*, 48:1575–1577, 1982.
11. C. Garrod and J. Hurley. Symmetry relations for the conductivity tensor. *Phys. Rev. A*, 27:1487–1490, 1983.
12. J. Verhás. An extension of the governing principle of dissipative processes to non-linear constitutive equations. *Ann. d. Physik*, 40:189–193, 1983.
13. V. Ciancio and J. Verhás. On the non-linear generalizations of Onsager's reciprocal relations. *J. Non-Equilib. Thermodyn.*, 19:184–194, 1994.
14. J. Meixner. Consistency of the Onsager-Casimir reciprocal relations. *Adv. Molecular Relax. Proc.*, 5:319–331, 1973.
15. J. Meixner. Macroscopic and microscopic reversibility. *Pap. on Math. Phys.*, 7:37–57, 1975.
16. V. Ciancio and J. Verhás. Reciprocal relations and dynamic degrees of freedom. *Physica A*, 189:665–675, 1992.
17. G. F. Smith and R. S. Rivlin. The anisotropic tensors. *Q. Appl. Math.*, 15:308–314, 1957.
18. M. Reiner. *Deformation, Strain and Flow.* Interscience, New York, 1960.
19. W. R. Schowalter. *Mechanics of Non-Newtonian Fluids.* Pergamon Press, New York, 1978.

20. H. A. Barnes, J. F. Hutton, and K. Walters. *An Introduction to Rheology*. Elsevier, Amsterdam, London, New York, Tokyo, 1989.
21. D. V. Boger and K. Walters. *Rheological Phenomena in Focus*. Elsevier, Amsterdam, London, New York, Tokyo, 1993.
22. I. Gyarmati. On the wave approach of thermodynamics and some problems of non-linear theories. *J. Non-Equilib. Thermodyn.*, 2:233–260, 1977.
23. M. Morse. Relation between the critical points of a real function of n independent variables. *Trans. Am. Math. Soc.*, 27:345–396, 1925.
24. V. Ciancio and J. Verhás. On the representation of dynamic degrees of freedom. *J. Non-Equilib. Thermodyn.*, 18:39–50, 1993.
25. E. P. Gyftopoulos and G. P. Beretta. *Thermodynamics: Foundation and Applications*. Macmillan, New York, 1991.
26. W. P. Cox and E. H. Merz. Correlation of dynamic and steady-flow viscosities. *J. Polym. Sci.*, 28:619–622, 1958.
27. W. P. Cox and E. H. Merz. Rheology of polimer melts. a correlation of dynamic and steady-flow measurement. *Am. Soc. Testing Material Spec. Techn. Publ.*, 247:178–188, 1958.
28. F. Greco. Prediction of shear-banding in flowing micellar systems. In I. Emri, editor, *Proceedings of the Fifth European Rheology Conference*, Progress and Trends in Rheology, Darmstadt, 1998. Steinkopff.
29. J. Verhás. A thermodynamic approach to viscoelasticity and plasticity. *Acta Mechanica*, 53:125–139, 1984.
30. V. Ciancio and J. Verhás. Plasticity as a consequence of the instability of viscoelastic flow. A thermodynamic description. *Le Matematiche (Catania)*, 46:77–83, 1991.
31. J. Verhás. The construction of dissipation potentials for non-linear problems and the application of Gyarmati's principle to plastic flow. *Z. Phys. Chemie*, 249:119–122, 1972.
32. F. C. Frank. On the theory of liquid crystals. *Disc. Faraday Soc.*, 25:19–28, 1958.
33. J. Verhás. The irreversible thermodynamics of nematic liquid crystals. *Acta Physica Hungarica*, 55:275–291, 1984.
34. C. Papenfuss, J. Verhás, and W. Muschik. A simplified thermodynamic theory for biaxial nematics. *Z. Naturforsch.*, 50a:795–804, 1995.
35. F. M. Leslie. Some constitutive equations for liquid crystals. *Arch. Rat. Mech. Anal.*, 28:265–283, 1968.

PHYSICS AND MECHANICS OF ICE COVER BREAK UP TECHNOLOGY

V. A. Zuev, E. M. Gramuzov, Yu. A. Dvoichenko, V. V. Knyazkov
Nizhnyj Novgorod State Technical University, Chair of Shipbuilding, ul. Minina 24, Nizhnyj Novgorod, 603600, GSP-41, Russia

Ice belongs to a number of well-known substances. However its properties have not been sufficiently investigated. This is accounted for a great diversity of its structure, composition, physico-mechanical properties of the ice cover and, to a great extent, by a variety of forces acting upon ice at different temperatures, with different intensity, durability and scope.

Ice and ice cover mechanics is one of the less developed areas of continuum mechanics. From the destruction point of view, ice reveals many properties which are very difficult, if possible at all, to observe in other materials.

Interest to the research of physics and mechanics of ice has risen dramatically for the last three decades, stimulated considerably by the necessity of developing natural resources in almost inaccessible regions of the globe with cold climate. Transport links in freezing seas and rivers including all-year-round navigation in the Arctic Ocean, are developing radically. Ice cover is used for transportation of goods and creating of cargo platforms.

Ice technology problems can be relatively divided into two large groups:

- research of physico-mechanical properties of ice and ice cover, modelling and criteria development of the ice cover break up;
- research of ice cover interaction with structures and developing of models of such interactions.

Practice demands for ice-techniques and navigation, especially long-term navigation, often result in contrary, opposed requirements as far as ice interaction is concerned. On the one hand, it's necessary to provide a maximum support capability of the ice cover, i.e. for running the ice passages, ice airfields, cargo-carrying areas etc, and on the other hand to create the best conditions of breaking up ice (e.g. by means of ice-breakers). In this connection there exist various approaches and understanding of the term "ice destruction", depending upon whether it is desirable or undesirable. Some investigators consider it as development of critical stresses and cracking of the ice cover, others - as absolute ice break up, when the outside loading is compensated only buoyancy of ice fragments.

Application of fracture mechanics methods for ice problems [1,2] has just started with no practical results yet obtained. In general all efforts here are concentrated in research of cracking growth characteristic features with "destruction resistance" and "stress intensity ratio" being involved. However, from our point of view, concepts existing in fracture mechanics make possible only to evaluate the possibility of ice cracking itself and at the same time can't provide an exhaustive solution to the problem of the absolute break up of the ice cover.

Observations of the ice cover deformation under loading have shown that the ice cover main cracking is not equal to the absolute ice destruction and failure of its carrying capability. This gave grounds for carrying out additional theoretic and experimental research.

To specify mechanism of the ice cover destruction its stress-strain state was studied at various stages of destruction. Ice cover was treated as a thin homogeneous isotropic plate of constant thickness lying on the hydraulic type elastic foundation. Behavior of such a plate under transverse loading is described by a rather simple differential equation in partial derivatives. However, the exact solution of this equation is possible to obtain by having rather simple boundary conditions, e.g. with infinitely large plate under central flexure. That is why this problem was solved by using numerical methods.

As a result a sequence of radial and concentric cracking was determined. First a mesh of radial cracking and then a concentric one is observed in a consecutive order beginning from the periphery to the place of loading applied. This entirely corresponds to the experimentally observed picture of ice destruction. It is also shown that due to the wedging effect of ice blocks formed by cracking appearance of cracks doesn't result in failure of the ice cover carrying capability. Deformation of the ice cover consisting of cracked blocks goes on. Interactions along crack sides are expressed in mathematical models by a hinge forming an elastics obstruction against rotation. Relationship between a transverse moment in crack M and vertical cracking angle in crack β is assumed to be liner $\beta=\alpha M/(aD)$, where a- non-dimensional coefficient characterizing interaction of crack sides and depending upon the wedging effect; D- plate cylindrical rigidity; $\alpha = \sqrt[4]{\rho_w g / D}$ - parameter of the plate flexure resting on elastic foundation; ρ_w- water density; g- acceleration of free fall.

Thus potential energy of the flexure is formed of the ice plate flexure energy and potential energy of the elastic foundation. Calculations show that elastic foundation deformation energy can be three times greater than ice plate deformation energy. This, for example, explains the effective operation of ice-breaking and ice cleaning attachments with transverse knives which result in ice failure localization and systematization.

To specify a complex mechanism of ice cover destruction and to verify some assumptions, field tests were carried out in the Gorky sea (storage lake) and model tests in the ice-breaking model basin of Nizhny Novgorod Technical University. Tests were carried out with "infinite" plates and canal cutted form plates being under kinematic loading. Kinematic loading has made possible to obtain complete flexure-loading diagrams. One of the diagrams obtained experimentally is outlined in Fig. 1. Point 1 corresponds to cracking and decrease of loading at moment of its creation. Shaded area of diagram is the amount of work spent for creation of new surfaces when cracking appears. Point 2 corresponds to the ice breaking maximum force P_p and flexure w_p at it. Point 3 (decline of the diagram) corresponds to absolute ice destruction and flexure w_R. Diagram entire area reflects destruction work that is several times greater than the amount of work spent on cracking. This is accounted for significant interaction energy of the crack sides.

Influence of the ice cover thickness upon the form (and area) of the destruction diagram is shown in Fig.2. Fig.3 shows influence of loading speed.

There are no analytic dependencies for evaluating parameters of absolute ice destruction (break up), therefore semiempiric ones are given which could be used in appropriate cases.

$$P_p = k_p \beta h^2, \tag{1}$$

where k_p - specific energy of ice cover destruction. (For infinite plate flexure k_p=1900 kPa with mean square deflection $\pm$ 100 kPa. For flexure of the canal cutted form plate k_p=(1100$\pm$100) kPa); $\beta = 1.0 + 1.34\alpha\sqrt{S}$ - loading area coefficient.

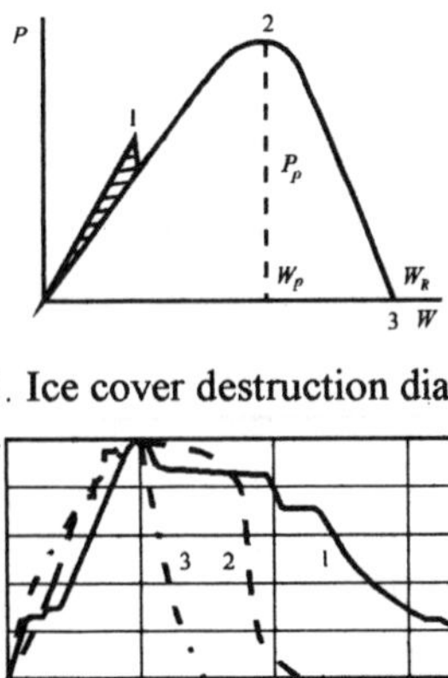

Fig. 1. Ice cover destruction diagram.

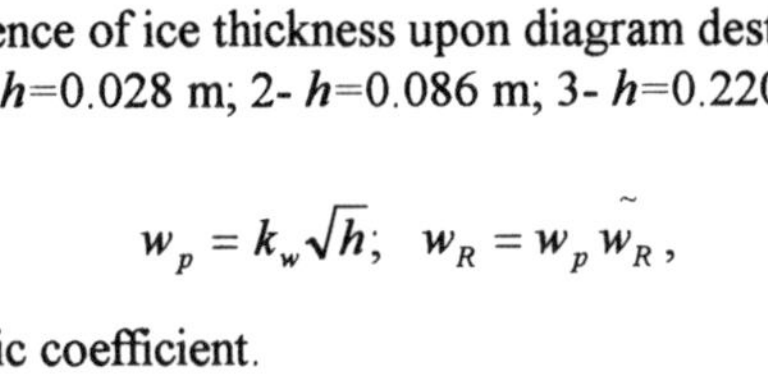

Fig. 2. Influence of ice thickness upon diagram destruction form (1- h=0.028 m; 2- h=0.086 m; 3- h=0.220 m)

$$w_p = k_w \sqrt{h}; \quad w_R = w_p \tilde{w}_R, \tag{2}$$

where k_w - empiric coefficient.

Amount of work for destruction

$$A_P = \int_0^{w_P} P dw = k_{AP} P_P w_P; \tag{3}$$

$$A_R = \int_0^{w_R} P dw = A_P + k_{AR} P_P (w_R - w_P). \tag{4}$$

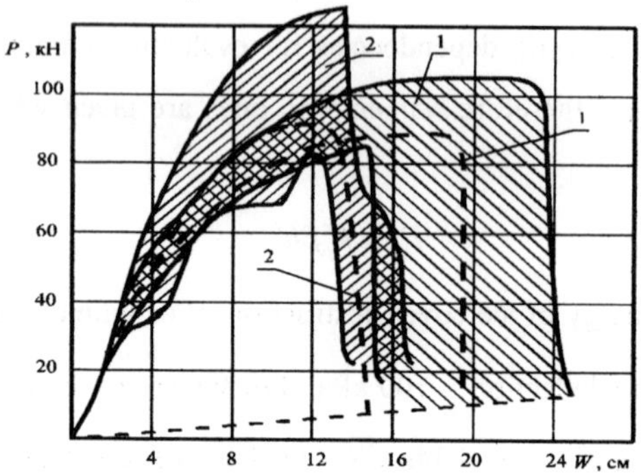

Fig. 3. Influence of loading speed upon destruction diagram form

(1 - v=0.01 m/s; 2 - v=0.14 m/s; 3 - v=0.22 m/s)

Ice cover destruction parameters under static loading in infinite plate (column A) and in canal cutted form plate (column B) are shown in the following table.

To define destruction parametrs

Destruction parameter	Calculation formula	Empiric coefficient		Coefficient value	
		Mark	Dimen-sionally	Column A	Coumn B
Break up force	$P_P = k_P \beta h^2$	k_P	MPa	1.9±0.1	1.1±0.1
Flexure under P_P	$w_P = k_w \dfrac{P_P}{D\alpha^2}$	k_w	$м^{0.5}$	0.28±0.04	0.41±0.04
	$w_P = k_w \sqrt{h}$	k_w	-	0.42±0.05	0.28±0.02
Flexure of absolute reaction slump, w_R	$w_R = w_P \tilde{w}_R$	$\tilde{w}_R$	-	1.0±0.01*	1.5±0.2*
Destruction work A_P	$A_P = k_P P_P w_P$	k_{AP}	-	0.64±0.08	0.69±0.03
Absolute destruction work A_R	$A_R = [k_{AP} + k_{AR}(\tilde{W}_R - 1)] P_P w_P$	k_{AP}		2.5±0.1*	0.87±0.07*

Note: * Coefficient values for $h > 0.2$ m.

References

1. Goldshtain P.V., Osipenco N.M. Ice cover destruction mechanics. - M.: IPM AN USSR, 71, (1982).
2. Miller K.J. The application of Fracture Mechanics to Ice Problems. - Phys. and Mech. Ice, Symp. Copenhagen, , Berlin, 1980, 265-277,(1979).
3. Zuev V.A., Gramuzov E.M., Dvoichenco Y.A. Ice cover destruction. - NTO im. Acad. A.N. Krylova, Gorky, 86, (1989).

LIST OF THE PARTICIPANTS

Prof. Vladimir I. Alshits
Institute of Crystallography
Academy of Sciences of Russia
Leninski pr. 59
117333 Moscow
Russia
alshits@mechan.incr.msk.su

Prof. dr hab. Jan Awrejcewicz
Łódź University of Technology
Institute of Fluid-Flow Machinery
ul. Wólczańska 219/223
93-005 Łódź
Poland
awrejcew@ck-sg.p.lodz.pl

Prof. dr hab. inż. Zbigniew Bilicki
Institute of Fluid-Flow Machinery
Polish Academy of Sciences
ul. Gen. J. Fiszera 14
80-952 Gdańsk
Poland
zb@imppan.imp.pg.gda.pl

Dr Jarosław Bośko
Gdańsk University of Technology
Department of Technical Physics and Applied Mathematics
ul. G. Narutowicza 11/12
80-952 Gdańsk
Poland
jbosko@kliper.task.gda.pl

Prof. Jaroslav Burak
Centre of Mathematical Modelling
NAS Ukraine
15 Dudaev str.
290005 Lviv
Ukraine
burak@cmm.lviv.ua

Prof. dr hab. inż. Michał Ciałkowski
Poznań University of Technology
Chair of Thermal Engineering
ul. Piotrowo 3
60-965 Poznań
Poland
cialkows@sol.put.poznan.pl.

Dr Mieczysław Cieszko
Pedagogical University in Bydgoszcz
Department of Environment Mechanics
ul. Mielżyńskiego 27/29
61-725 Poznań
Poland
ryban@rose.man.poznan.pl

Dr Elena V. Darinskaya
Institute of Crystallography
Academy of Sciences of Russia
Leninsky pr. 59
117333 Moscow
Russia
public@mechan.incr.msk.su

Dr Marina Dolfin
Universitá di Messina
Dipartimento di Matematica
Contrada Papardo, Salita Sperone 31
98166 Sant'Agata-Messina
Italy
dolfin@dipmat.unime.it

Dr Christine Drogomiretska
Lviv Polytechnic State University
Department of Applied Mathematics
Lviv
Ukraine
rashkev@polynet.lviv.ua

Mgr inż. Maciej Dutkiewicz
University of Technology and Agriculture
Faculty of Civil and Environmental Engineering
ul. Kaliskiego 7
85-796 Bydgoszcz
Poland
macdut@mail.atr.bydgoszcz.pl

Prof. Juri Engelbrecht
Estonian Academy of Sciences
Kohtu 6
EE-0001 Tallinn
Estonia
je@ioc.ee

Mgr Sylwia Futakiewicz
Kielce University of Technology
Chair of Mathematics
Al. 1000-lecia Państwa Polskiego 7
25-314 Kielce
Poland
grysa@eden.tu.kielce.pl

Prof. dr hab. Krzysztof Grysa
Kielce University of Technology
Chair of Mathematics
Al. 1000-lecia Państwa Polskiego 7
25-314 Kielce
Poland
grysa@eden.tu.kielce.pl.

Mr. Sebastian Gümbel
Technische Universität Berlin
Institut für Theoretische Physik
Hardenbergstrasse 36
D-10623 Berlin
Germany

Prof. Peter Haupt
Universität - Gesamthochschule Kassel
Institut für Mechanik
Monchebergstrasse 7
34109 Kassel
Germany
haupt@ifm.maschinenbau.uni-kassel.de

Mr. Heiko Herrmann
Technische Universität Berlin
Institut für Theoretische Physik
Hardenbergstrasse 36
D-10623 Berlin
Germany

Mgr Leszek Hożejowski
Kielce University of Technology
Chair of Mathematics
Al. 1000-lecia Państwa Polskiego 7
25-314 Kielce
Poland
grysa@eden.tu.kielce.pl.

Dr Janusz Jankowski
Poznań University of Technology
Institute of Applied Mechanics
ul. Piotrowo 3
60-965 Poznań
Poland
jankowsk@sol.put.poznan.pl

Prof. David Jou
Universitat Autonoma de Barcelona
Department de Fisica, Grup de Fisica Estadistica
Bellaterra 08193
Spain
jou@circe.uab.es

Doc. dr hab. inż. Jarosław Kaczmarek
Institute of Fluid-Flow Machinery
Polish Academy of Sciences
ul. J. Fiszera 14
80-952 Gdańsk
Poland
jarekk@imppan.imp.pg.gda.pl

Dr Mariusz Kaczmarek
Pedagogical University of Bydgoszcz
Department of Environment Mechanics
ul. Mielżyńskiego 27/29
61-725 Poznań
Poland
mkk@rose.man.poznan.pl

Miss Akiko Kato
Technische Universität Berlin
Institut für Theoretische Physik
Hardenbergstrasse 36
D-10623 Berlin
Germany

Dr. Garry Kerch
Centre of Science and Engineering
EKOTRA
Nicgales str. 4 - 235
LV-1035 Riga
Latvia
garry.k@usa.net

Mr. Vasyl P. Kopylchuk
Kiev National University
Physical Faculty of Taras Shevchenko
Lomonosova, 55
UA-252022 Kiev
Ukraine
baziv@hq.ups.kiev.ua

Dr Romuald Kotowski
Institute of Fundamental Technological Research
Polish Technology of Sciences
ul. Świętokrzyska 21
00-049 Warszawa
Poland
rkotow@ippt.gov.pl

Prof. dr hab. inż. Stefan J. Kowalski
Poznan University of Technology
Institute of Technology and Chemical Engineering
pl. M. Skłodowskiej-Curie 2
60-965 Poznań
Poland
kowal@rose.man.poznan.pl

Dr Wiesław Larecki
Institute of Fundamental Technological Research
Polisch Technology of Sciences
ul. Świętokrzyska 21
00-049 Warszawa
Poland
wlarecki@ippt.gov.pl

Mgr Artur Maciąg
Kielce University of Technology
Chair of Mathematics
Al. 1000-lecia Państwa Polskiego 7
25-314 Kielce
Poland
grysa@eden.tu.kielce.pl

Prof. dr hab. Bogdan T. Maruszewski
Poznań University of Technology
Institute of Applied Mechanics
ul. Piotrowo 3
60-965 Poznań
Poland
maruszew@sol.put.poznan.pl.

Dr Tadeusz Matuszek
Gdańsk University of Technology
Faculty of Mechanical Engineering
ul. Narutowicza 11/12
80-952 Gdańsk
Poland
tmatusze@pg.gda.pl

Prof. Gerard A. Maugin
Modelisation en Mecanique
Universite Pierre et Marie Curie (Paris VI)
Tour 66-4, Place Jussieu
75230 Paris Cedex 05
France
gam@ccr.jussieu.fr

Prof. Dr. Wolfgang Muschik
Institut für Theoretische Physik
Technische Universität Berlin
Hardenbergstrasse 36
D-10623 Berlin
Germany
muschik@physik.tu-berlin.de

Dr inż. Grzegorz Musielak
Poznań University of Technology
Institute of Technology and Chemical Engineering
pl. M. Skłodowskiej-Curie 2
60-965 Poznań
Poland
gmm@rose.man.poznan.pl

Dr. Christine Papenfuß
Technische Universität Berlin
Institut für Theoretische Physik
Hardenbergstrasse 36
D-10623 Berlin
Germany

Mr. Mindaugas Rackaitis
Kaunas University of Technology
Research Center VIBROTECHNIKA
Studentu 65
LT-3031 Kaunas
Lithuania
Mindaugas.Rackaitis@fmf.ktu.lt

Prof. dr hab. Andrzej Radowicz
Kielce University of Technology
Chair of Mechanics
Al. 1000-lecia Państwa Polskiego 7
25-314 Kielce
Poland
radowicz@tu.kielce.pl.

Prof. Liliana Restuccia
Universitá di Messina
Dipartimento di Matematica
Contrada Papardo, Salita Sperone 31
98166 Sant'Agata-Messina
Italy
lrest@dipmat.unime.it

Dr Patrizia Rogolino
Universitá di Messina
Dipartimento di Matematica
Contrada Papardo, Salita Sperone 31
98166 Sant'Agata-Messina
Italy
patrizia@dipmat.unime.it

Mr. Gunnar Rückner
Technische Universität Berlin
Institut für Theoretische Physik
Hardenbergstrasse 36
D-10623 Berlin
Germany

Prof. Jeremiah Rushchitsky
Institute of Mechanics
National Academy of Science of Ukraine
Nesterov str. 3
252680 Kiev
Ukraine
dynamic@imech.freenet.kiev.ua

Dr inż. Andrzej Rybicki
Poznań University of Technology
Institute of Technology and Chemical Engineering
pl. M. Skłodowskiej-Curie 2
60-965 Poznań
Poland
ryban@rose.man.poznan.pl

Dr Jarosław Rybicki
Gdańsk University of Technology
Department of Technical Physics and Applied Mathematics
ul. G. Narutowicza 11/12
80-952 Gdańsk
Poland
ryba@pg.gda.pl

Prof. dr hab. inż. Czesław Rymarz
Military University of Technology
ul. S. Kaliskiego 2
01-489 Warszawa
Poland
rymarz@glob.wic.wat.waw.pl

Prof. Igor Selezov
Department of Wave Processes
Institute of Hydrodynamics, Nat. Acad. Sci. of Ukraine
8/4, Sheliabov Str.
252057 Kiev
Ukraine
selezov@wave.pp.kiev.ua

Dr A. Mary Selvam
Indian Institute of Tropical Meteorology
Pune 411 008
India
selvam@tropmet.ernet.in

Prof. dr hab. Stanisław Sieniutycz
Warsaw University of Technology
Faculty of Chemical Engineering
ul. Waryńskiego 1
00-645 Warszawa
Poland
sieniutycz@ichip.pw.edu.pl.

Prof. Jozsef Verhás
Technical University of Budapest
H-1521 Budapest
Hungary
Verhas@phy.bme.hu

Prof. Valery Andreevich Zuev
Chair of Shipbuilding
Nizhnyj Novgorod State Technical University
ul. Minina 24
603600 Nizhnyj Novgorod,GSP-41
Russia
NSTU@nnpi.nnov.su